363.

1100.-

COMMUNICATION AND CONTROL IN ELECTRIC POWER SYSTEMS

IEEE Press
445 Hoes Lane
Piscataway, NJ 08854

IEEE Press Editorial Board
Stamatios V. Kartalopoulos, *Editor in Chief*

M. Akay	M. E. El-Hawary	M. Padgett
J. B. Anderson	R. J. Herrick	W. D. Reeve
R. J. Baker	D. Kirk	S. Tewksbury
J. E. Brewer	R. Leonardi	G. Zobrist
	M. S. Newman	

Kenneth Moore, *Director of IEEE Press*
Catherine Faduska, *Senior Acquisitions Editor*
John Griffin, *Acquisitions Editor*
Anthony VenGraitis, *Project Editor*

IEEE Power Engineering Society, Sponsor

PE-S Liaison to IEEE Press, Chanan Singh

COMMUNICATION AND CONTROL IN ELECTRIC POWER SYSTEMS

Applications of Parallel and Distributed Processing

MOHAMMAD SHAHIDEHPOUR
YAOYU WANG

IEEE Power Engineering Society, Sponsor

IEEE Press Power Engineering Series
Mohamed E. El-Hawary, *Series Editor*

IEEE PRESS

A JOHN WILEY & SONS, INC., PUBLICATION

Copyright © 2003 by John Wiley & Sons, Inc. All rights reserved.

Published by John Wiley & Sons, Inc., Hoboken, New Jersey.
Published simultaneously in Canada.

No part of this publication may be reproduced, stored in a retrieval system or transmitted in any form or by any means, electronic, mechanical, photocopying, recording, scanning or otherwise, except as permitted under Section 107 or 108 of the 1976 United States Copyright Act, without either the prior written permission of the Publisher, or authorization through payment of the appropriate per-copy fee to the Copyright Clearance Center, Inc., 222 Rosewood Drive, Danvers, MA 01923, (978) 750-8400, fax (978) 750-4470, or on the web at www.copyright.com. Requests to the Publisher for permission should be addressed to the Permissions Department, John Wiley & Sons, Inc., 111 River Street, Hoboken, NJ 07030, (201) 748-6011, fax (201) 748-6008, e-mail: permreq@wiley.com.

Limit of Liability/Disclaimer of Warranty: While the publisher and author have used their best efforts in preparing this book, they make no representation or warranties with respect to the accuracy or completeness of the contents of this book and specifically disclaim any implied warranties of merchantability or fitness for a particular purpose. No warranty may be created or extended by sales representatives or written sales materials. The advice and strategies contained herein may not be suitable for your situation. You should consult with a professional where appropriate. Neither the publisher nor author shall be liable for any loss of profit or any other commercial damages, including but not limited to special, incidental, consequential, or other damages.

For general information on our other products and services please contact our Customer Care Department within the U.S. at 877-762-2974, outside the U.S. at 317-572-3993 or fax 317-572-4002.

Wiley also publishes its books in a variety of electronic formats. Some content that appears in print, however, may not be available in electronic format.

Library of Congress Cataloging in Publication Data:

ISBN 0-471-45325-0

Printed in the United States of America.

10 9 8 7 6 5 4 3 2 1

Contents

Preface XIX

CHAPTER

1 Introduction 1
   1.1 The Role of Power Systems 2
   1.2 Complexity of Power Grids 4
   1.3 Communications System 6
       1.3.1 Fiber Optical Technique 7
       1.3.2 Fiber Optical Networks 8
       1.3.3 WAN Based on Fiber Optical Networks 9
       1.3.4 XML Technique 10
       1.3.5 IP-Based Real Time Data Transmission 12
   1.4 Interdependence of Infrastructures 13
   1.5 Catastrophic Failures 14
   1.6 Necessity for Distributed Processing 16
       1.6.1 Power System Control 16
       1.6.2 Distributed Implementation 17
       1.6.3 State Monitoring Based on GPS 19
   1.7 Vertically Integrated Power Systems 19
       1.7.1 Central Control Center 19
       1.7.2 Area Control Center 20
       1.7.3 SCADA/EMS 20

	1.7.4	Distributed Web-Based SCADA Systems	22
	1.7.5	Distributed Energy Management Systems	22
1.8	Restructured Power Systems	24
	1.8.1	GENCOs, TRANSCOs, and DISTCOs	24
	1.8.2	ISO	24
	1.8.3	OASIS	28
	1.8.4	RTO	29
1.9	Advantages of Distributed Processing	32
	1.9.1	Enhanced Reliability	32
	1.9.2	Enhanced Flexibility	33
	1.9.3	Economy	33
	1.9.4	Higher Efficiency	33
1.10	Foundations for Distributed Processing	34
	1.10.1	Agent Theory Applications	34
	1.10.2	Distributed Management and Control	39
	1.10.3	Adaptive Self-Healing	40
	1.10.4	Object-Oriented Modeling	41
	1.10.5	CIM	41
	1.10.6	Common Accounting Model	42
1.11	Organization of This Book	43

2 Parallel and Distributed Processing of Power Systems   47

2.1	Introduction	47
2.2	Parallel and Distributed Processing Systems	49
	2.2.1 Parallel Systems	49
	2.2.2 Distributed Systems	50
	2.2.3 Comparison of Parallel and Distributed Processing	50
	2.2.4 Representation of Parallel and Distributed Systems	54
2.3	Parallel and Distributed Algorithms	55
	2.3.1 Classifications	55
	2.3.2 Representation	56
2.4	Design of Parallel and Distributed Algorithms	57
	2.4.1 Task Allocation	58
	2.4.2 Data Communication	60
	2.4.3 Synchronization	62
2.5	Parallel and Distributed Computation Performance	65
	2.5.1 Speedup and Efficiency	65

CONTENTS

	2.5.2 Impacting Factors	66
2.6	Iterative Parallel and Distributed Algorithms	70
	2.6.1 Asynchronous Distributed Computation	72
2.7	Convergence Of Asynchronous Algorithms	73
2.8	Distributed Processing of Vertically Integrated Power Systems	74
	2.8.1 Distributed Processing of Transmission Systems	75
	2.8.2 Distributed Processing of Distribution Systems	76
2.9	Distributed Processing of Restructured Power Systems	79
	2.9.1 Distributed Processing System of an ISO	79
	2.9.2 Distributed Processing System of an RTO	80
	2.9.3 Distributed Processing of GENCOs, TRANSCOs, and DISTCOs	81
2.10	Distributed Energy Trading Systems	82
2.11	Computer Networks for Distributed Processing of Power Systems	83
	2.11.1 LAN	83
	2.11.2 ISO's Private WAN	83
	2.11.3 NERC ISN	84
	2.11.4 The Public Internet	84
2.12	Message-Passing Interface	87
	2.12.1 Development History and Features	87
	2.12.2 Data Communication	89
	2.12.3 Attributes of MPI	93
2.13	Other Forms and Techniques for Distributed Processing	96
	2.13.1 Client/server Architecture	96
	2.13.2 Network Programming	97
	2.13.3 Distributed Database	98
	2.13.4 Java Language	99
3	**Information System for Control Centers**	**101**
3.1	Introduction	101
3.2	ICCS in Power Systems	102
3.3	ICCS Configuration	103
	3.3.1 ICCS LAN	103
	3.3.2 Availability and Redundancy of ICCS	104

	3.4	Information System for ICCS	106
	3.5	CCAPI for ICCS	107
	3.6	Interfaces for ICCS Users	108
	3.7	ICCS Communication Networks	111
		3.7.1 Private WAN of the ISO	112
		3.7.2 NERC's Interregional Security Network	114
		3.7.3 Public Internet	115
	3.8	ICCS Time Synchronization	116
	3.9	Utility Communications Architecture	117
		3.9.1 Communication Scheme	118
		3.9.2 Fundamental Components	121
		3.9.3 Interoperability	123
	3.10	ICCS Communications Services	125
		3.10.1 Other Communication Services	127
	3.11	ICCS Data Exchange and Processing	127
		3.11.1 Real-Time Data Processing	128
		3.11.2 Transaction Scheduling and Processing	128
		3.11.3 Generation Unit Scheduling and Dispatch	129
	3.12	Electronic Tagging	129
		3.12.1 Tag Agent Service	130
		3.12.2 Tag Authority Service	131
		3.12.3 Tag Approval Service	132
	3.13	Information Storage and Retrieval	132
	3.14	ICCS Security	133
		3.14.1 Unauthorized Access	133
		3.14.2 Inadvertent Destruction of Data	134
4	Common Information Model and Middleware for Integration		135
	4.1	Introduction	135
		4.1.1 CCAPI/CIM	135
		4.1.2 What Is IEC?	136
		4.1.3 What Is CIM?	137
	4.2	CIM Packages	139
	4.3	CIM Classes	142

		4.3.1 Class Attributes 142

        4.3.1 Class Attributes 142
        4.3.2 Class Relationships 142

4.4 CIM Specifications 146

4.5 CIM Applications 147
        4.5.1 Example System 149

4.6 Illustration of CIM Applications 163
        4.6.1 Load Flow Computation 163
        4.6.2 Other Applications 164

4.7 Applications Integration 164
        4.7.1 Previous Schemes 164
        4.7.2 Middleware 165

4.8 Middleware Techniques 166
        4.8.1 Central Database 168
        4.8.2 Messaging 169
        4.8.3 Event Channels 169

4.9 CIM for Integration 172
        4.9.1 Integration Based on Wrappers and Messaging 172

4.10 Summary 173

5 Parallel and Distributed Load Flow Computation 177

5.1 Introduction 177
        5.1.1 Parallel and Distributed Load Flows 178

5.2 Mathematical Model of Load Flow 179

5.3 Component Solution Method 179

5.4 Parallel Load Flow Computation 180
        5.4.1 System Partitions 180
        5.4.2 Parallel Algorithm 181

5.5 Distributed Load Flow Computation 183
        5.5.1 System Partitioning 183
        5.5.2 Distributed Load Flow Algorithm 184
        5.5.3 Synchronous Algorithm of Load Flow 187
        5.5.4 Asynchronous Algorithm of Load Flow 188
        5.5.5 Boundary Power Compensation 191

5.6 Convergence Analysis 192

　　　　5.6.1　Convergence of Partially Asynchronous Distributed
　　　　　　　Algorithm 192
　　　　5.6.2　Convergence of Totally Asynchronous Distributed
　　　　　　　Algorithm 197
　　5.7　Case Studies 202
　　　　5.7.1　System Partition 202
　　　　5.7.2　Simulation Results 204
　　5.8　Conclusions 206

6　Parallel and Distributed Load Flow of Distribution
　Systems 209
　　6.1　Introduction 209
　　6.2　Mathematical Models of Load Flow 211
　　　　6.2.1　Model for One Feeder 212
　　　　6.2.2　Load flow Model for One Feeder with Multiple Laterals　214
　　6.3　Parallel Load Flow Computation 216
　　6.4　Distributed Computation of Load Flow 217
　　　　6.4.1　System Division 217
　　　　6.4.2　Synchronous Distributed Algorithm 217
　　　　6.4.3　Asynchronous Distributed Computation 218
　　6.5　Convergence Analysis 219
　　6.6　Distribution Networks with Coupling Loops 219
　　6.7　Load Flow Model with Distributed Generation 220
　　6.8　Joint Load Flow Computation of Transmission and
　　　　Distribution Systems 224
　　　　6.8.1　Problem Description 224
　　　　6.8.2　Joint Computation Based on Distributed Processing . . 225
　　　　6.8.3　Joint Computation Based on Separate Computations . . 225
　　6.9　Case Studies 228
　　　　6.9.1　The Test System 228
　　　　6.9.2　Simulation Results 232

7　Parallel and Distributed State Estimation 235
　　7.1　Introduction 235
　　7.2　Overview of State Estimation Methods 239

CONTENTS

		7.2.1	Applications of Parallel and Distributed Processing	244
	7.3	Components of State Estimation		245
	7.4	Mathematical Model for State Estimation		248
	7.5	Parallel State Estimation		249
		7.5.1	Data Partition	249
	7.6	Distributed State Estimation		253
		7.6.1	Distributed Topology Analysis	254
		7.6.2	Distributed Observability Analysis	256
		7.6.3	Computation of Distributed State Estimation	258
	7.7	Distributed Bad Data Detection and Identification		260
	7.8	Convergence Analysis of Parallel and Distributed State Estimation		260
	7.9	Case Studies		262
		7.9.1	Test System	262
		7.9.2	Synchronous Distributed State Estimation	263
		7.9.3	Partially Asynchronous Distributed State Estimation	263
		7.9.4	Totally Synchronous Distributed State Estimation	263
8	Distributed Power System Security Analysis			265
	8.1	Introduction		265
		8.1.1	Procedures for Power System Security Analysis	266
		8.1.2	Distributed Power System Security	267
		8.1.3	Role of the External Equivalent	269
	8.2	External Equivalence for Static Security		270
	8.3	Parallel and Distributed External Equivalent System		272
		8.3.1	Parallel External Equivalent	272
		8.3.2	Online Distributed External Equivalent	275
		8.3.3	External Equivalent of Interconnected External Systems	276
	8.4	Extension of Contingency Area		277
		8.4.1	Distribution Factors	278
		8.4.2	Extension of Subarea	283
		8.4.3	Case Study	286
		8.4.4	Contingency Ranking Using TDF	292
	8.5	Distributed Contingency Selection		293
		8.5.1	Distributed Computation of PI Based on Distributed Load Flow	295

		8.5.2	Distributed Computation of PI Based on Distributed External Equivalent 296
		8.5.3	Comparison of the Two Methods for PI Computation . . 297
	8.6	Distributed Static Security Analysis 298	
		8.6.1	Security Analysis Based on Distributed Load Flow . . . 298
		8.6.2	Security Analysis Based on Distributed External Equivalent 299
		8.6.3	Enhanced Online Distributed Static Security Analysis and Control 300
	8.7	Distributed Dynamic Security Analysis 302	
		8.7.1	External Equivalent 302
		8.7.2	Contingency Selection 303
		8.7.3	Contingency Evaluation 303
	8.8	Distributed Computation of Security-Constrained OPF . . 303	
		8.8.1	SCOPF Techniques 304
	8.9	Summary 305	

9 Hierarchical and Distributed Control of Voltage/VAR . . . 307

	9.1	Introduction 307	
	9.2	Hierarchies for Voltage/VAR Control 308	
		9.2.1	Two-Level Model 308
		9.2.2	Three-Level Model 309
		9.2.3	Hierarchical Control Issues 310
	9.3	System Partitioning 312	
		9.3.1	Partitioning Problem Description 313
		9.3.2	Electrical Distance 313
		9.3.3	Algorithm for System Partitioning 314
		9.3.4	Determination of the Number of Control Areas 315
		9.3.5	Choice of Pilot Bus 316
		9.3.6	Corrections Based on Expert Knowledge 317
		9.3.7	Algorithm Design 317
		9.3.8	Case Studies 318
	9.4	Decentralized Closed-Loop Primary Control 325	
	9.5	Distributed Secondary Voltage/VAR Control 326	
		9.5.1	Problem Description 326
		9.5.2	Distributed Control Model 327
		9.5.3	Closed-Loop Secondary Voltage/VAR Control 329

CONTENTS

        9.5.4   Case Studies 330
- 9.6 Distributed Secondary Voltage/VAR Control Based on Reactive Power Bids 332
  - 9.6.1 Introduction 332
  - 9.6.2 Optimization Based on Reactive Power Bidding 332
  - 9.6.3 Optimization Using Sensitivities 333
  - 9.6.4 Optimization and Control Considering Area Interactions 334
- 9.7 Centralized Tertiary Voltage/VAR Optimization 334
  - 9.7.1 Optimization Approaches 334
  - 9.7.2 Linear Optimization Approaches 336
  - 9.7.3 Mathematical Models 337
  - 9.7.4 Dantzig-Wolfe Decomposition Method 339
  - 9.7.5 Case Study 343
  - 9.7.6 Parallel Implementation of Dantzig-Wolfe Decomposition 345
- 9.8 Distributed Tertiary Voltage/VAR Optimization 346
  - 9.8.1 Negligible Interactions among Neighboring Control Areas 346
  - 9.8.2 Interactions among Neighboring Control Areas 347

10 Transmission Congestion Management Based on Multi-Agent Theory 349

- 10.1 Introduction 349
- 10.2 Agent-Based Modeling 352
- 10.3 Power System Modeling Based on Multi-Agents 354
- 10.4 Multi-Agent Based Congestion Management 357
  - 10.4.1 Congestion Management 357
  - 10.4.2 Application of Agents 360
  - 10.4.3 Agent Models of Market Participants 361
- 10.5 Multi-Agent Scheme for Congestion Mitigation 363
- 10.6 Application of PDF to Congestion Management 366
- 10.7 Objectives of Market Participants 369
  - 10.7.1 Objective of a GENCO 369
  - 10.7.2 Objective of a DISTCO Agent 371
  - 10.7.3 Objective of a TRANSCO Agent 372
  - 10.7.4 Objective of the ISO Agent 373

10.8	Decision-Making Process of Agents	374
	10.8.1 TRANSCO Agent	374
	10.8.2 GENCO and DISTCO Agents	376
10.9	First-Stage Adjustments	377
	10.9.1 Rescheduling of GENCOs	378
	10.9.2 Adjustment Process of Red Agents	380
10.10	Second Stage Adjustments	383
10.11	Case Studies	385
	10.11.1 Congestion Information	385
	10.11.2 The First-Stage Adjustment of Red Agents	386
	10.11.3 The Second-Stage Adjustment of Green Agents	389
10.12	Conclusions	389

11 Integration, Control, and Operation of Distributed Generation . . . 391

11.1	Introduction	391
11.2	DG Technologies	395
	11.2.1 Wind Turbines	395
	11.2.2 Photovoltaic Systems	396
	11.2.3 Fuel Cell	396
	11.2.4 Combustion Turbines	398
	11.2.5 Microturbines	398
	11.2.6 Internal Combustion Engines	400
	11.2.7 Comparison of DG Technologies	400
11.3	Benefits of DG Technologies	401
	11.3.1 Consumers	401
	11.3.2 Electric Utilities and ESPs	402
	11.3.3 Transmission System Operation	403
	11.3.4 Impact on Power Market	403
	11.3.5 Environmental Benefits	404
11.4	Barriers to DG Utilization	404
	11.4.1 Technical Barriers	404
	11.4.2 Barriers on Business Practices	406
	11.4.3 Regulatory Policies	406
11.5	DG Integration to Power Grid	408
	11.5.1 Integration	410
	11.5.2 Management of DG Units	411

CONTENTS

   11.5.3 Concerns for Utilizing DG in Distribution Systems . . 412

 11.6 Operation of Distribution System with DG 413
   11.6.1 Transition to a More Active System 414
   11.6.2 Enhanced SCADA/DMS 415
   11.6.3 Role of UDCs and ESPs 417
   11.6.4 Distributed Monitoring and Control 418

 11.7 Load Flow Computation 419

 11.8 State Estimation 422

 11.9 Frequency Stability Analysis 422
   11.9.1 System Models 423
   11.9.2 Stability Analysis 425

 11.10 Distributed Voltage Support 426

 11.11 DG in Power Market Competition 429
   11.11.1 Retail Wheeling 429
   11.11.2 Ancillary Services 431
   11.11.3 Role of Aggregators 433

 11.12 Congestion Elimination 436

12 Special Topics in Power System Information System . . . 439

 12.1 E-Commerce of Electricity 439

 12.2 Advantages of E-Commerce for Electricity 442

 12.3 Power Trading System 442
   12.3.1 Classifications 443
   12.3.2 Configuration Requirements 444
   12.3.3 Intelligent Power Trading System 446

 12.4 Transaction Security 447

 12.5 Power Auction Markets 448
   12.5.1 Day-ahead Market 448
   12.5.2 Hour-ahead Market 450
   12.5.3 Next Hour Market 451
   12.5.4 Real-time Operation 452

 12.6 Relationship between Power Markets 454
   12.6.1 Bidding Strategy 455

 12.7 Geographic Information System (GIS) 456
   12.7.1 GIS Architecture 457

        12.7.2 AM/FM/GIS ... 458
        12.7.3 Integrated SCADA/GIS ... 459
        12.7.4 GIS for Online Security Assessment ... 461
        12.7.5 GIS for Planning and Online Equipment Monitoring ... 462
        12.7.6 GIS for Distributed Processing ... 463
        12.7.7 GIS for Congestion Management ... 464
    12.8 Global Positioning System (GPS) ... 465
        12.8.1 Differential GPS ... 467
        12.8.2 Assisted GPS ... 468
        12.8.3 GPS for Phasor Measurement Synchronization ... 468
        12.8.4 GPS for Phasor Measurement ... 469
        12.8.5 GPS for Transmission Fault Analysis ... 470
        12.8.6 GPS for Transmission Capability Calculation ... 471
        12.8.7 GPS for Synchronizing Multiagent System ... 472
        12.8.8 GPS Synchronization for Distributed Computation and Control ... 474

APPENDIX

A   Example System Data ... 477
    A.1 Partitioning of the IEEE 118-Bus System ... 477
    A.2 Parameters of the IEEE 118-Bus System ... 478
    A.3 Bus Load and Injection Data of the IEEE 118-Bus System ... 480

B   Measurement Data for Distributed State Estimation ... 483
    B.1 Measurements of Subarea 1 ... 483
    B.2 Measurements of Subarea 2 ... 486
    B.3 Measurements of Subarea 3 ... 489
    B.4 Measurements on Tie Lines ... 492

C   IEEE-30 Bus System Data ... 493
    C.1 Bus Load and Injection Data of the IEEE 30-Bus System ... 493
    C.2 Reactive Power Limits of the IEEE 30-Bus System ... 494
    C.3 Line Parameters of the IEEE 30-Bus System ... 495

D Acronyms 497

Bibliography 503

Index 531

Preface

This book is intended to present the applications of parallel and distributed processing techniques to power systems control and operation. The book is written with several audiences in mind. First, it is organized as a tutorial for power engineering faculty and graduate students who share an interest in computer communication systems and control center design and operation. The book may also be used as a text for a graduate course on distributed power systems control and operation. Finally, it may serve as a reference book for engineers, consultants, planners, analysts, and others interested in the power systems communication and control.

The majority of the topics in this book are related to the restructuring of electricity which has necessitated a mechanism for various participants in an energy market to communicate efficiently and coordinate their technical and financial tasks in an optimal fashion. While the competition and control tasks in a vertically integrated electric utility company were loosely coordinated, such coordination is essential for the profit-oriented operation of today's energy markets. The competition in the restructured power industry and the volatility of energy market persuade electric power companies to become more vigilant in communicating their propriety data. However, the communication and control tasks cannot be accomplished without establishing a flexible information technology (IT) infrastructure in the restructured power industry. What is needed for managing a competitive electricity market is a framework that enables the pertinent data to be quickly communicated and transformed into usable information among concerned parties. The framework presented in this book is robust enough so that, as new situations arise, the up-to-the-second information can be exchanged and analyzed in ways not previously anticipated.

The chapters in this book are written in response to the migration of control centers to the state-of-the-art EMS architecture in restructured electric power systems. The migration is intended to integrate real-time, process-oriented applications with business-oriented applications. The new architecture will address the necessary attributes with respect to scalability, database access and user interface openness, flexibility, and conformance to industry standards. Today, major EMS providers offer a distributed workstation-based solution using the new standards for communications and control, user interfaces, and database management. The new EMS architecture provides a range of benefits including the reduction in the cost of EMS improvements, additional alternatives for software and hardware integration, and unlimited upgrading capabilities. The open EMS is designed according to the CIM-compliant relational database which facilitates the data communication in distributed power systems. The open EMS is further enhanced by the geographic information system (GIS) and the global positioning system (GPS) for the real-time monitoring and control of power systems in a volatile business environment.

The various topics presented in this book demonstrate that the competition among self-interested power companies in the electricity market and the availability of powerful parallel and distributed processing systems have created challenging communication and control problems in power systems. A common property of these problems is that they can be easily decomposed along a spatial dimension and can be solved more efficiently by parallelization, where each processor will be assigned the task of manipulating the variables associated with a small region in space. In such an environment, a communication system is an essential component of the power system real-time monitoring and control infrastructure. Since the control center design for the majority of ISOs and RTOs has a hierarchical and distributed structure, modern computer and telecommunication technologies are able to provide a substantial technical support for the communication and control in restructured power systems. When the communication system is mainly used to transfer metered data and other information such as the load forecast and electricity price and so on, the communication delay within a small range is usually allowed. When distributed processing is applied to the control of power systems, the communication overhead is a major factor that degrades the computation performance

The chapters in this book start with a review of the state-of-the-art technology in EMS applications and extend the topics to analyze the restructured environment in which power market participants handle their communication and control tasks independently while the entire power system remains integrated. The proposed style will help readers realize the migration from the traditional power system to the restructured system. The

PREFACE

tools that are reviewed in this book include distributed and parallel architectures, CIM, multi-agent systems, middleware and integration techniques, e-commerce, GPS, and GIS applications. The distributed problems that are studied in this book include control center technologies, security analyses, load flow, state estimation, external equivalence, voltage/var optimization and control, transmission congestion management, distributed generation, and ancillary services.

We would like to take this opportunity to thank several of the reviewers who read the first draft of this book and provided constructive criticisms. In particular, we appreciate the comments provided by Dr. Mariesa Crow (University of Missouri-Rolla), Dr. Daniel Kirschen (University of Manchester Institute of Science and Technology), Dr. Noel Schulz (University of Mississippi), Dr. Hasan Shanechi (New Mexico Institute of Technology), and Dr. Ebrahim Vaahedi (Perot Systems). Our communications with Dr. James Kavicky (Argonne National Laboratory) and Mr. Thomas Wiedman (Exelon Corporation) paved the way to understand some of the technical aspects of control center operations in an electric power company. The engineers and analysts at the Electric Power and Power Electronics Center (EPPEC) at IIT provided significant technical and editorial support to this project. We acknowledge the contributions of Dr. Zuyi Li, Bo Lu, Pradip Ganesan, Tao Li, Yong Fu, and Yuying Pan.

We offer our utmost gratitude to our respective families for supporting us throughout the completion of this book.

Mohammad Shahidehpour
Yaoyu Wang

Chicago, Illinois
May 2003

Chapter 1

Introduction

The focus of this book is on the design and the implementation of distributed power system optimization and control based on DEMS (distributed energy management systems) [Con02, Hor96, Lan93, Lan94, Kat92, Sch93, She91, Wan94]. Rapid developments in the computer hardware technology and computer communication have provided a solid foundation and necessary conditions for the parallel and distributed computation in various fields of engineering and science. Distributed computation and control is generally based on LAN (local area network), WAN (wide area network) and the Internet. The applications of parallel and distributed processing in vertically integrated electric power systems have already been presented as a viable option for modeling large-scale problems in real time [Alo96, Car98, Fal95], and it is envisioned that the applications will expand even further in restructured electric power systems.

Power industries around the world are experiencing a profound restructuring[1]. The purpose of restructuring is to provide electricity customers with better services at cheaper cost through the introduction of competition in power industries. The restructuring has initiated additional optimization and control tasks in the unbundled power systems, which necessitate parallel and distributed processing for analyzing the economic operation and enhancing the reliability of the system. For instance, each independent participant of a restructured power system has to optimize its functions to attain competitiveness. The global optimization in a

---

[1] It is often referred to as deregulation. However, we prefer to use the term 'restructuring" in place of deregulation as we believe that the current evolution in power system represents a restructuring process.

restructured power system, which can benefit all participants, would require a coordination among independent participants, which is devisable by parallel and distributed processing. Parallel and distributed computation and control techniques are specific to certain applications. Hence, we plan to discuss parallel and distributed computing systems in both vertically integrated and restructured power systems.

Parallel and distributed computing is represented by a cluster of computers at power system control centers. So we first introduce the structure and functions of control centers, such as CCC (central control center) and ACC (area control center), in a vertically integrated power system and discuss their operation strategies; next we discuss the changes in structures and functions of traditional control centers to form ISO (independent system operator) and RTO (regional transmission organizations) in restructured power systems. We discuss the functions of CCCs, ISOs, and RTOs to let readers comprehend the importance of parallel and distributed processing applications in power systems. We analyze the hierarchical and distributed computation and control structure of a vertically integrated power system that is evolved to form its restructured successor. Finally, we introduce the existing parallel and distributed processing theory and technology that can be employed in the power system optimization and control.

This chapter will provide readers with a brief background on structures and functions of power system control centers and a review of fundamentals of parallel and distributed computation and control.

1.1 THE ROLE OF POWER SYSTEMS

The role of electric power systems has grown steadily in both scope and importance with time, and electricity is increasingly recognized as a key to societal and economic progress in many developing countries. In a sense, reliable power systems constitute the foundation of all prospering societies. Since the key elements and operation principles of large-scale power systems were established prior to the emergence of extensive computer and communication networks, wide applications of advanced computer and communication technologies have greatly improved the operation and performance of modern power systems. As societies and economies develop further and faster, we believe that energy shortages and transmission bottleneck phenomena will persist, especially in places where demands grow faster than the available generation. Under these

INTRODUCTION

circumstances, rotating blackouts are usually practiced with certain regularity to avoid catastrophic failures of the entire system.

To better understand the issues that are now facing power systems, let us first have a look at how individual power systems evolved into interconnected power grids and how interconnections have progressively changed their roles. At the beginning, individual power systems were designed to serve as self-sufficient islands. Each power system was planned to match its generation with its load and reserve margins. The system planning criteria were based on the expected load growth, available generation sites, and adequate transmission and reactive power capabilities to provide an uninterrupted power supply to customers in the event of generation and transmission outages. However, it was soon realized that interconnections of power systems would have great advantages over isolated power systems [Sha00].

The primary requirement for interconnections was the sharing of responsibilities among utility companies, which included using compatible control and protection systems, helping neighboring systems in emergencies, and coordinating maintenance outages for the entire interconnection. For instance, in an emergency, a system could draw upon the reserve generation from its neighboring systems. The burden of frequency regulation in individual systems would be greatly reduced by sharing the responsibility among all generators in the interconnection. In addition, if the marginal cost of generation in one system was less than that in some other systems, a transaction interchange could possibly be scheduled between the systems to minimize the total generation cost of the interconnection.

To some extent, power system restructuring has exacerbated the operations of interconnected grid. In a restructured environment, generators could be installed at any place in the system without much restriction; hence, transmission bottlenecks could become more common in restructured power systems. It is envisioned that new criteria for planning, design, simulation, and optimization at all levels of restructured power systems have to be coordinated by appropriate regulating authorities. The restructuring could also result in a slower coordination across the interconnected power network because some aspects of coordination might need to go through several intermediate procedures and be endorsed by different power market participants.

As shown in Figure 1.1, a power system has two major infrastructures: one is the power grid and the other is the communications system. Most coordination activities among control centers can be enhanced by making use of advanced computer tools, although much of the coordination is still based on conventional telephone calls between system operators. An intimate interaction of extensive computer-based communication network with the power grid will certainly facilitate the operation and control of power systems. To realize this, we need to change our view of power systems and better understand power systems as complex interacting systems.

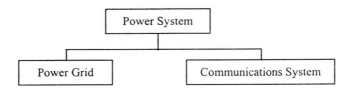

Figure 1.1 Power System Infrastructures

1.2 COMPLEXITY OF POWER GRIDS

The power grid is a complex network involving a range of energy sources including nuclear, fossil fuel, and renewable resources, with many operational levels and layers including control centers, power plants, and transmission, distribution, and corporate networks [Ami01]. Many participants, including system operators, power producers, and consumers, can affect the operational state of the power grid at any time. The interactions of these power grid elements, including various physical components and human beings, also increase the complexity of the power grid. On the other hand, the diversity of the time scale at which the power grid elements operate contributes to this complexity. The time scales for various control and operation tasks can be as short as several microseconds and as long as several years, which greatly complicates the modeling, analysis, simulation, control, and operation of a power grid.

The restructuring of the power industry has had profound effects on the operation of the power grid. The power grid was previously operated by vertically integrated electric utilities for delivering the bulk power reliably and economically from power plants to load areas. As a noncompetitive, regulated monopoly, these utilities put much emphasis on reliability at the expense of economy. Now, with the restructuring, intensive market competition, stressed transmission grid, and increased demand for high-quality and reliability of power supply challenge unbundled electric utilities additionally in satisfying the basic objectives [Sha01].

As various functionalities in the world are being further automated and digitized, higher power quality becomes a new requirement in the power grid complexity. Further improvements in power supply reliability will mean that the expected duration of energy not served will drop from 8 hours to 32 seconds a year.

The following factors could make it difficult to operate and control restructured power systems more reliably and efficiently:

- A large number of power system components are tightly interconnected and distributed in a vast area.

- Power system components are operated in multiple hierarchical layers.

- Power system components in different hierarchical layers have different spatial and temporal requirements for operation. For instance, the time frame of a protective device that is adaptive to lightning should be within several milliseconds, while the black start of a generating unit could span over several hours. The breakdown of a transformer at a user side could only affect a small area of the distribution system, but a short circuit fault on a backbone transmission line could affect a large part of the system.

- A variety of participants, including system operators, power producers, and energy customers, act on the system at different places simultaneously.

- The requirements for time are rigid and the requirements for reliability are extremely high.

- Random disturbances, including natural disasters such as lightning, unusually high power demands, and operational faults, can lead to widespread failure almost instantaneously.

- No single centralized entity can evaluate, monitor, and control all the interactions in real time.

- The relationships and interdependencies among various power components are so complex that conventional mathematical theories and control methods are hard to apply to many issues.

1.3 COMMUNICATIONS SYSTEM

The reliable and economic operation of power systems relies heavily on its efficient communications system, which is growing fast and becoming increasingly complicated with the rapid development of modern communications technologies. The communications system should also be operated with high reliability while its operations rely on secure and high-quality power supplies. Generally, the communications system of a power system is composed of the following three kinds of networks:

- Fixed networks including public switched telephone and data networks
- Wireless networks including cellular phones and wireless ATM
- Computer networks including various dedicated LANs, WANs, and the Internet

The satellite network is another segment of the communications system that can provide important services that are hard to attain from regular communication techniques. These services include detailed earth imaging, remote monitoring of dispersed locations, highly accurate tracking, and time synchronization using the continuous signals of GPS (global positioning system). The Internet is rapidly expanding the range of applications for satellite-based data communications. Some satellite systems allow users to browse Web pages and download data through a roof-mounted dish receiver connected to a personal computer at a speed that is as high as 400 kbps. This capability could become a valuable tool for expanding an enterprise network to remote offices around the world.

Electric utilities have diversified their businesses by investing in telecommunications and creating innovative communications networks that cope with industry trends toward distributed resources, two-way customer communications, business expansion, as well as addressing the measurement of complex and data-intensive energy systems via wide-area monitoring and control [Ada99]. Electric utilities use communications media for different purposes such as real-time monitoring, control and protection. Network services such as real-time monitoring are expanding the use of broad bandwidth communications networks as new remote real-time protection and control techniques become more pervasive.

Although some operations such as isolating or restoring a damaged portion of the power grid could take several hours or days, the high-speed communication is desirable especially for the real-time operation of power

INTRODUCTION

systems. To improve services and reduce operating costs, broad bandwidth communications are used for monitoring and control, which would benefit both the utility and customers. Some typical applications of broad bandwidth communications are as follows:

- Data acquisition from generation, transmission, distribution and customer facilities
- Communication among different sites, substations, control centers, and various utilities
- Real-time information provided by power markets and weather service
- Database information exchange among control centers
- Relay setting adjustments
- LFC and generation control
- Load shedding based on contingency analysis
- Control of devices such as FACTS facilities

With recent advancements in IT techniques such as multi-channel WDM (wavelength division multiplexed) connection and XML (eXtensible markup language), a dedicated fiber optical communication network could be built for power systems based on IP over WDM. Next, we discuss the architectures and protocols used for such a fiber optical communication network

1.3.1 Fiber Optical Technique

The particular characteristics of optical fibers such as low attenuation, high bandwidth, small physical cross section, electromagnetic interface immunity, and security make them most suitable for information communication in power system monitoring and control. The manifestation of fiber optical networks such as WDM is attributed to advancements in a host of key component technologies such as fibers, amplifiers, lasers, filters, and switching devices [Kwo92].

A fiber optical network has two major sets of components: switching components and optical linking components. The switching components include tunable transmitter/receivers; the optical linking components include WDM multiplexier/demultiplexers, WDM passive star coupler, and the like. Multiplexers aggregate multiple wavelengths onto a single fiber, and demultiplexers perform the reverse function. OADMs (optical add/drop multiplexers) are programmable devices configured to add or drop different wavelengths.

OXC (optical cross-connects) is a large photonic switch with N-full duplex ports as each port can connect to any other device. OADM is a 2×2 degenerate form of the $N \times N$ OXC that extracts and reinserts certain light-paths for local use. The OXC cross-connects sometimes are also referred to as wavelength routers or wavelength cross-connects. New amplifier technologies have increased the distances between the signal re-generators. Two basic optical amplifiers include SOA (semiconductor optical amplifier), which can be integrated with other silicon components for improved packaging, and EDFA (erbium doped fiber amplifier), which can achieve high gains. Optical packet switches are nodes with an optical buffering capability that perform packet header processing functions of packet switches [Gow95, Vee01].

1.3.2 Fiber Optical Networks

Currently most WDM deployments adopt the point-to-point scheme and use SONET/SDH (synchronous optical network/synchronous digital hierarchy) as the standard layer to interface to higher layers of the protocol stacks. Different protocol stacks provide different communication functionalities. The SONET/SDH layer mainly interfaces with electrical and optical layers, delivers highly reliable ring-based topologies, performs mapping of TDM (time-division multiplexing) time slots from digital hierarchical levels, and defines strict jitter bounds. The ATM (asynchronous transfer mode) layer mainly performs segmentation and reassembly of data, creates class of service, and sets up connections from source to destination. The WDM layer multiplexes electrical signals onto specific wavelengths in a point-to-point topology, which constructs the backbone of power system communications [Ada98].

The communication networks with multi-layer sometimes have problems with time delays and function overlaps. Increasing efforts have been recently devoted to the development of prototypes for transmitting IP packets in the optical domain. The best choice is the IP over WDM, which has inherent advantages due to the absence of many layers. The IP over WDM has the property of virtual fibers where each wavelength is considered as a dedicated connection. The signals are not converted into an electrical domain for performing control operations. Hence, the latency in the IP/WDM system is smaller compared to that encountered in the

INTRODUCTION

SONET system. Besides, the absence of vendor specific components makes the system services transparent.

The ATM function of traffic engineering is being absorbed into IP by using IP over WDM, and transport capabilities of SONET/SDH are being absorbed by the optical layer. Therefore the four-layer architecture is converted into a two-layer architecture. The MPLS (multiprotocol label switching) or in the case of the optical layer, a slightly modified version, MPλS (multiprotocol lambda switching) are chosen to control both layers [Awd01]. However, the IP over WDM becomes a reality only when all end-to-end services are offered optically.

With the application of fiber optics, the trunk capacity of traditional wire-based telephone networks has been increased significantly. A fiber trunk can carry at least 1000 times the communication traffic of a copper pair, and more than one hundred optical fiber pairs are laid on one single route.

1.3.3 WAN Based on Fiber Optical Networks

WDM allows multiple network protocols to coexist on one network. WDM is utilized and further optimized by wavelength routing because of the increasing operational cost to deploy fiber rings. The overall fiber optical network has a mesh architecture. The nodes of this mesh are electric utilities with the IP protocol for data communications. New technologies such as all-optical cross-connects and all-optical add-drop multiplexers made it possible for simple point-to-point WDM links to evolve to a complicated network. Figure 1.2 depicts the architecture of such a fiber optical network.

Similar to other communication protocols, the communication protocol for fiber optical communication networks has the following three basic layers:

- **Physical layer.** The physical layer is the layer at which signals are exchanged. Transmitting optics are based on laser and bit rates are as high as OC-48 (2.5 Gbps) to even OC-192 (10 Gbps).

- **Data link layer.** The data link layer, underneath the network layer, is responsible for delimiting data fields, acknowledgment of receipt of data, and error control, such as a parity check. In most communication systems, receipt of information that passes the error check is acknowledged to the sending station. In addition, the data link layer may contain a flow control mechanism to prevent problems when two devices with different speeds try to communicate with each other. By not employing a retransmission-request procedure, first-time transmission messages can be ensured to all traffic in the fiber optical network. No link will be made busy by nodes trying to overcome neighbor-to-neighbor communications problems. In combination with the antibody algorithm, this helps guarantee that there will be no network congestion.

- **Application layer.** Power system applications are allocated to the application layer. The wide area fiber optical network is used for information exchange among utilities, substations, and other entities. The system information, such as rate schedules, operating constraints, and available transmission capacity, is shared among different utilities. Data communication is used for various purposes including power system control, protection, monitoring, and scheduling.

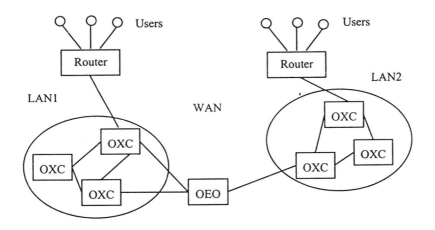

Figure 1.2 Fiber Optical Network

When combined with computer networks in the system, such a high-speed and reliable fiber optical communication network has formed the foundation for distributed processing of power systems [Yos00].

1.3.4 XML Technique

WAN provides a low-cost, easily accessible infrastructure for data communication and exchange. However, the diversity of data format and operating platform makes it difficult to exchange information efficiently.

INTRODUCTION

To further facilitate information exchange especially for distributed processing, XML was created by the World Wide Web consortium [Web15]. All organizations can set personal and corporate agendas to cooperate in the construction of a genuinely open standard, driven entirely by user needs.

XML is a markup language for structured data, and is viewed as a language for data modeling and exchange. A well-formed XML document is a database document with associated DTD (document type definition) as its schema. XML addresses the critical need for a universal interchange format that enables the data exchange especially among heterogeneous platforms [kir94, Cle00, Nam02, Sin01, Web16, Wid99]:

- XML specifies a rigorous, text-based way to represent the structure that is inherent to data so that it can be authored, searched, and interpreted unambiguously.

- XML provides a human-readable format for defining data object names, attributes, and methods. It also provides a means for an application to find additional information about data objects embedded in the DOM and send to the server for information access.

- XML is platform, vendor and language neutral, and ideal to act as the common media among the numerous proprietary standards that currently exist. It can facilitate seamless data sharing and exchange among different utility systems and applications.

Power system operation requires a seamless information exchange among heterogeneous databases with transparency of any internal data formats and system platforms. One feasible way to resolve this problem is to adopt a single data interchange format that serves as the single output format for all data exporting systems and the single input format for all importing systems. Fortunately, this objective is realized by using XML. With XML, proprietary formats and platforms are utilized within each utility and substation, and XML performs necessary translations for information exchange.

XML is utilized for many purposes including transparent metadata exchange, distributed processing, online transactions, and data presentation. Many power system applications involve intensive computations, and may need to retrieve the metadata from a number of distributed databases. Each database at a utility or substation could have several terabytes of data on event recorders. To use this information, a flexible platform and vendor-independent protocol is needed. In using XML, a computation-intensive distributed processing process is changed into a brief interaction with database servers.

An example of the XML application is the OASIS (open access the same time system) for on-line transmission services transactions. Customers buy electricity according to a vendors' price. The transaction data on the client site are represented as XML-tagged data sent to the OASIS server. The OASIS server will then perform the required real-time authentication and send the results to the user. In some applications, the customer can present different views of a group of data to cope with multiple users. Users can have multiple choices of data presentations by using XML. For instance, different kinds of graphic displays of one substation can be easily achieved by using XML style sheets.

Although XML is ideal for man-to-machine interface or machine-to-machine information exchange, security is a problem because it is implemented in a text form. Certain security measures such as firewalls and encryption are to be implemented when using XML. Firewalls are used for the server to minimize the possibility that unauthorized users to access any critical information. Access through the Internet is permitted only when designated security requirements are met. Sophisticated data encryption techniques such as 128-bit encryption algorithm are used to transfer sensitive data across the Internet.

1.3.5 IP-Based Real Time Data Transmission

Currently data communication is mainly based on TCP/UDP/RTP/IP/HTTP (transmission control protocol/user datagram protocol/real-time protocol/internet protocol/HTTP), which is commonly referred to as IP. The IP protocol belongs to the very basic layer in data communication. IP-based protocols such as TCP/UDP are used for real-time data communication [Qiu99]. As was mentioned before, XML is utilized to establish a common standard format for exchanging data among entities. Figure 1.3 shows the XML-based client/server architecture for data exchange.

The XML server acts as the mediator between different databases. The server receives and processes XML-tagged requests from clients, and then converts the processing results into DOM format and sends it to the associated clients. The architecture supports on-line power market transactions, seamless data exchanges among utility databases, and distributed processing.

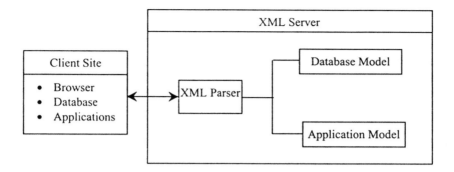

Figure 1.3 XML Client/Server Architecture

    Utilities among other entities usually have different protocols for information exchange. Their DTD/XML schema definitions and XML tags could be different. When they need to retrieve the historical data stored in several utility databases, a distributed processing application will send an XML file to each utility database. After receiving this request, each utility server will convert the XML-tagged file into SQL statements, run the queries, and get the results. The query result will be packed as an XML document. Usually this distributed processing application does not have a specific requirement for the data format of utility databases and can parse data in the XML format.

1.4 INTERDEPENDENCE OF INFRASTRUCTURES

Modern power systems are becoming increasingly complex. A major source of complexity is the interdependence of the power grid and the communications system. These two infrastructures have become progressively interdependent on issues that range from the highest command and control level (i.e., power system information system) to the individual power stations and substations at the middle level, and then to devices and equipment at the lowest level. The Internet/intranet connecting middle-level stations is an Ethernet network with individual gateways. The dedicated communications link is a fiber optic cable or microwave system. The satellite systems are used for a range of utility business applications including direct-to-home interactive services, wide area monitoring and control, and time synchronization.

The restructuring of power industry has further enhanced the interdependence of these two infrastructures, since restructuring requires more information exchange among system operators and other entities in a power system. This interdependence not only challenges the reliable operation and control of the power system, it also presents new venues in modeling, forecasting, analysis, and optimization of both infrastructures. A failure in one infrastructure can often lead to failures or even breakdowns of the other.

The intensive interdependence of these two infrastructures requires a more powerful decision-making system for the operation of a power system. DEMS based on WAN, which is composed of multiple computers at power system control centers, is viewed as a powerful distributed processing system. Further, in some sense, there is a need for self-healing mechanisms of these infrastructures at the local level to mitigate the effect of random disturbances.

1.5 CATASTROPHIC FAILURES

Modern power systems have become more vulnerable at the presence of various outages such as power grid or communications system failures, random disturbances, and human errors [Hey01]. Catastrophic failures in the past merely took into account the customers affected by the incidence and the duration for which their power was interrupted. However, the spectrum of events that are described as catastrophic has become much wider as modern technological industries are getting extremely sensitive to voltage variations and disruptions.

Power supply and demand should be kept at balance in real time for the secure operation of power grids. However, as the supply becomes limited, the system reserve margin becomes lower. Factors contributing to catastrophic failures include the stressed state of power systems when facilities are operated closer to their operational limits, generation reserves are minimal, and reactive power supply is insufficient. In a stressed state, a minor fault can become a triggering event. There are hidden failures in the protection system that can surface only when some triggering events occur, and can lead to false tripping of other grid facilities. Electromechanical instability or voltage instability may develop because of these events, which can lead to the separation of a grid into islands of excess load or generation and the eventual collapse of a load-rich region into a blackout.

INTRODUCTION

On the other hand, if it lacks sufficient real-time measurements and analytical tools, the power system operator can make incorrect assessments of the system operation state. For example, ATC (available transfer capability) is an index used to measure the operating state of transmission grids. Inaccurate computation of ATC can endanger the power grid operation especially as catastrophic failures become imminent.

In the past a number of cascading failures in power systems have caused huge economic losses to electric utilities, so system operators now pay more attention to the prevention of cascading failures. According to NERC, outages affected nearly seven million customers per decade within the last two decades of the twentieth century, and most of these outages were exacerbated by cascading effects. The well-publicized blackout of 1996 in the western grid of the United States that led to islanding and blackouts in eleven U.S. states and two Canadian provinces was estimated to cost $1.5 to $2 billion. The analyses that followed showed that the cascading blackout could have been prevented if 0.4% of the total load had been shed for about 30 minutes. This argument further reinforces the importance of distributed monitoring and control in efficiently processing local disturbances within a very tight time limit.

To understand the origin and the nature of catastrophic failures and to develop defense strategies and technologies that significantly reduce the vulnerability of the power system infrastructure, SPID (Strategic Power Infrastructure Defense) was launched by EPRI and DoD as a research initiative. The most important concept of the SPID project was to provide self-healing and adaptive reconfiguration capabilities for power systems based on the contingency assessment. To achieve the objectives of SPID, power systems would need to set up a broad bandwidth communications infrastructure to support their operational requirements. The communications system for the self-healing purpose could include the substation automation network, wide area information network, and advanced IT technologies such as WDM and XML [Mak96]. An all-fiber network for the communications infrastructure is built based on IP over WDM. To achieve high reliability, innovative control and protection measures are provided for defense against various disturbances, and most of these measures require wide area distributed monitoring and control for power systems.

1.6 NECESSITY FOR DISTRIBUTED PROCESSING

As we explained earlier, complex power system infrastructures are composed of highly interactive, nonlinear, dynamic entities that spread across vast areas. In any situation, including disturbances caused by natural disasters, purposeful attack, or unusually high demands, centralized control requires multiple, high-speed, two-way communication links, powerful central computing facilities, and an elaborate operations control center to give rapid response. However, a centralized control might not be practical under certain circumstances because a failure in a remote part of the power system can spread unpredictably and instantaneously if the remedial response is delayed. The lack of response from the control system can cripple the entire power system including the centralized control system, so the centralized control can very likely suffer from the very problem it is supposed to fix. For instance, in clearing a short-circuit fault, the centralized control may require the pertinent information transmitted by the faulted transmission line.

In this situation a proper question is: What effective approaches can be employed to monitor and control a complex power system composed of dynamic, interactive, and nonlinear entities with unscheduled discontinuities? A pertinent issue is managing and robustly operating power systems with hierarchical and distributed layers. An effective approach in such cases is to have some way of intervening in the faulted system locally at places where disturbances have originated in order to stop the problems from propagating through the network. This approach is the essence of distributed processing. Distributed processing can greatly enhance the reliability and improve the flexibility and efficiency of power system monitoring and control. The attributes of distributed processing in power systems are discussed next.

1.6.1 Power System Control

The operation of modern power systems could be further optimized with innovative monitoring, protection, and control concepts. Generally, a power system operator tries to achieve the highest level of reliability with limited available control facilities. To realize this objective, effective tools should be developed and include the assessment of the system vulnerability, mechanisms for the prevention of catastrophic failures, and distributed control schemes for large-scale power systems. Synchronized phasor measurements are very useful for monitoring the real-time state of power systems. The wide area measurements together with advanced analytical tools using multi-agent theory, adaptive protection, and adaptive self-healing strategies could lead to major improvements in monitoring, contingency assessment, protection, and control of power systems [Hey01].

With the aid of GPS (global positioning system), IEDs (intelligent electronic devices) are widely spread in a power system and synchronized to capture the system data; such synchronization gives an instantaneous snapshot of the real-time state of a power system. Suppose that the system is observable for control purposes, and that the system operator must find a means of optimal control for the system. If a potentially unstable power swing is recognized before its occurrence, corrective control actions must be taken within a very short time period to bring the system back to a stable state. The WAN-based communications network provides the foundation for implementing these functionalities.

1.6.2 Distributed Implementation

Substation automation is the foundation of the power system's distributed processing and control. Let us start with the distributed communications network for substation automation. Advanced computer and network techniques facilitate the substation automation. A substation automation system requires a large amount of equipment supported by an efficient communications system. Specifically, substation automation includes the following aspects:

- Automation of data (i.e., analog and digital data) acquisition from various elements. Consolidated metering, alarm, and status information can greatly facilitate local operations.

- Automation of substation monitoring and control. The control hierarchy exists among control centers, local substations and IEDs.

- High availability and redundancy. Substation automation must meet the "no single point of failure" requirement.

Substation automation demands high-speed, real-time communication links between field and station units. Current and voltage signals sampled by data acquisition units at the field are transmitted to the station within a few milliseconds for the online protection and control. Data communication via fiber optical media has robustness against electrical magnetic interference.

As shown in Figure 1.4, the substation communications network is divided into three levels.

- Level 1 involves IEDs connected to equipment. Protection IEDs report equipment status and execute protection algorithms. Control IEDs function as gateway between substation servers and protection IEDs.

- Level 2 consists of a substation server. All monitoring and control operations are performed through this server. Besides, the substation server also communicates with the control center for the information exchange.

- Level 3 is the utility control server. This utility control server monitors and controls the whole substation. Different LAN systems are adopted for substation automation.

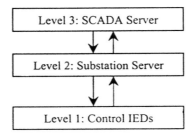

Figure 1.4 Substation Communication Network

When implemented in a distributed manner, the operations of IEDs, communications network, and host computers are separated so that a failure in the host computers would have no effect on the operation of other devices. All data available on the communication network are shared by all IEDS. A legacy SCADA can still function by using gateways that act as protocol translators. Some legacy IEDs may still be useful in such an integrated environment.

Some protection schemes use hard wires and wired logics to implement their protection schemes; the current peer-to-peer communications allow information transfers using virtual inputs and outputs. Any devices can define a virtual interface linked to an object in another IED. The linkage would be specified by an IED address, the object's name, the object's type, and security. The requesting device gets access to the desired object either by a request, as states change, or periodically. Since plant control requires a high degree of reliability, provision is made to implement redundant communications from IEDs and provide support for a redundant LAN.

1.6.3 State Monitoring Based on GPS

In recent years, synchronized phasor measurements based on GPS have received wide applications in power systems. Phasor measurement units can compute positive sequence voltages as frequently as once per cycle of the power frequency. These measurements, when collected to a central location, can provide a coherent picture of the system's state and this is the starting point for most system monitoring, contingency assessment, protection, and control processes.

Power grids critically depend on information and sensing technologies for system reliability, operation, protection, and many other applications. In a restructured power system, this dependence has increased with the appearance of new information infrastructures, including OASIS, wide area measurement systems, and low earth orbit-based satellite communications systems. With regard to its functions, this infrastructure is a double-edged sword. On the one hand, it provides critical support for system monitoring, measurement, control, operation, and reliability. On the other hand, it also provides a window of accessibility into power grids for external agents with malicious intents. These external agents could purposely sabotage the grids to cause catastrophic failures.

1.7 VERTICALLY INTEGRATED POWER SYSTEMS

1.7.1 Central Control Center

The CCC (central control center)[2] is like the eye and the brain of a vertically integrated power system for online monitoring and control of the system operation and reliability. To implement the functions of CCC, all online measured data, including voltages and active and reactive load flow, are required to be transmitted to CCC in a very tiny time limit. Because large amounts of data are to be processed in real time, such a mechanism for data processing needs a high-performance computing machine at CCC, except for a reliable communications system. With more and more functionalities deployed to support the operation of power systems at CCC, the computational burden of CCC becomes heavier, and many CCCs have

---

[2] Also called CDC (central dispatching center)

to expand their hardware and software facilities to meet online requirements.

1.7.2 Area Control Center

The ACC (area control center) is charged with monitoring and control of the regional power system. The function of ACC is quite similar to that of a CCC. The difference between these two is that a CCC monitors and controls the entire power system whereas an ACC is only responsible for the monitoring and control of a regional system. With the aid of a number of ACCs, the computational burden of CCC can be greatly reduced. In addition, this structure can provide a large saving on capital investment for the construction of expensive long-distance communication links. Actually, in most power systems, ACCs appeared prior to CCCs because most CCCs of large-scale power systems are the results of a merger of several regional power systems that were previously monitored and controlled by ACCs. Most ACCs became the satellite control centers of CCCs after CCCs were formed. Currently the definition of a control area and the functions of an ACC are being reexamined by the NERC Control Area Criteria Task Force.

1.7.3 SCADA/EMS

Both CCC and ACCs perform their monitoring and control through SCADA/EMS (supervisory control and data acquisition/energy management system) that are installed at these centers. SCADA is mainly responsible for remote measurements and control; EMS represents a set of senior online applications software for power system optimization and control.

Rapid developments in computer networks and communications technologies have profoundly affected the power system operation in the last two decades. With the aid of high performance computers, senior EMS applications – such as online state estimation (SE), load flow (LF), optimal power flow (OPF), load forecast (LF), economical dispatch (ED), dynamic security assessment (DSA), and restoration strategy (RS) – have become practicably applicable, and the utilization of these applications has greatly improved the performance of power system operation and finally brought along huge economic benefits to electric utilities.

INTRODUCTION

Figure 1.5 depicts specific functions of the SCADA/EMS system. SCADA collects measurement data of power system via dispersed RTUs (remote terminal units).

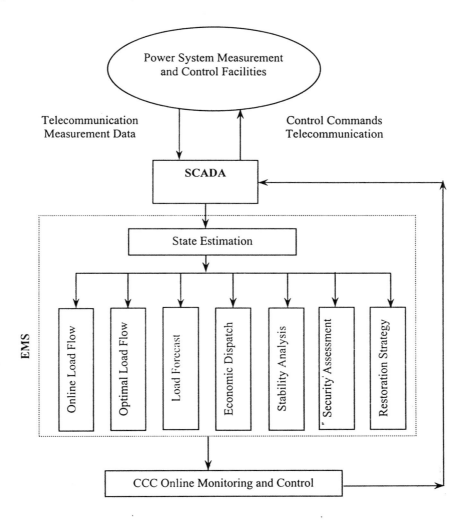

Figure 1.5 Function Structures of SCADA/EMS

There are two kinds of measurement data: one is the digital data that describe the status of breakers and switches, and the other is the analog data that represent power injections and the voltage level at a certain bus. These measurements data are "raw" because they are not creditable enough to be applied directly, meaning there could be errors in them caused by modulation, demodulation of communication signals, or by some random disturbances. State estimation is the software that can turn these "raw" data into "credible" data through error correction. According to Figure 1.5, state estimation is the basis of other senior applications of EMS. We will study the state estimation problem in detail in Chapter 7.

The traditional EMS exclusively had a centralized structure, but since 1980s, with advancements in the automation of distribution, the trend has been toward extending the computer control to the customer level and to provide other services such as energy price and load management for customers. In order to realize these control functions, a complicated and expensive communications system had to be constructed. CCC had to be equipped with a high-performance computing machine because the data transmitted to CCC are processed in real time. Even so, some EMS applications such as state estimation would have difficulties in meeting rigid time requirements for online power systems computation and control.

1.7.4 Distributed Web-Based SCADA Systems

The SCADA system employs a wide range of computer and communication techniques. The SCADA system collects real-time operation data that are geographically distributed in power systems. Most commands for controlling the system are also issued through the SCADA system. Besides its organizational hierarchy, the SCADA system is mostly characterized by its geographical spread. The system operator would need to exchange information with other entities, while some of the exchanges would mostly require data security but are not quite time critical. Because www browsers integrate various communication services into one user interface, www provides a convenient, low cost, and effective way for information exchange, including access to SCADA information. This scheme can be implemented on the client/server architecture.

1.7.5 Distributed Energy Management Systems

Since the rapid advancement of computer and communication technologies in the late 1980s, there has been a trend to optimize and control power systems in a distributed manner. The strict requirements for modern power system operations are promoting the EMS design toward a DEMS (distributed energy management systems) architecture. DEMS is a large and complex computation and control system based on WAN that is composed of computers at CCC and ACCs. In DEMS, CCC and ACCs are geographically distributed in a wide area that can be hundreds and even thousands of miles long and wide. The computers at CCC and ACCs could be heterogeneous and connected through various communication media.

INTRODUCTION

The hierarchical structure of DEMS is depicted in Figure 1.6. All ACCs are under the control of CCC. An ACC will communicate with CCC and its neighboring ACCs when necessary. In this regard, the communications system is most important to DEMS. The characteristics of data communication between CCC and ACCs will be discussed in Chapter 2.

Each EMS component of DEMS functions independently to monitor and control its own local system, and in some cases, they may be required by CCC to work together to solve a common task, for instance online load flow or state estimation computation for the entire system.

The optimization and control functions of an ACC will be fewer and less complicated than those of CCC because the size of a control area is usually much smaller than the entire system. For instance, most ACCs execute a simple state estimation and load flow computation but are not responsible for the stability analysis. Furthermore, an ACC could probably have some SACCs (subarea control centers).

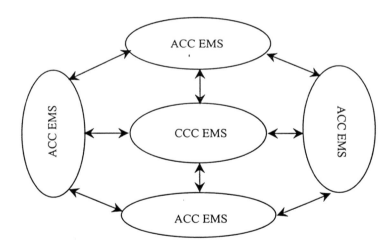

Figure 1.6 DEMS Structures

1.8 RESTRUCTURED POWER SYSTEMS

1.8.1 GENCOs, TRANSCOs, and DISTCOs

Since 1990s, power industries around the world have been experiencing a period of restructuring. Traditional vertically integrated utilities have been required by law to dissect themselves into three independent entities:

- Power producers (GENCOS), which are responsible for the power supply in a restructured power system.

- Transmission providers (TRANSCOs), which are responsible for transmitting and wheeling the bulk energy across power grids of a restructured power system.

- Bulk power consumers (DISTCOs), which buy power in bulk from a power marketplace and distribute power to customers.

GENCOS, TRANSCOs, and DISTCOs can get useful information from OASIS to help them make decisions or even adjust market behavior according to spot prices. A special form of power trading in restructured power systems is that transactions are made through schedule coordinators (SCs) outside the power marketplace. This scenario happens when GENCOs and DISTCOs sign bilateral contracts for energy transactions. The activities of individual participants of restructured power systems would require the approval of the regional ISO before their activities can be practically realized. The relationships of market participants of a restructured power system are shown in Figure 1.7.

1.8.2 ISO

The centralized optimization and control of traditional CCCs cannot meet many new functions and requirements arising with the restructuring of power systems. Hence, as required by FERC Order 889, CCCs are replaced by an ISO in restructured power systems. An ISO is a nonprofit organization with a primary responsibility for maintaining the reliability of grid operation through the coordination of participants' related activities. An ISO is established on a regional basis according to specific operating and structural characteristics of regional transmission grids.

There could be some differences among the responsibilities of ISOs. However, their common goal is to establish a competitive and nondiscriminatory access mechanism for all participants. In some systems, the ISO is also responsible to ensure efficient power trading. An ISO has the control over the operation of transmission facilities within its region and has the responsibility to coordinate its functions with neighboring control areas. Any ISO should comply with all the standards set by NERC and the regional reliability council.

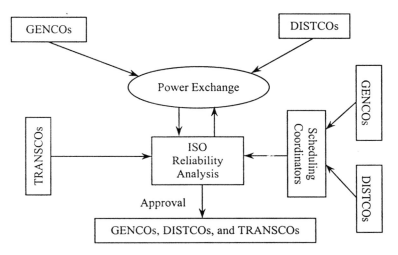

Figure 1.7 Energy Transaction in Restructured Power System

A critical task for an ISO is the transmission congestion management including the collection and distribution of congestion revenues. When transmission congestion occurs, the ISO calls for adjustments of generation and demand to maintain the system reliability. During the process of mitigating transmission constraints, an ISO is used to ensure that proper economic signals are sent to all participants and to encourage an efficient use of resources capable of alleviating congestion. Under emergency circumstances, an ISO is authorized to dispatch any available system resources or components.

In some restructured power systems, like the California power system, the ISO does not have a responsibility to execute optimal power flow (OPF), economic dispatch (ED), and so on. However, in other restructured power systems, like the PJM power system or the New England power pool, the ISO will perform some of these tasks. Compared with a CCC, the ISO has many new functions, including the congestion management and the ancillary services auction, in order to meet restructuring requirements. The typical EMS monitoring and control functions of an ISO in a restructured power system are depicted in Figure 1.8.

The ISO is not responsible for generation dispatch, but for matching electricity supply with demand. An ISO should control the generation to the extent necessary to enhance reliability, optimize transmission efficiency, and maintain power system stability.

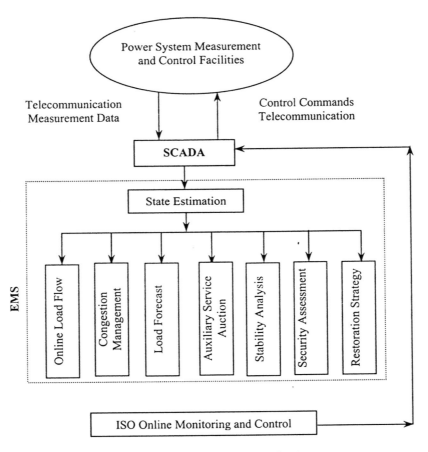

Figure 1.8 EMS Functions of ISO

To maintain the system integrity, an ISO is responsible for acquiring necessary resources to remove transmission violations, balance the system in a second-to-second manner and maintain the system frequency at an acceptable level.

INTRODUCTION

To comply with the FERC Order 889, an ISO is responsible for maintaining an electronic information network, that is, OASIS, through which the current information on transmission systems is made available to all transmission system providers and customers. Some capacity services should be contracted by the ISO with market participants in case they are not provided on OASIS. These services include operating reserves, reactive power and frequency response capability, and capacity to cover losses and balancing energy. To make these services available, the ISO contracts with service providers so that these services are available under the ISO's request. Usually, the ISO chooses successful providers based on a least-cost bid.

An ISO will face the same problem as a traditional CCC for monitoring and control of a large-scale power system. Inspired by the DEMS architecture in vertically integrated power systems, the ISO's optimization and control could also be implemented in a distributed manner. In the New England Power Pool, the ISO has SCCs (satellite control centers) for monitoring and controlling local areas of the interconnection. Figure 1.9 shows the architecture for distributed optimization and control of an ISO, which is quite similar to the DEMS architecture depicted in Figure 1.6.

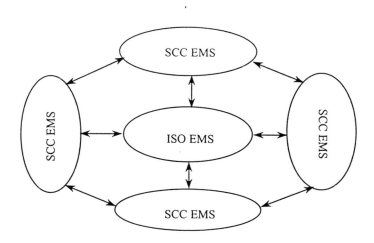

Figure 1.9 Architecture of Distributed ISO

1.8.3 OASIS

In its Order 889, FERC mandated the establishment of OASIS. The ISO is responsible to post certain metering data and market clearing information such as ATC (Available Transmission Capacity) or TTC (Total Transmission Capacity) on the OASIS. Each TRANSCO is also required to establish an OASIS node alone or with other TRANSCOS, and update its information about ATC for further commercial activities on its OASIS node. As shown in Figure 1.10, all customers can access the OASIS, using Web browser to reserve transmission capacity, purchase ancillary services, re-sell transmission services to others, and buy ancillary services from third party suppliers. Services offered by the OASIS could differ from one market to another.

To a certain extent, the success of the restructuring of power industries is attributed to the rapid development and wide application of communication and computer network techniques. The Internet has made it possible for participants to access necessary data almost instantly from OASIS. Besides, energy trading relies heavily on the Internet; for instance, GENCOs and DISTCOs submit bidding prices to the PX (power exchange) through e-mails. Likewise, they can reserve transmission rights from TRANSCOs. All these behaviors assume that communications among participants are fast and reliable.

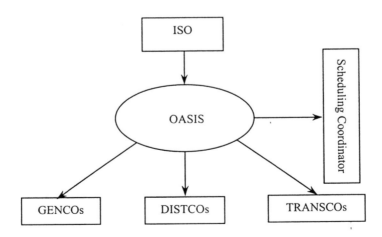

Figure 1.10 OASIS

INTRODUCTION

1.8.4 RTO

According to the FERC Order 2000, published in November 1999, energy transmission organizations of power systems are required to participate in RTOs. A region is a collection of zones, and there can be more than one ISO in one region. The purpose of establishing RTOs is to get rid of discriminatory transmission practices and eliminate pancaked transmission prices. The formation of an RTO should make energy transmission more fluid in the region.

According to the FERC Order 2000, an RTO has the following four minimum characteristics:

- Independence: All RTOs must be independent of market participants.

- Scope and Regional Configuration: RTOs must identify a region of appropriate scope and configuration.

- Operational Authority: An RTO must have an operational authority for all transmission facilities under its control and be the NERC security coordinator for its region.

- Short-Term Reliability: RTOs must have exclusive authority for maintaining the short-term reliability of its corresponding grid.

An RTO is also required by FERC to have the following eight fundamental functions.

- Tariff Administration and Design: An RTO is the sole provider of transmission services and sole administrator of its own open access tariff, and as such must be solely responsible for approving requests for transmission services, including interconnection.

- Congestion Management: It is the RTO's responsibility to develop market mechanisms to manage congestion.

- Parallel Path Flow: An RTO should develop and implement procedures to address parallel path flow issues. It directs RTOs to work closely with NERC, or its successor organization, to resolve this issue.

- Ancillary Services: An RTO must be the provider of last resort of all ancillary services. This does not mean the RTO is a single control area operator.

- OASIS, TTC and ATC: An RTO must be the single OASIS site administrator for all transmission facilities under its control. FERC requires RTOs to calculate TTC and ATC based on data developed partially or totally by the RTO.

- Market Monitoring: An RTO must perform a market monitoring function so that FERC can evaluate and regulate the market, despite many concerns about the RTO intruding into the markets.

- Planning and Expansion: An RTO must encourage market-motivated operating and investment actions for preventing and relieving congestion.

- Interregional Coordination: An RTO must develop mechanisms to coordinate its activities with other regions whether or not an RTO exists in other regions. The Commission does not mandate that all RTOs have a uniform practice but that RTO reliability and market interface practices must be compatible with each other, especially at the "seams."

According to the FERC proposal, RTOs should have an open architecture so that RTOs can evolve over time. This open architecture will allow basic changes in the organizational form of the RTO to reflect changes in facility ownerships and corporate strategies. As to the organizational form of an RTO, FERC does not believe that the requirements for forming an RTO favor any particular structure and will accept a TRANSCO, ISO, hybrid form, or other forms as long as the RTO meets the minimum characteristics and functions and other requirements.

The initial incentive to establish RTOs was to promote regional energy transactions, however, it is also clearly specified in the Order 2000 that an RTO should be responsible for the cooperation with its adjacent RTOs. In this sense, RTOs should also act as bridges connecting adjacent regions. Obviously, RTOs can cooperate to monitor and control the security operation of the "seams" tie lines that are interconnecting regions. The ultimate or ideal control form of these "seams" is transactions without difference.

It is assumed that there will be only one RTO in a region. However, there can be more than one ISO in each region. There can be overlaps in the RTO's functions and those of its subsidiary ISOs, and this is why ISOs are expected to function as TRANSCOs. An RTO can also take the form of a TRANSCO or the combinatory form of an ISO and RTO. Figure 1.11 depicts the control hierarchy for an RTO.

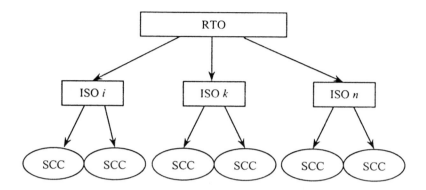

Figure 1.11 Control Hierarchy of RTO

As shown in Figure 1.11, when an RTO is set up in a control area that is composed of a number of ISOs, it has the responsibility to coordinate all subordinated ISOs. The RTO is responsible for monitoring and controlling the entire system, and must focus on the security monitoring of tie lines between zones. In such a structure, an ISO's satellite control center monitors and controls a designated zone of the system.

Regardless of the locations of RTO boundaries, it is important for neighboring RTOs to cooperate for reliability/operations purposes. There are two kinds of seam issues: reliability practices across seams and market practices across seams. For reliability/operations purposes, there can be a set of regions within a larger region for rates and scheduling. Theoretically, a super RTO may be required to operate and coordinate inter-RTO activities. Practically, it is not of much significance to set up such a super RTO because, in most cases, the coordination of neighboring RTOs can be realized through reciprocity agreements, which may be easier to achieve than having all RTOs in an interconnection, agree to form a super RTO. In some cases, determining RTO boundaries is less important than ensuring that seams do not interfere with the market operation. Accordingly, adjacent RTOs may be required to adopt consistent methods for pricing and congestion management to encourage seamless trading.

1.9 ADVANTAGES OF DISTRIBUTED PROCESSING

As contemporary power systems become more complex, distributed processing (i.e., distributed computation, monitoring, and control) will provide a feasible means for power system operators to manage the system efficiently within a limited period. Usually, the first step in implementing a distributed optimization and control is to divide the entire power system into a number of control areas, and then set up a lower level control center for each control area. This hierarchical and distributed monitoring and control scheme has many advantages over the centralized control, especially when the system is in an emergency state. Compared with the centralized processing, distributed processing will not only enable large capital investment for communications systems but also be more reliable, flexible, economical, and efficient. These attributes are discussed next.

1.9.1 Enhanced Reliability

Centralized control requires all data to be transmitted to the control center within a stringent period. The implication is that communication components of the system should function properly; otherwise, CCC would fail. For instance, if a major communication linkage to CCC is malfunctioning and a major power plant data cannot be transmitted to CCC, the EMS/SCADA at CCC would not be able to function properly. Should this occur, the distributed control would enhance the reliability of power system monitoring and control. The distributed control would first divide the entire system into several control areas, each with one control center that is responsible for the real-time monitoring and control of its control area. The disturbances in individual control areas then would be processed locally in order to minimize its impact on other control areas.

When an area control center is malfunctioning due to a purposeful attack, or damaged by earthquake, its neighboring control centers in the same hierarchical layer can take over the functions of the faulted control center. This scheme was the original incentive for promoting distributed processing. There are a few control centers in distributed areas for the monitoring and control of the entire system that function as coordinators of control centers. However, as most random disturbances are processed locally, the burden of upper level control centers is greatly reduced and the reliability of the monitoring and control for the entire system is largely enhanced when less information is required by the upper level control centers.

INTRODUCTION

1.9.2 Enhanced Flexibility

A good strategy for the monitoring and control of large-scale power systems is to process the regional problems locally. If every problem, no matter how serious, needs to be processed by the centralized control center, the centralized control center can be overwhelmed by trivial affairs and may overlook the most serious issues. What is more, stringent time requirements of power system operation do not allow a centralized control center to function properly in many occasions. With distributed control, disturbances occurring in a control area will be processed by the local area control center. Correspondingly, the upper level control center intervenes in handling disturbances that may affect more than one control area. With most of the disturbances solved locally, the upper level control center will have better chances of focusing on serious disturbances.

1.9.3 Economy

An outstanding drawback of centralized control is that it needs to build a complex communications system to reach the entire power system. For that reason, entire data are sent to the centralized control center, no matter how far the location of the data collection is from the centralized control center. This process requires a massive amount of long distance communication links and a large capital investment on communication systems. A critical shortcoming of such complicated communications systems for centralized control is that a faulty communication link can cripple the entire EMS/SCADA system.

With distributed control, the data in each control area are sent to the local area control center. So the communications system of an area control center is much simpler and the data communication distances are much shorter. Since there are a limited number of communication links between area control centers and the upper level control centers, large investments on communication systems can be saved. A partial saving can be used on establishing distributed control centers. However, we should bear in mind that the main purpose of distributed control is not to save money in building communication links but to enhance the reliability, flexibility, and efficiency of the power system monitoring and control.

1.9.4 Higher Efficiency

The centralized control center can easily become overwhelmed by a large number of tasks. Even if powerful computers are employed in centralized control centers, it may be difficult to meet the rigid processing time requirements, especially in emergencies.

The efficiency of power system monitoring and control should be greatly improved with distributed control. In distributed control, the task of power system monitoring and control is divided into a number of subtasks, accordingly to the number of control areas. These subtasks are processed concurrently in a distributed manner with the same number of computers at area control centers. With more computers to solve the same task, the efficiency is greatly improved. With the utilization of reliable, high-data-rate communication links and accurate time synchronization functions of GPS, a distributed computing system can reach the performance and efficiency of a parallel machine, even though a distributed computing system is geographically distributed in a vast area.

1.10 FOUNDATIONS FOR DISTRIBUTED PROCESSING

In addition to the advancement of computer and communication technologies, other techniques and theories have provided a substantial support for advancing the distributed processing in power systems. These techniques and theories include the agent theory, distributed management and control, adaptive self-healing, object-oriented modeling, and common information and accounting models. We discuss these techniques briefly next.

1.10.1 Agent Theory Applications

Agent Modeling

Power systems are extremely complex and highly interactive. They can be modeled as a collection of intelligent entities, that is, as agents that can adapt to their surroundings and act both competitively and cooperatively in representing the power system. These agents can range in sophistication from simple threshold detectors, which choose from a few intelligent systems based on a single measurement, to highly intelligent systems. The North American power grid is represented by thousands of such agents, and agent theory has already been applied to the decision-making and control of power systems.

By mapping each component to an adaptive agent, a realistic representation of power system can be achieved. In practice, disturbances like lightning may last for a few microseconds, but the network's ability to

INTRODUCTION

communicate data globally may take much longer. Thus, each agent must be charged with real-time decisions based on local information. The adaptive agents could manage the power system using multilevel distributed control. These agents are to be designed with relatively simple decision rules based on thresholds that give the most appropriate responses to a collection of offline studies. Through its sensors, each agent receives messages from other agents continuously. If an agent senses any anomalies in its surroundings, several agents can work together, in a distributed manner, to keep the problem locally by reconfiguring the power system. In this sense, agents help prevent the cascading of power system disturbances. On the other hand, by simulating agent-based models, the power system operators can better understand the dynamics of complex inter-component and inter-system actions.

The object-oriented method and hierarchies of simpler components can be used to model more complex components. For instance, a power plant agent is composed of a number of generator agents, thus creating a hierarchy of adaptive agents. The agent theory and technique is well suited for power system operation and control. Agents can assess the situation based on measurements from sensing devices and the information from other entities, and can influence the system behavior through sending commands to actuating devices and other entities.

Agent Evolution

Intelligent agents can be designed to have the capability to evolve; that is, they can gradually adapt to their changing environment and improve their performance as conditions change. For instance, a bus agent strives to stay within its voltage and load flow limits while still operating in the context of the voltages and flows that power system managers and other agents impose on it. In order to be aware of context and evolutions, an agent is usually represented as an autonomous active object that is equipped with appropriate intelligence functions. Evolutions are enabled through combining evolutionary techniques such as genetic algorithms and genetic programming. This way object classes are treated as an analogy of biological genotypes and objects are instantiated from them as an analogy of their phenotypes. When instantiating objects form individual agents, operations typical of genetic algorithms, such as crossover and mutation, can select and recombine their class attributes, which -define all the potential characteristics, capabilities, limitations, or strategies these agents might possess. The physics specific to each component will determine the object-agent's allowable strategies and behaviors.

Context-Dependent Agents

Adaptive agents are known to take actions that drive the system into undesirable operating states. When the system is disturbed, agents could act as previously programmed, but the problem is that the pre-designed actions cannot be the best responses to a particular situation. It is better if for some instances the agents are cognizant of the context and recognize that the preprogrammed action is not be appropriate.

Context-dependent agents are different from adaptive agents. In a context-dependent agent-based power systems, agents cooperate in their local operations while competing with each other in pursuing the global goal set by their supervisors. In a power system, local intelligent controllers represent a distributed computer that communicates via microwaves, optical cables, or power lines and limits their messages to the required information for optimizing the power grid that is recovering from a failure. This way, controllers become context-dependent intelligent agents that cooperate to ensure successful overall operation but act independently to ensure adequate performance.

The application of context-dependent agents requires a dynamic and real-time computing system. This system of agents will provide timely and consistent contexts for distributed agents. For instance, in response to random disturbances that affect the dynamics of power systems, the security of power systems requires coordination among distributed agents. Correspondingly, event-driven real-time communication architecture will assemble relevant distributed agents into task-driven teams and provide the teams with timely and consistent information to carry out coordinated actions for maintaining the power system's security.

Multi-agent System

MAS (multi-agent system) is a distributed network of intelligent hardware and software agents that work together to achieve a global goal. Although each agent is intelligent, it is difficult for individual agents to achieve a global goal for complex large-scale power systems. MAS will model a power system as a group of geographically distributed, autonomous, and adaptive intelligent agents. Each agent will only have a local view of the power system, but the team of agents can perform wide area control schemes through both autonomous and cooperative actions of agents. Certain kinds of coordination are necessary due to the autonomy of each

INTRODUCTION

agent, since conflicts are possible among decisions and actions from a number of agents.

The MAS of power grids is classified into cognitive and reactive agents. A cognitive agent has a complete knowledge base that comprises all the data and know-how to carry out its tasks and to handle interactions with other agents. The reactive agent, in contrast, is not necessarily intelligent but can respond to stimuli with a fast speed. As depicted in Figure 1.12, a MAS usually has three layers. The bottom layer (i.e., the reactive layer) performs preprogrammed self-healing actions that need immediate responses. This layer is distributed at every local subsystem and interfaces with the middle layer. The middle layer (i.e., the coordination layer) identifies which triggering event from the reactive layer is urgent based on heuristics. A triggering event will be allowed to go to the deliberative layer only if it exceeds a preset threshold. This layer also analyzes the commands to the top layer and decomposes them into actual control signals to the agents of the bottom layer. The top layer (i.e., the deliberative layer) prepares higher-level commands, such as contingency assessment and self-healing by keeping current with information from the coordination layer.

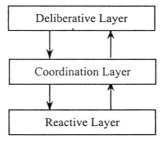

Figure 1.12 Layers of MAS

The coordination layer has to compare the system models continuously between the deliberative and reactive layers because agents in the deliberative layer do not always respond to the current situation of the power system. Besides, it will update the current system model and check if the commands from the deliberative layer match the status of the system. When a command does not align with the real-world model, the coordination layer will ask the deliberative layer to modify the command. Sometimes events from the reactive layer contain too much information for the deliberative layer to process, or the commands from the deliberative layer might be too condensed for the agents in the reactive layer to implement. Then deliberative layer must at the same time send more than a few control commands to the reactive layer.

In the MAS of a power grid, the lowest layer consists of a number of reactive agents that can handle a power system from a local point of view to achieve fast and adaptive control. A number of cognitive agents are placed on the deliberative layer that can analyze and monitor the entire system from a wide area point of view. The agents in the reactive layer perform short-term planning, while the agents in the deliberative layer can plan for the long term. The agents in the deliberative layer can inhibit control actions and decisions initiated by the reactive layer for wide area control purposes. For instance, a generation agent may decide, based on its local view, to trip the generator. However, if reconfiguration agents in the deliberative layer based on the global view decide to block the tripping action, then the action of the generation agent will be inhibited. The deliberative layer does not always respond to the current state of the power system. Thus, decisions made by the deliberative layer might be inconsistent with current power system conditions. The coordination layer, therefore, continuously updates and stores the current state of the power system and verifies the plans from the deliberative layer with the current state of the power system. This coordination layer also examines the importance of events and alarms received from the reactive layer.

Agents on different layers will communicate with each other. Agents on the same layer can further interchange their information. Each agent is endowed with a certain communication capability. The communication among agents is at a knowledge level that is quite similar to human communication. This kind of agent communication guarantees the openness and flexibility of MAS. For instance, the hidden-failure-monitoring agent on the deliberative layer identifies the sources of hidden failures and sends the results to the contingency assessment agent on the same layer. The contingency assessment agent calculates the contingency index of the system and broadcasts this index to other agents. Once the load-shedding agent receives the contingency index, the agent initiates a decision making process and provides load-shedding control actions. All communications among agents can be implemented through a dedicated intranet with potential access through the Internet.

The innovative multi-agent approach makes use of the real-time information from diverse sources, and it has the potential to prevent catastrophic failures of large power systems. Based on the understanding of the origin and nature of catastrophic failures of power systems, a novel

INTRODUCTION

multi-agent-based platform can be used to evaluate system vulnerability to catastrophic events by taking into account various environmental factors. Several new concepts associated with wide area measurements and controls, networked sensors, and adaptive self-healing are being used to reconfigure the network to minimize the system contingency. The objective is to provide useful solutions to interconnected power system networks.

1.10.2 Distributed Management and Control

Agent-based modeling is only a part of what is involved in capturing the level of complexity of the power system infrastructure. It is more important to know what will be happening based on the available measurements and to develop distributed management and control systems to keep the infrastructure robust and operational. The distributed management and control actions include the following:

- Robust control: Manages the power system to avoid cascading failures in the face of destabilizing disturbances such as enemy threats or lightning strikes.

- Disturbance propagation: Predicts and detects system failures at both local and global levels. Thresholds for identifying events that trigger failures are established.

- Complex systems: Develop theoretical foundations for modeling complex interactive power systems.

- Dynamic interactions of interdependent network layers: Create models that capture network layers at various levels of complexity.

- Modeling in general: Develops efficient simulation techniques and ways to create generic models for power systems; develops a modeling framework and analytical tools to study the dynamics and failure modes for the interaction of economic markets with power systems.

- Forecasting of network behavior and handling uncertainty and risk: Characterizes uncertainties in large distributed power networks; analyzes network performances using stochastic methods; investigates the mechanism for handling rare events through large-deviations theory.

1.10.3 Adaptive Self-Healing

Adaptive self-healing is a novel idea that is used to protect power systems against catastrophes. To show this, we use an analogy that classifies the principles of the adaptive self-healing approach for power systems.

When a certain disturbance occurs in a power system, its severity is assessed according to certain criteria. For instance, whether the disturbance can result in failures of the power system or communications system. In general, these criteria include three assessments:

- Degree of danger
- Degree of damage
- Speed of expansion

If the disturbance causes failure of certain elements of a power system, it is determined whether the disturbance will affect only a small portion or a wide area of the system based on the operating conditions of the system. Just as the system determined the severity of the disturbance, it can determine whether the disturbance will damage equipment and result in a widespread blackout. After the extent of damage is specified, it is determined whether the effect of the disturbance is progressing or has ended. If the effect of the disturbance is progressing, the speed at which it would spread is determined. Once it is determined that the disturbance is affecting a wide area of the system, the self-healing idea will be invoked to contain the damage, which can be realized by breaking up the system into smaller parts and isolating the effect of the disturbance. The smaller parts can operate at a slightly degraded level. The entire system is restored once the effect of the disturbance is removed.

This self-healing approach considers several aspects of a competitive restructuring environment. The self-healing includes abilities to reconfigure the system based on contingency analysis, to identify appropriate restorative actions to minimize the impact of an outage or contingency, and to perform important sampling that helps determine weak links in a power system. The self-healing features are developed because agents tend to be competitive. A number of analytical tools have been used to study the behavior of competing agents, and these analyses range from conventional time-domain analysis to artificial intelligence.

INTRODUCTION

Global self-healing approaches for the entire system are realized using a DDET (dynamic decision event tree). The DDET is an extension of an event tree. An event tree is a horizontally built structure that models the initiating event as the root. Each path from the root to the end nodes of an event tree represents a sequence or scenario with an associated consequence. The DDET can trigger various self-healing actions, such as the controlled islanding, with boundaries changing with changes of operating conditions. The islanding can be triggered by using conventional out-of-step relaying or by using synchronized phasor measurements. Analytical techniques to determine appropriate islands based on the system's topological characteristics can also be derived. When a disturbance occurs in the system, DDET will determine what self-healing action should be initiated

1.10.4 Object-Oriented Modeling

The popular object-oriented method has been widely utilized in information processing and computation programming, especially in the modeling of large-scale CASs (complex adaptive systems). An object is a physical entity or abstract concept in a general sense. An object model creates an independent representation of data, which makes is easy to visualize its local environment as well as its interaction with other outside elements. The physical representation is standardized by applying UML (unified modeling language); hence, object models can easily be shared with others.

Object-oriented modeling also provides a way to standardize the information exchange between devices; for instance, the GOMSFE (general object model for substation and field equipment) standardized in UCA, contains models for metering, protection elements, control, security, and a host of other items. These models are based on what is perceived to be the most common elements found in IEDs. Hence, based on the GOMSFE, users can issue requests to IEDs for any standard object value; for example, a utility SCADA system can automatically request volts, amps, watts, vars, and status from IEDs without having any knowledge of the manufacturers of IEDs.

1.10.5 CIM

To reduce the cost and time needed to add new applications to EMS, and to protect the resources invested in existing EMS applications that are working effectively, in 1990 EPRI launched a project, named EPRI Control

Center Application Program Interface (CCAPI) research project. The main goal of the CCAPI project was to develop a set of guidelines and specifications to enable the creation of plug-in applications in a control center environment. Essentially CCAPI is a standardized interface that enables users to integrate applications from various sources by specifying the data that applications could share and the procedure for sharing it. The essential structure of a power system model is defined as the Common Information Model (CIM) that encompasses the central part of CCAPI.

CIM provides a common language for information sharing among power system applications. CIM has been translated into an industry standard protocol, called Extensible Markup Language (XML), which permits the exchange of power system models in a standard format that any EMS can understand using standard Internet and/or Microsoft technologies. The North American Electric Reliability Council (NERC) mandates the use of this standard by security coordination centers to exchange models.

The CIM compliance offers control center personnel a flexibility to combine, on one or more integrated platforms, software products for managing power system economy and reliability. This compatibility allows the personnel to upgrade, or migrate, their EMS systems incrementally and quickly, while preserving prior utility investments in customized software packages. It is perceived that migration could reduce upgrade costs by 40 percent or more, and enable energy companies to gain strategic advantages by using new applications as they become available.

1.10.6 Common Accounting Model

The companies in a restructured power system have to process a large amount of electronic transactions as electricity is traded in the power market. Because most of these transactions span over multiple companies, it is hard to track electronic transactions within a single company. In order to achieve interoperability, the accounting object model, which is quite similar to power system elements models, is proposed and adopted. This model has a three-layer architecture that is widely used in IT (information technology) industries. The top layer is the external interface through which users view the system. The middle layer is the service layer, which performs computations and other services. The bottom layer is the actual data model. This model provides a focal point for all services and is especially useful for e-commerce of electricity.

1.11 ORGANIZATION OF THIS BOOK

So far we have discussed both the conventional and the distributed structures of CCC, ISO, and RTO. Many optimization and control problems of power systems can be solved more efficiently in parallel and distributed manner. As mentioned in Section 1.1, most CCC, ISO, and RTO have a hierarchical and distributed structure. Modern computer and telecommunication technologies have provided substantial technical supports for hierarchical and distributed computation and control of power systems. Generally speaking, parallel and distributed processing can be utilized to solve the following problems:

- Hierarchical and distributed optimization and control of CCC, ISO, and RTO

- Parallel computation and control of large-scale power systems at CCC, ISO and RTO levels

- Distributed computation of large-scale power systems based on a WAN composed of computers at ACCs, SCCs, or ISOs

- Distributed information exchange and energy transaction processing.

This book is intended to introduce parallel and distributed processing techniques that are applied to power system optimization and control. Power engineering professionals who are interested in this subject should find it a useful reference book. The problems that are studied in this book include load flow, state estimation, external equivalence, voltage/var optimization and control, transmission congestion management, and ancillary services management. The chapters of this book are organized as follows:

In Chapter 2, pertinent issues on parallel and distributed processing are reviewed briefly. Since the topic of parallel and distributed processing is very broad, this chapter is written with an assumption that readers will have some basic knowledge of parallel and distributed processing.

In Chapter 3, the information system architecture in modern power system control centers is discussed. In particular, the discussion focuses on the real-time distributed processing of the ISO and the RTO, which are based on the modern information system.

In Chapter 4, the infrastructure for applying parallel and distributed processing to power system is discussed. The main discussion is on the application of EPRI CIM (common information model). The applications of CIM to power system integration and control are presented and several examples are discussed.

In Chapter 5, parallel and distributed load flow algorithms based on MPI (message passing interface) and the application of COW (cluster of workstations) is discussed. Random data communication time delay is considered in the asynchronous distributed load flow computation. The mathematical analyses for the convergence of asynchronous load flow computations are discussed.

In Chapter 6, parallel and distributed load flow algorithms for distribution systems are discussed. The applications of distributed generation in the modeling and the calculation of load flow are presented. Distributed distribution management systems (DDMSs) are analyzed with a few examples.

In Chapter 7, parallel and distributed state estimation algorithms are discussed. State estimation is the most basic and the core function of an EMS, and its application to a restructured and highly distributed power system is presented. The convergence properties of asynchronous distributed state estimation are discussed.

In Chapter 8, the distributed security analysis of power systems based on DEMS is discussed. The chapter presents a distributed model for the calculation of external equivalence. The applications of distributed security analysis in improving the operational performance of restructured power systems are discussed.

In Chapter 9, the hierarchical and distributed voltage optimization and control are presented. The local properties of voltage control in a distributed power system are highlighted. The chapter analyzes the role of the ancillary services market in the distributed processing of power system operation and control.

In Chapter 10, the agent theory is introduced. The applications of multi-agent theory to power system transmission management are discussed. It is shown that the MAS model provides a more flexible and efficient approach to congestion mitigation and management.

INTRODUCTION 45

In Chapter 11, various techniques (e.g., PV, wind, and fuel cells) for distributed power generation and utilization are discussed. It is shown that the existence of multiple distributed generation units affect the operation characteristics and control of distribution systems.

In Chapter 12, several advanced topics are presented, including the electronic commerce (e-commerce) of electricity, geographic information system (GIS) and global positioning system (GPS), for distributed processing of power systems. It is shown that the utilization of GIS and GPS techniques can greatly improve the operational performance of power systems.

Figure 1.13 depicts an overview of the organization of topics in the book.

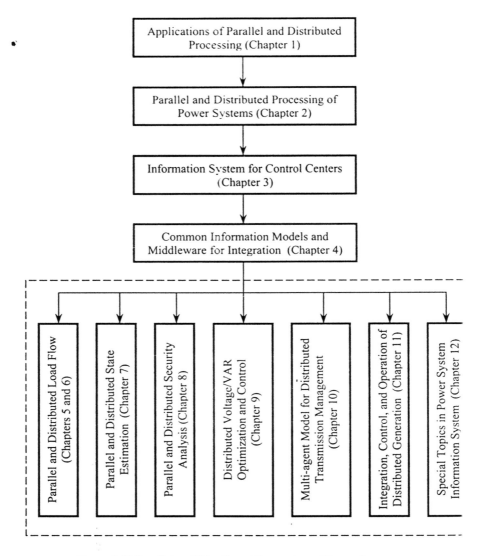

Figure 1.13 Parallel and Distributed Processing of Power Systems

Chapter 2

Parallel and Distributed Processing of Power Systems

2.1 INTRODUCTION

In this chapter we review a few pertinent topics on parallel and distributed algorithms and apply those topics to the parallel and distributed processing of electric power systems. Because the topic of parallel and distributed processing is very broad, we have written this chapter with the assumption that readers have some basic knowledge of parallel and distributed computation.

There has always been a need for fast and reliable solutions of various large-scale engineering problems [Alo96, Car98, Fal95, Le96, WuF92]. In recent years, rapid developments in computer and communication technologies have provided substantial technical supports for parallel and distributed processing of many problems that were very time-consuming when they were processed in the traditional sequential manner. Parallel computation is usually implemented on a parallel machine or a SAN (System area network); distributed computation, and control is generally based on LAN, WAN, or the Internet.

The current restructuring of electric power systems provides a proper forum for parallel processing of various tasks. As designated by FERC, it is the ISO's responsibility to monitor the system reliability[1], and a

---

[1] Reliability of a system is interpreted as satisfying two major functions: adequacy and security. An adequate amount of capacity resources should be available to

large amount of real-time information would need to be processed for performing related tasks. However, much of these data are controlled by self-interested agents (GENCO, TRANSCO, DISTCOS, etc.), and only a minute amount of time is available for the ISO to perform the online monitoring and control of large-scale power systems.

Traditional approaches to the centralized monitoring and control of power systems would not only need high-performance computers to meet rigid time requirements but would also need to build up expensive and complicated communication networks to transmit locally measured data to CCC (central control center). Another drawback of the centralized processing is that local faults or system malfunctions can affect the computation and control of the entire system. For example, online centralized computation cannot fully proceed if a data communication link is suddenly broken. In comparison, distributed processing not only can save a large amount of capital investment in communication networks but also limit the effect of local faults within the related local area. Furthermore, the rapid development of the Internet technology has created a credible mechanism for distributed processing. We have reason to expect that the Internet technology will thrive in power systems and provide a flexible and versatile foundation for distributed processing.

The reliability of communication networks is crucial to the real-time monitoring and control of power systems. Hence, for some important communication links, in addition to a very strict routine inspection and maintenance, back-up communication facilities should be provided. Sometimes, however, even if the reliability of communication networks can be guaranteed, communication delays among control centers are difficult to predict. These time delays may be aggravated by many factors such as the imbalance in task allocations and differences in the computer performances of a distributed system. Communication delays may be mitigated when some communication systems are specifically designed for distributed processing. However, when communication delays between the ISO and its satellite control centers are hard to predict, asynchronous distributed computation can be used instead of synchronous computation, although in most cases the latter exhibits better convergence performance.

---

meet the peak demand (adequacy), and the system should be able to withstand changes or contingencies on a daily and hourly basis (security).

2.2 PARALLEL AND DISTRIBUTED PROCESSING SYSTEMS

Parallel and distributed processing has been an intense research area for several decades. Although traditional supercomputers with a single processor are fast, they are extremely expensive and their performance depends on their memory bandwidth. With rapid advancements in computer and communication technologies, traditional supercomputers with a single processor are being gradually replaced by less expensive and more powerful parallel and distributed processing resources.

Over the past several decades, parallel and distributed processing techniques have evolved as a coherent field of computation. The main process for solving large-scale problems with these techniques can be outlined as follows: First, the problems are formulated for parallel and distributed processing. Second, specific parallel and distributed algorithms are developed for the solution of these problems. Third, the complexity and the performance of the algorithms are evaluated for measuring the accuracy and the reliability of the proposed solutions. Parallel and distributed systems are discussed next.

2.2.1 Parallel Systems

A parallel system represents the physical arrangement for parallel processing. There are two types of parallel system. One is a parallel machine, and the other is a computer network such as a SAN dedicated to parallel processing. Both types of parallel system comprise a number of processors that are closely coupled within a small physical space. There are many types of parallel machines available in the market. They include mainframes like Cray, generic parallel computers like SGI parallel machines, and specially constructed computers that are embedded with multiple processors. When a computer network is used for parallel computation, it represents a virtual parallel machine with all the computers on the network as processors of this virtual machine.

Communication links between the processors of a parallel system are usually very short; for instance, the processors of a parallel machine are located on the same motherboard. The data communication between processors of a parallel system is assumed to be very reliable, and communication delays, if considered, are predictable.

The main intention of a parallel system is to speed up the computation by employing more than one processor at a time. In other words, the sole purpose of employing a parallel system is to obtain a fast solution by allowing several processors to function concurrently on a common task.

2.2.2 Distributed Systems

Quite similar to a parallel system, a distributed system is the physical arrangement for distributed processing. But unlike a parallel system, a distributed system is usually a computer network that is geographically distributed over a larger area. The computers of a distributed system are not necessarily the same and can be heterogeneous. A distributed system can be used for information acquisition; for instance, a distributed system could be a network of sensors for environmental measurements, where a set of geographically distributed sensors would obtain the information on the state of the environment and may also process it cooperatively. A distributed system can be used for the computation and control of large-scale systems such as airline reservation and weather prediction systems.

The correct and timely routing of messages traveling in the data communication network of a distributed system are controlled in a distributed manner through the cooperation of computers that reside on the distributed system. The communication links among computers of a distributed system are usually very long and data communications over a distributed system are relayed several times and can be disturbed by various communication noises. In general, a distributed system is designed to be able to operate properly in the presence of limited, sometimes unreliable, communication links, and often in the absence of a central control mechanism. The time delays of data communication among distributed computers can be very difficult to predict, and this is especially true with a distributed processing, which has rigid time requirements such as voltage/VAR control in a power system.

2.2.3 Comparison of Parallel and Distributed Processing

Parallel processing employs more than one processor concurrently to solve a common problem. Historically the only purpose for parallel processing is to obtain a faster solution. Parallel computation can be performed in a parallel mainframe such as Cray, or in a parallel computer like Silicon Graphic Systems, or in a personal computer with multiple processors. Most vendors of parallel machines will provide parallel compilers for Fortran and C/C++ with customers. Parallel languages such as OCAM are specially provided for some parallel computers like Transputers, although in most cases other computer languages such as high-performance Fortran and C/C++ can also be used for parallel computation in these machines.

Because the data communication links among parallel computing systems are very short, data communication is usually assumed to be very reliable. Therefore, unless asynchronization is purposely introduced, parallel algorithms are generally designed to be totally synchronized. To guarantee that each iteration step is completely synchronized, each processor should not proceed for the next iteration until it has received all the data that are supposed to be transmitted by other relevant processors. Theoretically, when no approximation is introduced in parallelization, parallel algorithms should have the same convergence property as their serial counterparts.

Although parallel computation solves problems in parallel, it still has a centralized manner when seen from outside the parallel system. This is because the entire system data have to be sent to the location of the parallel system. For instance, to implement the parallel state estimation, measurement data from all over the system would need to be sent to SCC for parallel processing.

There are many approaches to parallel computation, among which the following two are currently the most popular:

- SMP (Share memory processing): SMP is usually used for parallel computation on parallel machines with multiple processors. There is no real data communication in SMP. All the data that need to be exchanged among participating processors will be stored in a designated memory block to which all participating processors have the right to access. The operating system of most parallel machines support the SMP application and the only requirement for SMP is that these parallel machines should have large enough memories to solve the problems.

- MPI (Message-passing interface): MPI is a protocol for data communication among processors. With MPI, the processors participating in parallel computation can store their data in local memories and then exchange them with other processors. See Section 2.12 for further information.

Both parallel and distributed processing are based on the concurrency principle. Hence, in most cases, except when the parallel processing is implemented on a real parallel machine, these two terms can be applied interchangeably and generally denoted as *parallel processing*. A widely accepted distinction between these two processing techniques is that the parallel processing employs a number of closely coupled processors while distributed processing employs a number of loosely coupled and geographically distributed computers. In this sense, the distinction between these two techniques mainly depends on how system components are physically organized. Although distributed processing can be regarded as an extension of parallel processing, to avoid confusions, we use parallel processing and distributed processing distinctly in this book

Parallel processing deals with problems in a centralized manner, which means the information to be processed needs be transmitted to the place where the parallel system is located. In contrast, distributed processing deals with the information in a distributed manner. For example, as depicted in Figure 2.1, generation and demand data of system buses will be gathered at the CCC when the system operator executes a real time parallel load flow program, which requires a complicated telecommunication network to transmit the data. As depicted in Figure 2.2, however, if the operator executes a distributed load flow program, generation and demand data at each bus are sent to the respective ACC rather then the CCC.

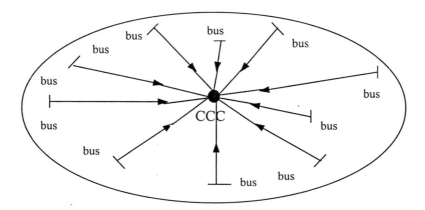

Figure 2.1 Data Communication for Parallel Processing

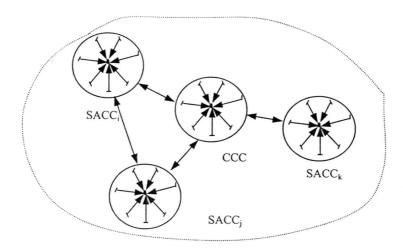

Figure 2.2 Data Communication for Distributed Processing

Unlike parallel processing which is performed on one machine, distributed processing is performance on a computer network. Usually a distributed system occupies a larger physical space than a parallel system. Some special distributed systems, for instance a DEMS, are distributed in an area, which could be hundreds, even thousands of miles wide and long. Some can even be based on the Internet. Under this situation the time for data communication among computers needs to be taken into account for some distributed algorithms, such as the distributed load flow based on DEMS. In some distributed systems the time for computation can even be shorter than the time for data communication among computers.

The following two schemes are widely applied to distributed processing:

- **File sever.** A file sever can provide a data-sharing place for all computers on the same computer network. The data exchanges among these computers are realized by sharing a file on the server, which is quite similar to the memory block sharing of the SMP method. However, this file server scheme will have a low efficiency because the shared data file cannot be written by more than one user at a time.

- **MPI.** It is suitable not only for parallel computation but also for distributed computation [Ong02, Pac97, Sod02, Sta02, Vet02]. See Section 2.12 for further information. There are two versions of MPI for distributed computation. One is the MPI for LAN composed of personal computers and the other is for COW or WAN composed of Unix workstations. The performance-enhanced MPI protocol can be applied to the Internet.

As we will see later in Chapter 3, the purpose of the parallel load flow computation is to acquire a fast solution. However, compared with the parallel load flow computation, the distributed load flow computation offers additional incentives such as an investment saving in building expensive communication networks as well as improvements in the performance of power system computation and control.

2.2.4 Representation of Parallel and Distributed Systems

Generally, the topology of a parallel or a distributed system can be represented by a graph $G = (V, E)$, where a node i of the graph corresponds to a processor of the system, and an edge (i, j) indicates that there is a direct communication link between nodes i and j in both directions. It is theoretically assumed that the communication can take place simultaneously on incident links of a node and in both directions. Let distance (i, j) denote the length of the shortest directed path from node i to node j in G; otherwise distance $(i, j) = 0$. The diameter of a parallel or a distributed system is defined as the maximum distance (i, j), taken over all (i, j). The diameter of a system can reflect the size of the system, and a large diameter of a distributed power system indicates that the distributed system spans over a large area.

Parallel and distributed systems can take a variety of network topologies. For instance, parallel machines whose principal function is numerical computation could have a symmetrical topology as shown in Figure 2.3; while geographically distributed systems composed of CCC and ACCs could have an asymmetrical topology as shown in Figure 2.4.

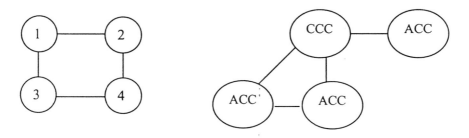

Figure 2.3 Topology of a Four-Processor Parallel Machine

Figure 2.4 Distributed Power System

In general, the parallel and distributed algorithms design for an asymmetrical system is more difficult than that for a symmetrical system since the asymmetrical topology could add communication complexity to the algorithm design.

2.3 PARALLEL AND DISTRIBUTED ALGORITHMS

2.3.1 Classifications

Parallel and distributed algorithms are classified according to the following attributes:

- **Inter-process communication method.** Parallel and distributed algorithms run on a collection of processors that need to communicate with each other effectively in order to accomplish a common task with a high efficiency. The generic communication methods for parallel and distributed processing include shared memory, message passing, and RPCs (remote procedure calls). The choice of the inter-process communication method depends mostly on the architecture of the processing system. The parallel computation implemented on a parallel machine at a power system control center can use shared memory and message-passing techniques. However, for the distributed processing of CCC, message passing and RPCs will be used for communications among computers.

- **Timing model.** A timing model is used to coordinate the activities of different processors for parallel and distributed processing. At one extreme of this model, all processors of the system are completely synchronous and thus perform communications and computations in a perfect lock-step pattern. At the other extreme, they are totally asynchronous and thus take steps at arbitrary speeds and in arbitrary orders. In between, a wide range of options can be considered under the designation of partially synchronous, where processors have only partial information about the timing of events. For instance, processors may be given bounds on their processing speeds.

Generally, synchronization is easy to implement in a processing environment where the communication delay is predictable, while asynchronization is usually applied to scenarios where communication delays are more difficult to predict. Parallel processing of power system are designed to be synchronous because data communications among the processors of a parallel machine or SAN are assumed to be reliable. However, the distributed processing of power system is usually designed as asynchronous because data communications delays between power system control centers are hard to predict.

Although parallel and distributed systems are assumed to be reliable, distributed systems could face more uncertainties. These uncertainties could include independent inputs to different computers, several software programs executed simultaneously though starting at different periods and speeds, and possible computer and communication failures. In some cases, even if the software code for an algorithm is very short when it is executed in an asynchronous manner, there could be various modes in which the algorithm can behave. In this sense, it could be very hard to predict the process for executing an algorithm. This is quite different with the synchronous parallel processing for which we can often predict the behavior of the algorithm at every point in time.

2.3.2 Representation

Parallel and distributed algorithms can be represented by a DAG (directed acyclic graph) like $G = (N, A)$, which is a directed graph that has no positive cycles, that is, no cycles with forward arcs exclusively. Suppose that $N = \{1,..., |N|\}$ is the set of nodes and A is the set of directed arcs. Each node represents an operation performed by an algorithm, and a directed arc represents the relationship of data dependency. An arc $(i, j) \in A$ indicates that the operation corresponding to node j uses the results of the operation corresponding to node i. An operation can be either elementary or high level like the execution of a function. The DAG of a parallel or distributed algorithm can give the designer a very clear picture about the algorithm; it indicates the tasks that a processor performs and the data dependency of a processor at each step of processing; it is also very useful in an algorithm analysis and optimization. The DAG of a parallel or distributed algorithm is optimized before the corresponding program code is developed. To specify this point, we illustrate in Figure 2.5 the parallel computation of (x1*x2 + x1*x3 + x2*x3).

In Figure 2.5, three processors are employed to compute the polynomial (x1 *x2 + x1*x3 + x2*x3) in parallel. In the first step of the parallel computation, values of x1, x2, and x3, are assigned to processors 1, 2, and 3 respectively; in the second step, processor 1 sends x1 to processor 2 and receives x2 from processor 2. Similarly, processor 2 sends x2 to processor 3 and receives x3 from processor 3, and processor 3 sends x3 to processor 2 and receives x2 from processor 3. Finally, the three processors perform the multiplication operation of x1*x2, x1*x3, and x2*x3 simultaneously. In the third step, processor 1 and processor 3 send their computation results of x1*x2 and x2*x3 to processor 2, and the latter performs the summation of (x1*x2 + x1*x3 + x2*x3).

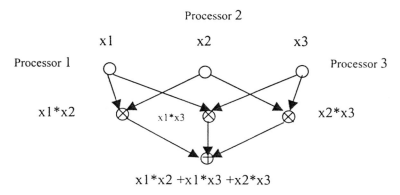

Figure 2.5 Parallel Computation of (x1*x2 + x1*x3 +x2*x3)

The representation of distributed algorithms can become very complex when they are executed on a distributed and irregular network topology like that of a power system. This issue will be pursued further in Chapters 5, 6, and 7, where parallel and distributed load flow and state estimation of power systems are discussed in detail.

2.4 DESIGN OF PARALLEL AND DISTRIBUTED ALGORITHMS

Parallel and distributed algorithms are designed on the principle of concurrency and parallelism. As a strategy for performing large and complex tasks faster, concurrency and parallelism is done by first breaking up the task into smaller tasks, then assigning the smaller tasks to multiple processors to work on those smaller tasks simultaneously, and finally coordinating the processors. Parallel and distributed algorithms follow design methodologies that are quite similar. In general, the design of parallel and distributed algorithms is based on the following three steps:

- Task allocation
- Data communication
- Synchronization

2.4.1 Task Allocation

The first step for the parallel and distributed processing is to partition the initial task into smaller tasks and then allocate them to processors of the parallel or distributed system. For a given task, there are numerous task allocation options. For example, the load flow computation is to solve a group of nonlinear equations composed of n separate equations. The task allocation in parallel load flow algorithm design is to partition these n nonlinear equations into a number of smaller groups of equations, and then assign each group to one processor of the parallel machine. The number of tasks could be between 1 and n. There are as many as $n(n+1)/2$ partitioned solutions for the parallel load flow computation.

In the distributed load flow computation, the task allocation mainly depends on the results of system partitioning, rather than the partitioning of load flow equations. The number of distributed algorithm tasks should be equal to the number of partitioned subareas, and subarea load flow computation tasks are assigned to the computers at each subarea. The task allocations for parallel and distributed load flow computation are depicted in Figures 2.6 and 2.7, respectively.

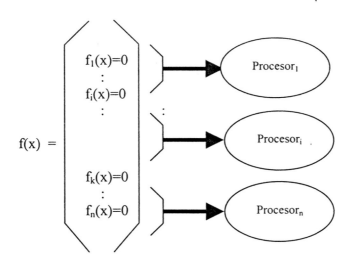

Figure 2.6 Task Allocation for Parallel Load Flow Computation

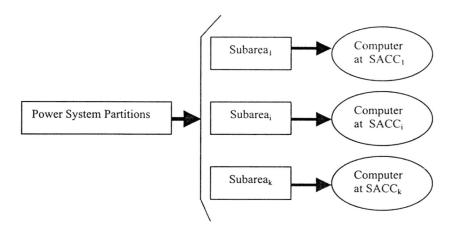

Figure 2.7 Task Allocation for Distributed Load Flow Computation

Theoretically, we could group load flow equations arbitrarily for parallel computation; likewise, we could partition a power system arbitrarily into subareas for the distributed load flow computation. Practically, however, the number of tasks cannot exceed the number of available processors of the parallel machine at system control centers, and the number of subarea computers of the distributed system should represent CCC and SACCs of a power system.

One important concern in the task allocation for parallel and distributed computation is to balance tasks among participating processors. As we will show later, load balancing is a major factor that affects the performance of parallel and distributed computation. To achieve the best computation performance, each processor's load ought to be commensurate with its performance. For example, when processors of the parallel and distributed systems are homogeneous, an equal amount of task should be allocated to each processor, but when the processors of the parallel and distributed systems are heterogeneous, a heavier task should be allocated to the processors with higher performance.

Compared with a parallel system, a distributed system usually has less freedom in its task allocation because most often the task allocation is restricted by other factors. For instance, in a parallel state estimation, a system operator could divide the system into several subareas arbitrarily. However, once the subarea control jurisdiction is set, the computation task will be allocated to each subarea at the same time.

2.4.2 Data Communication

Data communication among processors becomes critical when a task is completed through the cooperation of a number of individual processors. Data communication in parallel and distributed processing facilitates the processors' exchange of intermediate results. As was mentioned earlier, unlike the data communication among processors of a parallel machine, which is predictable, data communications in distributed processing is unpredictable. In this regard the data communication among computers becomes more problematic, which must be taken into account in the design of distributed algorithms.

2.4.2.1 Data Communication Mechanisms

The following two mechanisms are commonly used for data communication in parallel and distributed processing.

- **Shared memory.** Figure 2.8 shows that the data, which are to be exchanged among processors, in a shared memory block, will be stored in a designated common memory block, which will be accessed by all processors. Data communications are realized instantly and reliably by accessing this common memory block. Shared memory is mostly used for parallel processing as there are technical difficulties with its implementation in a distributed computer network.

- **Message passing.** Message passing is a communication protocol for transferring data from one processor (sender) to another (receiver). A message is transferred as packets of information over communication links or networks. Message passing requires the cooperation of sending and receiving processes. A send operation of message passing requires the sending processor to specify the location, size, type, and the destination of data. In addition, the associated receive operation should match the corresponding send operation.

As shown in Figure 2.9, the sender can communicate directly with the receiver by using point-to-point message passing. A sender can also broadcast a message to all other members by using a one-to-all broadcast. In addition, any number of members can broadcast messages to all members at the same time by using an all-to-all broadcast call.

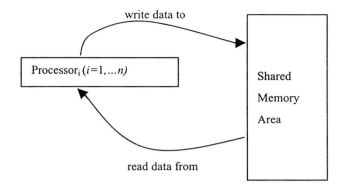

Figure 2.8 Shared Memory Communication Scheme

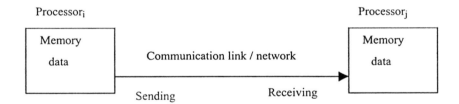

Figure 2.9 Point-to-Point Message Passing

The message passing offers two types of communication: blocking and non-blocking. A communication routine is blocking if the completion of the call is dependent on certain events. For senders, data must be successfully sent or safely copied to the system buffer space so that the application buffer that contains the data can be reused. For receivers, data must be safely stored in a receive buffer so that it can be used. A communication routine is non-blocking if a call is returned without waiting for the completion of any communications events.

It is not safe to modify or use the application buffer after the completion of a non-blocking send. It is the programmer's responsibility to ensure that the application buffer is free for reuse. Non-blocking communications are primarily used to overlap computations with communications for gaining performance. Based on the blocking mechanism, the send operation of the message passing is classified into two types: synchronous and asynchronous sends. A synchronous send is completed only after the acknowledgment is received that the message was safely received by the receiving process. An asynchronous send may be completed even though the receiving process has not actually received the message. To allow for an asynchronous communication, the data in the application buffer may be copied to/from the system buffer space.

Message passing has been widely used for parallel and distributed processing. A message-passing application package includes a message-passing library, which is a set of routines embedded in the application code to perform send, receive, and other message-passing operations. Mature message-passing techniques include MPI (message-passing interface), PVM (parallel virtual machine), and MPL (message-passing library).

Message passing mostly suits the distributed processing system where each processor has its own local memory accessed directly by its own CPU. By message passing, a distributed processor sends data to or receives data from all or some of the other processors. The data transfer is performed through data communications, which differ from a shared memory system that permits multiple processors to directly access the same memory resource via a memory bus. Message passing can also be used in place of a shared memory on a parallel machine. The choice of message passing or shared memory depends on the system architecture and requirements for the design of algorithms. Because the distributed processing of power system is essentially based on WAN, we use message passing for the distributed computation and control of power systems.

2.4.3 Synchronization

The coordination becomes essential when more than one processor works jointly on a common task. Synchronization is a mechanism that allows processors to wait at some predetermined time points for the completion of other processors' computation. Synchronization provides an opportunity for cooperation, which can influence the solution and performance of parallel and distributed algorithms. Synchronization is considered based on the following three models:

2.4.3.1 Synchronous model. In synchronous model, processors perform the computation steps simultaneously; in other words, processors complete computation and communication tasks in synchronous rounds. Suppose that two processors are employed to accomplish a common task and their synchronous model is as depicted in Figure 2.10. Synchronous computation can be applied in circumstances where communications are reliable and communication time delays among processors are predictable, if not negligible. For instance, parallel algorithms such as the parallel load flow for a parallel machine is usually designed to be synchronous. One merit of the synchronous computation is that it can guarantee the convergence of computation, and generally, synchronous parallel and distributed algorithms have the same convergence characteristics as their serial counterparts.

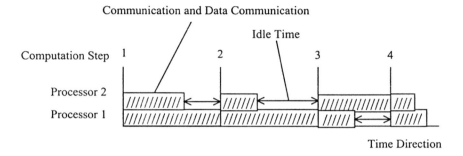

Figure 2.10 Synchronous Computation

2.4.3.2 Totally asynchronous model. A totally asynchronous model assumes that processors can exit certain steps arbitrarily and without any bounds on individual processors speed; in other words, a processor would not need to wait for any data transfer from other processors and can continue its computation with the existing information, though it might be outdated. The updated information will be used for computation immediately after it becomes available. As illustrated in Figure 2.11, there is no idle period in this model, and every processor proceeds with the computation until the common task is accomplished.

The totally asynchronous algorithms are usually harder to program than their synchronous counterpart because of the chaotic behavior of the participating processors. Totally asynchronous algorithms are more general and portable because the totally asynchronous computation does not involve synchronization. In some cases, however, totally asynchronous computation cannot guarantee an efficient or even a correct solution of the problem.

The totally synchronous model is usually utilized in circumstances where communication time delays among processors are hard to predict. If the synchronous model is employed in these circumstances, some processors may have to wait for a long time for data communication, which could result in a very low even impractical computation efficiency. The distributed computation based on CCC and SACCs should be designed with an asynchronous model because time delays of data communication among power system control centers are usually unpredictable.

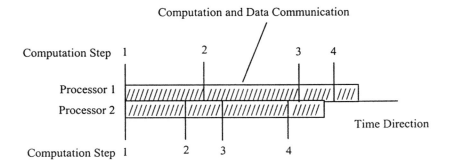

Figure 2.11 Asynchronous Computation

2.4.3.3 Partially synchronous model. A partially synchronous model imposes time restrictions on the synchronization of events. However, the execution of every component is not as lock-step as in the synchronous model. In a partially synchronous model it is assumed that a processor's time step and message delivery time are between upper and lower bounds. Thus a partially synchronous model lies properly between synchronous and totally asynchronous models. A partially synchronous model represents one of the most realistic communication characteristics of distributed systems. The communication delays among power system control centers are hard to predict. However, in most cases, a communication delay has an upper bound, especially when data communications are performed on a dedicated WAN of a power system.

A partially synchronous model has less uncertainty than a totally asynchronous model. However, this does not mean that a partially synchronous model is easier to program. On the contrary, a partially synchronous model is usually more difficult to program due to extra complications that may arise from the timing. Partially synchronous algorithms are often designed so that their performances depend on timing assumptions. Algorithms designed with information on the timing of events can be efficient, although they can be fragile and will not run correctly if timing assumptions are violated.

A simple and practical way to design parallel and distributed algorithms is to parallelize their serial counterparts. Parallel algorithms are usually designed to be synchronous. This is because data communications among processors of a parallel system are fast and reliable. It is more difficult to design distributed algorithms because of the existing uncertainties. However, distributed algorithms are supposed to work correctly even as certain components fail.

2.5 PARALLEL AND DISTRIBUTED COMPUTATION PERFORMANCE

2.5.1 Speedup and Efficiency

The performance of a serial algorithm is usually measured by the computation time, but the performances of parallel and distributed algorithms are measured by their *speedup* and *efficiency*. To see this, suppose a particular parallel or distributed system with p processors or computers is chosen to solve a problem that is parameterized by an integer variable n. Further suppose that the algorithm terminates in time $T_p(n)$, and $T_s(n)$ denotes the time required by the best possible serial algorithm to solve the problem. *Speedup* describes the speed advantage of parallel and distributed algorithms, compared to the best serial algorithm, while *efficiency* indicates the fraction of time that a typical processor or computer is usefully employed. The *speedup* and *efficiency* are respectively defined as

$$S_p(n) = \frac{T_s(n)}{T_p(n)}$$

$$E_p(n) = \frac{S_p(n)}{p}$$

An ideal case is $S_p(n) = 1$ and $E_p(n) = 1$. However, this ideal situation is practically unattainable because it requires that no processors or computers ever remain idle or do any unnecessary work during the entire computation process. A practical objective is to aim at an efficiency that keeps bounded away from zero as n and p increase.

As we noted earlier, the main idea behind parallel computation is to obtain a solution at faster speed. For distributed computation, a faster computation speed represents, in most cases, a partial objective as distributed computation can emphasize other issues such as the full use of distributed computing resources while the environment is more suitable for distributed computation. Nevertheless, it is a common goal of parallel and distributed computations to achieve a faster computation speed, and the speed up depends mainly on how parallel and distributed algorithms are designed.

2.5.2 Impacting Factors

The performance of the components of a parallel or distributed system can affect the performance of the entire system. To achieve a satisfactory computation performance, the behavior of the system components should be coordinated optimally in the algorithm design process. The computation performance of a parallel or a distributed algorithm can be affected by many factors, and among these, the following five factors are most critical.

2.5.2.1 Load balancing. Load balancing means every processor participating in parallel and distributed computation has almost the same amount of computation task. This condition can cause each processor to arrive simultaneously at the preset synchronization point at every iteration and thus can greatly reduce the time waiting for data communication.

Theoretically, load balancing refers to the distribution of tasks in a way that ensures the most time efficient parallel execution. If tasks were not distributed in a well-balanced manner, certain processors or computers would still be busy with their heavy tasks while others would be idling after completing their light tasks. According to the Amdahl's law, the performance of such parallelization could not be the best because the entire computation process could be slowed down by those components with heavy computation burdens.

To reach load balancing, processors are assigned to subtasks of similar size when they are homogeneous. For instance, in the design of the parallel load flow algorithm, shown in Figure 2.6, the subgroups of load flow equations should approximately have the same size. If all processors can finish their computation tasks designated for one synchronous round at the preset time for synchronization, the waiting time for synchronization could be largely reduced. In a system with heterogeneous processors, for instance, where most distributed systems are composed of heterogeneous computers whose computing power is quite different, the load allocation should include the computation of power differences for load balancing.

For instance, the subarea partitioning in Figure 2.7 should allow more powerful computers to take on larger subareas such that the distributed computation can be balanced algorithmically and better computational performance is achieved.

2.5.2.2 Granularity. Granularity is defined as the ratio of the time of computation to that of communication for a parallel or distributed algorithm. Because parallelism can be pursued at different levels, parallel and distributed algorithms fall into two categories: fine-grain parallelism and coarse-grain parallelism. In fine-grain parallelism, the common task is divided into very small subtasks that are then allocated to many processors. Fine-grain parallelization is also called massively parallel processing, and can be implemented when sufficient computing resources are available. In coarse-grain parallelism, the common task is partitioned into fewer subtasks that are then allocated to several processors that are more powerful. Coarse-grain parallelism applies when only a limited number of computing resources are available. For instance, in a fine-grain parallel load flow algorithm, load flow equations are divided into many small sub-groups and each is assigned to a processor. In a coarse-grain parallel load flow algorithm, load flow equations are divided into a few sub-groups and each is assigned to a processor.

Generally, the more processors that participate in parallel computation, the higher the communication overhead will be. Accordingly, fine-grain parallelism refers to a small granularity, while coarse-grain parallelism represents a large granularity. Fine-grain parallelism facilitates load balancing in a more precise way. However, when granularity is very fine the overhead required for communications and synchronization between tasks could take longer than the computation. Compared with the coarse-grain parallelism, the fine-grain parallelism has less opportunity to enhance the computation performance. The coarse-grain parallelism is typified by long computations between synchronization cycles, which implies more opportunities for enhancing performance. However, the coarse-grain parallelism will not support the load balancing precisely.

In sum, the most efficient granularity depends on the algorithm requirements and the hardware environment in which the algorithm runs.

2.5.2.3 Data dependency. Data dependency corresponds to the sequence of information usage. Data dependency exists when a processor needs additional information from other processors to proceed with its computation. Any data dependency inhibits the implementation of parallelism and thus degrades the system computation performance. In most cases data dependency requires communication at synchronization points. For instance, in the parallel load flow computation, if processor i needs the information from processor j for its next step computation, but processor j is not able to provide this information as required, the component i will have to cease its computation momentarily. A proper way of reducing the redundant data dependency in parallel load flow computation is to allow nodes in each sub-group of equations to represent physically interconnected buses of a subarea of the power system. In this case the only information that needs to be exchanged is the state variables of boundary buses, which is analogous to the distributed computation of load flow.

2.5.2.4 Communication features. The communication overhead of parallel and distributed computation depends largely on the given system's communication features. In this sense, communication features of the processing system affect the computation performance. As shown in Figure 2.12, the communication overhead generally represents the time for the following four communication steps: data processing, queuing, transmission, and propagation. The data processing time is the required time for preparing the information for transmission, such as assembling information in packets and appending, addressing, and controlling the information to the packets. The queuing time is the time that a packet spends while waiting to be transmitted. The transmission time is the time used for the transmission of the information. The propagation time is the time between the end of transmission of the information at the sending end and the reception of the information at the receiving end.

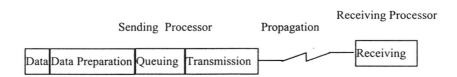

Figure 2.12 Data Communication Process

Besides the latency, which can be thought of as the time required to send a zero byte message, communication overhead in parallel and distributed computation depends on two factors, one is the amount of data to be transmitted, and the other is the communication rate or bandwidth of communication links. For instance, the time t needed to send Z words between two components could be calculated as $t = L + Z/W$, where L is the communication latency and W is bandwidth expressed in bits per second.

The communication features of a given system are presumably set and so are difficult to change. However, we can decrease the communication overheads by reducing the size of the communication data. In addition, with the rapid development of computer and communication technology, broad bandwidth communication techniques have become available, and they provide a substantial level of support to distributed computation based on WAN in power systems.

In some special cases where there are many data to be transmitted in parallel algorithms, it is generally assumed that data communication in parallel commutation are short and accurate; there is no fault in commutation among processors and communication expenses are negligible. The case is very different for distributed computation. In a distributed computing system, communication lines can become hundreds even thousands of miles long, and there are usually several communication relays in between. Thus the communication time cannot be neglected and should be taken into consideration, especially when the communication time is larger than the computation time in an iteration.

2.5.2.5 Computer performance. The processors in parallel machines or computers with multiple processors are most often the same. If the processors are assigned well-balanced subtasks, they can obtain almost ideal synchronization and therefore a faster parallel computation speed. For distributed computation, the situation is quite different. Computers connected to a LAN or a WAN are usually heterogeneous, not homogeneous. Hence, even if they are assigned to well-balanced subtasks in synchronous distributed algorithms, many of the higher performance computers still have to wait for computation results of the lower performance computers, which generally lead to low computation efficiency.

In sum, the computation performance can be affected by many factors that interact in a parallel or distributed algorithm. To achieve an ideal computation performance, the parallel and distributed algorithms design should aim at specific parallel or distributed systems, and take load balancing, granularity, data dependency, and communication features of the given system into account. For instance, to further reduce the synchronization penalty, the processors performance and communication delays would need to be considered in the task allocation so that the entire system would be "algorithmically" balanced. To obtain an ideal synchronization, the communication time delay should be taken into consideration in the design of an algorithm.

2.6 ITERATIVE PARALLEL AND DISTRIBUTED ALGORITHMS

The iterative approach is the most general solution approach to many linear and nonlinear equations and is regarded as an efficient approach to parallel and distributed processing of a wide range of engineering problems. Many power system problems, such as load flow and state estimation, are well suited for an iterative solution. To lay a foundation for the parallel and distributed processing of power systems that will be discussed later in this book, we here describe a general iterative process for parallel and distributed computations.

Suppose we intend to solve the following fixed point problem with either a parallel or a distributed system

$$x = f(x) \qquad (2.1)$$

where $x \in X \subset R^n$, $f: X \ R^n$. Let $X_1, X_2, \ldots, X_m$ be the given sets, where $m \leq n$ and X is the Cartesian product $X = X_1 \times X_2 \times \cdots \times X_m$. For any $x \in X$, we write $x = (x_1, x_2, \cdots, x_m)$ where x_i is the corresponding element of X_i, i=1,..., m. If $f_i : X \to X_i$ is a given function, we define function $f : X \to X$ as $f(x) = (f_1(x), f_2(x), \cdots, f_m(x)), \forall x \in X$. The problem here is to find a fixed point on f such that $x^* \in X$ and $x^* = f(x^*)$. Hence,

$$x_i = f_i(x_1, \cdots, x_m), \quad \forall i = 1, \cdots, m. \qquad (2.2)$$

Equation (2.2) represents the component solution method of (2.1), which provides a basis for the distributed iterative computation proposed below.

Suppose that there is an iterative method $x_i = f_i(x_1, \cdots, x_m), \forall i = 1, \cdots, m$. Let $x_i(t)$ represent the value of x_i at time t. Suppose that at a set of times $T = \{0,1,2,\ldots\}$ one or more components x_i of x are updated by some of the processors of the distributed system. Let T^i be the time at which x_i is updated. We assume that the updated x_i would not have an access to the most recent value of other components of x, then

$$x_i = f_i(x_1(\tau_1^i(t)),\cdots,x_m(\tau_m^i(t))), \quad \forall t \in T^i, i = 1,\cdots,m. \tag{2.3}$$

where $\tau_j^i(t)$ represents the time when the jth component becomes available at the updated xi(t) at time t. In addition $\tau_j^i(t)$ satisfies $0 \leq \tau_j^i(t) \leq t$, $\forall t \in T$. For all times $t \notin T^i$, x_i is fixed and $x_i(t+1) = x_i(t)$, $\forall t \notin T^i$. In such a distributed environment there is no requirement for a shared global clock or synchronized local clocks of processors. The time difference $(t - \tau_j^i(t))$ is viewed as a communication delay. To give a specific description on synchronization we introduce the following assumptions.

- **Synchronous assumption:** For any i, there is $\tau_j^i(t) = t$, $\forall j$.

- **Totally asynchronous assumption:** Suppose T^i is indefinite and $\{t_k\}$ is an definite sequence of T^i elements we have $\lim_{k \to \infty} \tau_j^i(t_k) = \infty$, $\forall j$.

- **Partially asynchronous assumption:** Suppose there is a positive integer M, and for each i and $t = 0$, $\{t, t+1, \ldots, t+M-1\}$ will have at least one element which belongs to T^i; for all i, and $\forall t \in T^i$, there is

$$t - M < \tau_j^i(t) \leq t \tag{2.4}$$

Then for all i and $\forall t \notin T^i$, there is

$$\tau_j^i(t) = t \tag{2.5}$$

The distributed iterative method (2.3) could be a synchronous, totally asynchronous, or partially asynchronous iterative algorithm using the respective synchronous, totally asynchronous, or partially asynchronous categories discussed earlier. Some asynchronous algorithms have a faster computation speed than their synchronous counterparts because they are not required to cease for synchronization [Ber89].

The iterative parallel and distributed load flow and state estimation algorithms, which will be discussed in detail in Chapters 5, 6 and 7, are designed according to the iterative model above. When we substitute (2.1) with load flow and state estimation equations, the parallel and distributed iterative algorithms discussed above become the corresponding parallel and distributed load flow and state estimation algorithms.

2.6.1 Asynchronous Distributed Computation

The communications system is essential to power system operations, and its reliability is critical for fulfilling the objectives of CCC, ISO, and RTO. Many interconnections around the world have defined their minimum operating reliability criteria, which require written operating instructions and procedures for each control area to enable a continued operation of the system during the loss of telecommunication facilities. In the event of a total or partial failure or corruption of the communications network, it is required that there further be an alternative communication system that is tested periodically.

Area control centers are usually hundreds of miles apart, and data transmission must pass through a number of converting procedures, with some even relayed a few times. Therefore, time delays for the data communication between area control centers are often very hard to predict, although they are normally supposed to be within a certain limit. Accordingly, if totally synchronous algorithms were used for the distributed processing of power systems, the efficiency would be low. To solve this problem, asynchronous algorithms are specially designed for the distributed processing of DEMS.

Asynchronous distributed processing does not have to wait for the data communication at every computation step, and the local computation can proceed by using the information at hand. However, convergence conditions of asynchronous algorithms are more stringent than their synchronous counterparts because asynchronous algorithms require convergence without waiting for the communicated data. In addition, since there will be no need to wait for data communication, the design of asynchronous algorithms will become quite different from their synchronous counterparts.

Asynchronous algorithms can certainly be utilized under synchronous circumstances. Moreover, in many cases, asynchronous algorithms can obtain a faster solution speed than their synchronous counterparts without waiting for data communication; they can even get a faster speed than their serial counterparts because of the possibility of acceleration in asynchronization However, the asynchronous convergence analysis becomes more complicated [Ber89]. Since strict convergence conditions are to be satisfied, some of the problems cannot be solved by asynchronous computation.

2.7 CONVERGENCE OF ASYNCHRONOUS ALGORITHMS

Parallel iteration, such as the Newton method for load flow, updates variables once at every iterative step. The synchronization scheme inside parallel algorithms guarantees the proper convergence of the algorithms. However, for some distributed systems it is difficult to realize the synchronization because communication delays between processors are hard to predict due to long-distance data communications. Even if the synchronization can be realized, the computation efficiency could be very low because processors have to spend a substantial amount of time for synchronization. Asynchronous iterations can replace synchronous iterations in this circumstance. However, the convergence of asynchronous algorithms becomes problematic because the asynchronous iteration will not wait for synchronization.

In most cases, asynchronous algorithms pose more stringent convergence conditions than their synchronous counterparts. Generally, the convergence analysis of asynchronous algorithms is very complicated. Later we will explore in depth the convergence conditions for the asynchronous parallel and distributed load flow and state estimation algorithms in Chapters 5, 6, and 7. It is imperative to present a general pattern for proving the convergence of asynchronous algorithms and provide a set of the sufficient conditions that can guarantee the convergence of asynchronous iterative algorithms, especially when applied to some fixed-point problems as discussed in Section 2.6.

Suppose that there is a sequence of nonempty set $\{X(k)\}$, with $\cdots \subset X(k+1) \subset X(k) \subset \cdots \subset X$ which satisfies the following two conditions:

- **Synchronous convergence condition** For any k and any $x \in X(k)$, there is $f(x) \in X(k+1)$. Moreover, if $\{y^k\}$ is a sequence such that for every k there is $y^k \in X(k)$, then every limiting point of $\{y^k\}$ is a fixed point of f. The synchronous convergence condition implies that if x belongs to $X(0)$, the limiting points of sequences generated by synchronous iteration are fixed points of f.

- **Box condition** For every k, there are sets $X_i(k) \subset X_i$ so that $X(k) = \subset X_1(k) \times X_2(k) \times \cdots \times \subset X_m(k)$. Then this Box condition implies that, by combining components of vectors in $X(k)$, we obtain a new element of $X(k)$. For instance, if $x \in X(k)$ and $\bar{x} \in X(k)$, then we get an element of $X(k)$ by replacing the ith element of x with the ith element of $\bar{x}$.

Accordingly, we present the following sufficient conditions for the asynchronous iteration.

Asynchronous Convergence Theorem If synchronous convergence and Box conditions hold, and the initial solution $x(0) = (x_1(0), \cdots, x_m(0))$ belongs to the set $X(0)$, then every limiting point of $\{x(t)\}$ is a fixed point of f.

This asynchronous convergence theorem provides a foundation for the convergence analysis of asynchronous load flow and state estimation algorithms.

2.8 DISTRIBUTED PROCESSING OF VERTICALLY INTEGRATED POWER SYSTEMS

A power system is conventionally divided into transmission and distribution systems, which are distinguished by their voltage levels and network structures. The transmission system is generally composed of power plants and transmission lines ranging above 110 kv. The distribution system is composed of distribution networks extending from the transformers in substations to transformers at customer sides, with voltages ranging below 35 kv. The transmission system usually has a meshed network topology, while the distribution system usually has a radial structure with a high resistance. One consideration for the separation of the distribution system from the transmission system is that these two systems have different concerns for the real-time monitoring and control. The real-time monitoring and control of transmission system focuses on the system reliability and security, while the real-time monitoring and control of distribution system focuses on service qualities such as the optimal reconfiguration of distribution networks.

2.8.1 Distributed Processing of Transmission Systems

As we saw in Chapter 1, a CCC and a number of SACCs are responsible for the secure operation of the entire vertically integrated power system. The CCC is not necessarily located in a central place, but it is responsible for the security monitoring and control of the whole system. A SACC can be taken as an agent of CCC with limited rights and responsibility and is only responsible for the monitoring and control of a local area assigned by CCC. Figure 2.13 depicts the hierarchical and distributed processing of transmission system.

As shown in Figure 2.13, together CCC and SACCs form a hierarchical and distributed system. During the daily operation of the system, CCC bypasses trivial events with minute effect on the entire system and leaves them to associated SACCs, which are better equipped to deal with local events. If coordination is needed for the optimization and control of the entire system, CCC will send corresponding instructions to individual SACCs according to their respective boundary conditions.

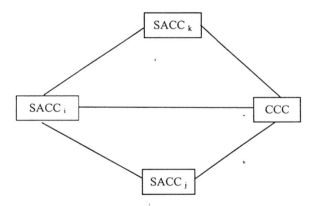

Figure 2.13 Hierarchical and Distributed Processing of Transmission System

The hierarchical and distributed transmission system may have an irregular network topology as the system is irregularly partitioned with different geographical distances among control centers. Besides, control areas of SACCs may vary in size: computers at SACCs may be heterogeneous, and communication links among SACCs may constitute different communication media. The communication links between CCC and SACCs are usually very long, and communication networks are a crucial component of this distributed system. The hierarchical and distributed transmission system is shown in Figure 2.14, which is extracted from Figure 2.13.

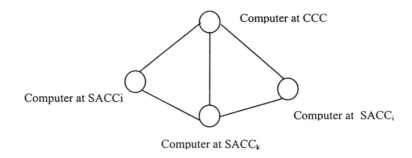

Figure 2.14 Distributed Transmission System

Like CCC, a SACC is equipped with SCADA/EMS with the following basic functions:

- Real-time data acquisition
- Real-time monitoring and control based on state estimation, load flow, and security analysis
- Data communication with CCC

SACCs have greatly reduced the real-time computation and control burden of CCC, since they deal with most events locally. The CCC, in turn, focuses on monitoring and control of tie lines among control areas and acts as a coordinator for the distributed computation and control of SACCs. As CCC and SACCs complete a common task such as the load flow computation in a distributed manner, the EMS at each control center constitutes a distributed energy management system (DEMS) that performs its functions based on the distributed system shown in Figure 2.14.

2.8.2 Distributed Processing of Distribution Systems

In the vertically integrated power system, the distribution system operator (DSO) at the control center is responsible for the real-time monitoring and control of the distribution system. A large amount of real-time data needs to be processed centrally for this purpose. An alternative way to the real-time monitoring and control of a distribution system is to adopt distributed processing.

As shown in Figure 2.15, the distribution system is comprised of a number of distribution networks that are separate and independent regardless of their interactions with the transmission system. A distribution network operator (DNO) at a local control center operates a distribution network, and the DSO and all DNOs constitute a distributed system as depicted in Figure 2.16.

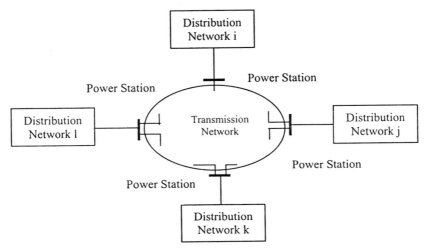

Figure 2.15 Networks of Distribution System

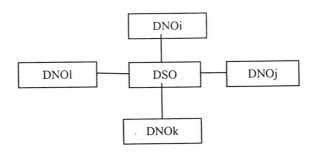

Figure 2.16 Distributed System of Distribution System

Rarely is there any physical connection among distribution networks. Almost no direct communication occurs between DNOs, and DNOs primarily communicates with the DSO.

A distribution network has feeders that are stemming out of substations; there are several laterals along a feeder with branches that are connected to loads. Therefore, a distribution network would have a tree-like radial topology. Each feeder is monitored and controlled by a computer at DNO and the participating computers form a distributed system as shown in Figure 2.17. Because every feeder tends to operate in an open-loop state, there would be no direct communication among feeder computers.

The distributed processing of distribution networks can be further refined. A feeder usually would have a number of laterals, and these laterals could be monitored and controlled by dedicated computers. The distributed system has a radial topology, as shown in Figure 2.18, and the representation of such distributed system is shown in Figure 2.19.

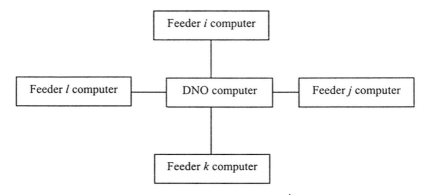

Figure 2.17 Distributed System for Distribution Network

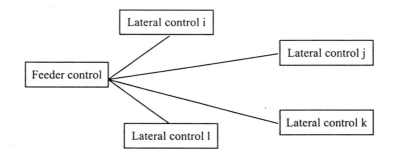

Figure 2.18 Distributed System for Distribution Network

The DSO and DNOs use SCADA/DMS to fulfill their missions. One main function of SCADA is to collect the real-time information from customers and provide value-added information such as electricity price to customers. Similar to the DEMS of transmission system, DEM can also adopt the distributed processing to attain management flexibility and save investments in communication networks.

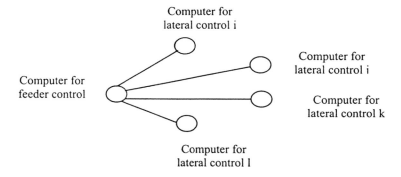

Figure 2.19 Distributed System for Distribution Feeder

2.9 DISTRIBUTED PROCESSING OF RESTRUCTURED POWER SYSTEMS

2.9.1 Distributed Processing System of an ISO

With the restructuring of the power system, both CCC and ACCs are being replaced by the ISO CC (control center) and the ISO SCCs (satellite control centers). However, the computer system architecture for the distributed processing system of a power system would almost remain unchanged. The distributed processing system corresponding to Figure 1.5 is shown in Figure 2.20.

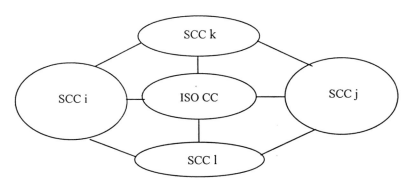

Figure 2.20 Distributed System for ISO

As required by the FERC Order 888, an ISO should establish a cooperation mechanism with its neighboring control areas, or ISOs, in the case of emergency. An RTO will act as the coordinator when geographically distributed ISOs cooperate in real time for the optimization and control of a power system.

2.9.2 Distributed Processing System of an RTO

The purpose of establishing RTOs in the restructured power system is to overcome pancaked transmission tariffs and to provide transmission customers with better and non-discriminatory services. The creation of an RTO signifies the unification of ISOs in restructured power systems. In this sense, it would be essential for an RTO to adopt the distributed processing for the real-time monitoring and control of a large-scale power system. This is why we put more emphases on the RTOs' hierarchical and distributed computation and control in this book. The distributed processing system of a RTO is depicted as in Figure 2.21.

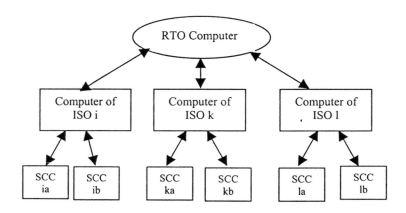

Figure 2.21 Distributed System for RTO

In both the ISO and RTO control centers, EMS is the core tool for the real-time monitoring and control of a power system. Therefore, the implementation of a distributed computation and control of an RTO refers mainly to the distributed implementation of EMS functions. In such a complex hierarchical and distributed processing system, the hierarchical optimization and control methodologies will be necessary for most applications. Based on DEMS, the hierarchical and distributed pyramid of an RTO is shown in Figure 2.22.

In restructured power systems, RTOs appear to be interconnected, though their interactions are usually weak. However, with the rapid development in power system, restructuring, weak interconnections among

PARALLEL AND DISTRIBUTED POWER SYSTEMS

RTOs could become strong and the cooperation among RTOs could become indispensable. This strong interconnection among complicated and dynamical energy systems will prove essential in emergency conditions.

As required by the FERC Order 2000, an RTO should establish a cooperating mechanism with its adjacent control areas or RTOs in the case of an emergency. Similar to the cooperation among adjacent ISOs, when multiple RTOs cooperate in the real-time analysis and control of a power system, their cooperation can be based on the distributed processing system of RTOs, as shown in Figure 2.23. To better coordinate the RTOs in an interconnected power system, one RTO would need to be the designated coordinator of the participating RTOs. The coordinator could be regarded as a super RTO.

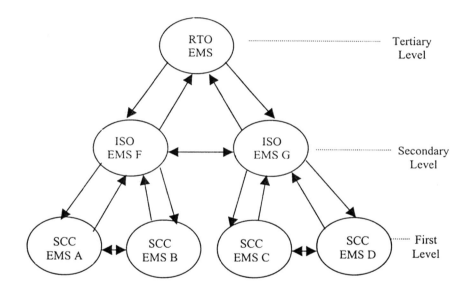

Figure 2.22 Hierarchical and Distributed System for RTO

2.9.3 Distributed Processing of GENCOs, TRANSCOs, and DISTCOs

GENCOs, TRANSCOs, and DISTCOs may individually possess their own computer networks for business purposes. For instance, a GENCO may have several power plants that are geographically distributed in the power system; a network of computers located at these power plants could be instrumental to the operation and control of these plants as one entity and to the information exchange among these member power plants. The establishment of such a computer network could even rely on the public Internet for data communication. Likewise, when a TRANSCO has several transmission companies, it may build up a dedicated computer network for its own purposes. The distributed processing system of a GENCO or TRANSCO would have an irregular network topology because power plants of a GENCO and transmission companies of a TRANSCO may be distributed irregularly in the power system.

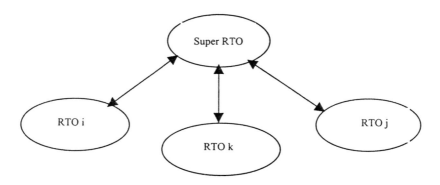

Figure 2.23 Distributed System for RTOs

Unlike distribution utilities in vertically integrated power systems which were responsible for the operational reliability of distribution network, DISTCOs in restructured power systems may not be responsible for the monitoring and control of the distribution system. Rather, the distribution system is monitored and controlled by DNO (distribution network operator). The distributed processing system shown in Figure 2.16 is suitable for the distributed computation and control of distribution systems. Therefore, a DNO will execute the distributed computation and control similar to that shown in Figure 2.23. A DISTCO would build a distributed computer network to provide customers with more value-added information and better services. A DISTCO could also execute a distributed computation based on its own distributed system, which would represent a new trend in the electric utility service management.

2.10 DISTRIBUTED ENERGY TRADING SYSTEMS

In restructured power systems, energy trading is realized through auction in the power market. Energy producers and consumers submit their bids to the power exchange for competition. Ancillary services are also auctioned on OASIS at the ISO control center. Information exchanges between market participants and the power exchange is implemented mostly based on computer networks, that is either the public Internet or the dedicated computer network of the ISO. With the advancement of restructuring, additional energy-trading brokers will appear in the power system, and this will require extended energy trading systems for interfacing with market participants.

2.11 COMPUTER NETWORKS FOR DISTRIBUTED PROCESSING OF POWER SYSTEMS

In a restructured power system, an RTO/ISO could use the following four types of computer networks for distributed processing:

- LAN
- ISO's private WAN
- NERC ISN
- The public Internet

These four computer networks exhibit different performances when they are used for distributed computation and control of power systems. The main concern in selecting any of these networks is whether the data can be transmitted with a satisfactory speed and level of security.

2.11.1 LAN

Since LAN can greatly facilitate resources sharing, most computers at power system control centers are connected to a LAN. When used for distributed processing, LAN could exhibit better performance and efficiency than serial processing. For instance, when conventional EMS is reorganized into DEMS, which is based on a LAN at the control center, the performance and efficiency of EMS can be greatly improved. Because LAN is an intranet, in most cases there will only be a limited data communication on the network. It is perceived that the data communication among individual computers on LAN is quite reliable. For instance, when MPI is used for the distributed load flow computation on LAN, the performance would be as good as a parallel machine using the same communication scheme.

2.11.2 ISO's Private WAN

The ISO's private WAN is particularly built to interconnect the ISO's control center and all its subarea control centers for data communication.

The major communication protocols for this intranet include OSI (open system interconnection), ICCP (Inter-control center communications protocol), and ELCOM (electricity utilities communication). OSI works more effectively than TCP/IP for the online and distributed computation and control because it does not use time tags. ICCP and ELCOM support communications of various data such as measurements/analog values, digital values, binary commands, analog and digital set points, and text messages. The distributed computation and control applications such as distributed load flow, distributed state estimation, and distributed voltage control can also be implemented on this WAN since the ISO's private WAN uses wideband communication links, and data communications are mostly limited to that between control centers.

2.11.3 NERC ISN

As required by NERC, an ISO must be able to quickly and reliably communicate with neighboring security coordinators. The NERC ISN (Inter-regional Security Network) is used for the ISO's communication with other NERC security coordinators, external control areas, and other ISOs. An ISO has access to the NERC ISN for the purpose of meeting the NERC security coordinator requirements. The protocol used for ISN access is ICCP which is known as IEC60870-6 TASE.2[2] and specifies database-oriented communications methods. ICCP only specifies methods for data transfer between utility control centers and will not be used for peer-to-peer data communication. The ICCP could be used to support the distributed processing of system security applications, based on the NERC ISN, such as the distributed external equivalent and security analysis.

2.11.4 The Public Internet

Because of its ability to reach every site within power systems, the Internet becomes the most convenient and useful facility for an RTO/ISO's distributed processing. TCP/IP provides the possibility for distributed processing based on the Internet; the major issues here are the efficiency, the reliability, and the security of data communication over the Internet. The data transmission speed becomes the key issue because the communication congestion often occurs randomly on the Internet. TCP/IP was originally designed to transport data streams from one point to another or from one application to another. Because TCP/IP streams use timers to

---

[2] TASE.2 stands for Telecontrol Application Service Element Number 2. See Chapter 4 for more details.

detect the end of the stream of data, TCP/IP may sit idle for a moment while it is waiting to see if more data will come through. However, for applications that are not time-critical, the Internet is the best option for data communications.

The process of data communications over the Internet is described more specifically as follows: Suppose that the SCADA data are to be exchanged between two subarea control centers, namely SACC A and SACC B. The SCADA data of SACC A will be first attached to a company-owned router, which then sends out a single packet containing the name of the Web site of SACC B to the nearest domain name server that will translate the address. This packet will enter the Internet backbone and core routers, which will figure out the optimal network paths; these are paths that offer the least number of links and are least congested. The packet leaves the backbone at the distant Internet network router and reaches the FTP (File Transfer Protocol) server of SACC B. When this FTP server acknowledges the data communication request of SACC A, the data exchange between the SACC A and SACC B starts. SACC A may send many packets of real-time SCADA data to SACC B that will acknowledge the receipt of the data. This process is depicted in Figure 2.24.

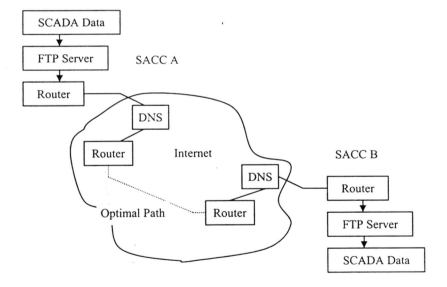

Figure 2.24 Data Communication Over Internet

As indicated in Figure 2.24, the key issue for efficient data transmission over the Internet is to find the optimal communication path. However, the problem is that for many applications the optimal path may not meet the time requirement. For this reason, time-critical functions, such as the online distributed load flow, are difficult to implement over the Internet.

To improve the performance of data communications over the Internet, ISPs (Internet service providers) offer many value-added services for time-sensitive applications. First, the security of data transfer is guaranteed by adding AH (authentication header) and ESP (encapsulating security payload) in communication protocols. Thus, the integrity of data is ensured. Second, the reliability of data connection is guaranteed by using at least three backup paths in each switching node. Third, the time delay experienced by a packet is guaranteed not to exceed a certain value that depends on the service class and the features of communication links. However, the issue is difficult to solve by only the IPSs. The data transmission congestion on the Internet is random and mostly user-related. As many users visit a Web site at the same time, the paths to this Web site can become congested. For this reason, it is difficult to guarantee that time-critical data will be transmitted continuously with satisfactory speed over the Internet.

Certain security measures are taken to protect the data exchange over the Internet. Generally, the Internet security infrastructure includes the following three basic components

- A firewall to enforce the access policy
- Proxies to regulate services through the firewall
- Packet gateways to pass permitted packets through the firewall

The encryption and authentication are effective measures for the data exchange security. However, these measures increase the complexity of data exchange, so more processing time is used in exchanging the data over the Internet.

The Internet can be used for many purposes: from generic information sharing to the proposed distributed computation. Internet interfacing with the SCADA system at the system's control center might be a favorable way to monitor the system, and the ATM (asynchronous transfer mode) based data communication might be used for priority controls of power systems [Sumi98, Meul98, Seri99, Doi99]. Since the time delay in data communication over the Internet is acceptable for some applications, studies of data transmission via the public Internet may help determine which applications are suitable in terms of their time requirements. Once these applications have been determined, some applications could be moved from the ISO private network to the Internet, and thus the data communications burden of the ISO private network would be mitigated.

2.12 MESSAGE-PASSING INTERFACE

MPI provides specifications for message-passing libraries, and thus forms a widely used message-passing standard for parallel and distributed processing. MPI in fact has become the standard and most powerful message-passing library for parallel and distributed systems including massively parallel machines, SMP clusters, workstation clusters, and heterogeneous networks. There are various implementations available from vendors as well as in public domain. Some parallel machine vendors have created their own MPI implementations that are capable of exploiting native hardware features to optimize performance. MPI programs are portable; there is no need for modifying programs when applications are migrated to a different platform.

All processes in a message-passing system communicate by sending or receiving messages. Programs designed using MPI must explicitly identify all parallelism and implement the algorithms in terms of MPI constructs.

In the following, we introduce the history and main functions of MPI as we demonstrate the use of MPI for parallel and distributed computation of power systems. We also discuss MPI attributes in order to help readers comprehend the application of MPI to parallel and distributed programming.

2.12.1 Development History and Features

MPI was developed through the efforts of many individuals and groups. The objective behind MPI was to establish a practical, portable, efficient, and flexible standard for message passing. Parallel and distributed computing began to develop quickly in 1980s and early 1990s. When a number of incompatible software tools for parallel and distributed computing appeared with trade-offs between portability, performance, functionality, and price, a need for a standard arose.

In April 1992 the Workshop on Standards for Message Passing in DEM (distributed memory environment) was held, which was sponsored by the Center for Research on Parallel Computing in Williamsburg, Virginia. The workshop discussed the basic features essential to a standard MPI, and a working group was established to begin the standardization process. A preliminary draft proposal was developed soon after, and the working group met during November 1992 in Minneapolis. The MPI draft proposal (MPI-1) from Oak Ridge National Laboratory (ORNL) was presented at that meeting and the group adopted procedures and organization to form the MPI Forum (MPIF). MPIF eventually included 175 individuals from 40 organizations, among these parallel computer vendors, software writers, academics, and application scientists. A draft of the MPI standard was presented at the Supercomputing 93 Conference and the final version of MPI-1 was released at http://www.mcs.anl.gov/ Projects/ mpi/ standard.html.

MPI-1 can be used with C and Fortran languages, and the number of tasks dedicated to run a parallel program is static, which means new tasks cannot be dynamically spawned during run time. MPIF began discussing the enhancements to the MPI standard in March 1995. In December 1995, at the Supercomputing 95 conference, the Birds of a Feather meeting discussed the MPI-2 extension issues. In November 1996, at the Supercomputing 96 conference, an MPI-2 draft was made available and comments were solicited. The MPI-2 document is available at www.ERC.MsState.Edu/mpi/mpi2.html.

The key enhancements that have now become a part of the MPI-2 standard included the following issues:

- Dynamic processes: Removes the static process model of MPI and provides routines to create new processes.

- One-sided communications: Provides routines for one-directional communications; includes shared memory operations and remote accumulate operations.

- Extended collective operations: Allows for non-blocking collective operations and application of collective operations to inter-communicators

- External interfaces: Defines routines that allow developers to create a layer on the top of MPI, such as for debuggers and profilers.

- Additional language bindings: Describes C++ and Fortran90 bindings.

- Parallel I/O: Discusses MPI support for parallel I/O.

2.12.2 Data Communication

In MPI, data communication is realized through the execution of various communication routines. Most MPI communication routines require programmers to specify a communicator as an argument. A communicator or group is an ordered set of processes. Therefore, in the initialization process, MPI first uses communicators and groups to define which collection of processes may communicate with others. Each process in a communicator or group is associated with a unique integer rank, which is also referred to as a process ID. All ranks of a communicator or group are contiguous and begin at zero. In the MPI communication, ranks are mainly used by programmers to specify the source and destination of messages. In parallel and distributed processing, ranks can also be used by applications to conditionally control the program execution. For example, in a parallel or distributed program, we can specify a code as "if rank=x do this" or " if rank=y do that."

MPI communications routines are specified in two categories: point-to-point communication routines and collective communication routines. These routines are discussed as follows:

Point-to-Point Communication Routines

MPI point-to-point communication routines are utilized for data communications between a sending processor and a receiving processor. The argument list of these routines takes one of the following formats:

Blocking communication

>MPI_Send (*buffer, count, type, dest, tag, comm*)

>MPI_Recv (*buffer, count, type, source, tag, comm, status*)

Non-blocking communication

>MPI_Isend (*buffer, count, type, dest, tag, comm, request*)

>MPI_Irecv (*buffer, count, type, source, tag, comm, request*)

The arguments of these communication routines are explained as follows:

A *buffer* is the application address space that references the data to be sent or received. In most cases, this is simply the variable name that is to be sent/received. For C programs this argument is passed by the reference and usually must be pre-pended with an ampersand.

A data *count* indicates the number of data elements of a particular type to be sent or received.

A data *type* must use the data types predefined by MPI, though programmers can also create their derived types. We should note that the MPI types, MPI_BYTE, and MPI_PACKED do not correspond to standard C or Fortran types.

A *dest* indicates the send process where a message should be delivered, it is specified as the rank of the receiving process.

A *source* indicates the originating process of the message; as the counterpart of the *dest*, it also specifies the rank of the sending process. The *source* may be set to the wild card MPI_ANY_SOURCE to receive a message from any task.

The *tag* is an arbitrary non-negative integer assigned by the programmer to uniquely identify a message. The send and receive operations should match message tags. For a receive operation, the wild card MPI_ANY_TAG can be used to receive any message regardless of its tag. The MPI standard guarantees that integers 0-32767 can be used as tags, but most implementations allow much wider range than this.

A *comm* represents the designated communicator and indicates the communication context, or the set of processes for which the source or destination fields are valid. The predefined communicator MPI_COMM_WORLD is usually used, unless a programmer is explicitly creating new communicators.

A *status* indicates the source of the message and the tag of the message for a receive operation. In C, this argument is a pointer to a predefined structure MPI_Status. In Fortran, it is an integer array of size MPI_STATUS_SIZE. Additionally the actual number of bytes received is obtainable from *status* via the MPI_GET_COUNT routine.

A *request* is used by non-blocking send and receive operations. Since non-blocking operations may return before the requested system buffer space is obtained, the system issues a unique request number. The programmer uses this system assigned handle later to determine completion of the non-blocking operation. In C, this argument is a pointer to a predefined structure MPI_Request. In Fortran, it is an integer.

In the following, we present an example of point-to-point communication. Suppose that in the distributed load flow computation, $SACC_i$ and $SACC_j$ will exchange boundary state variables. For simplicity but without losing generality, we assume that $SACC_i$ will only send the magnitude of one boundary voltage variable. If the ranks of $SACC_i$ and $SACC_j$ are assigned by the distributed system are x and y, respectively, then $SACC_i$ and $SACC_j$ can use the following function calls to realize synchronous communication:

$SACC_i$ MPI_Send (v_k, 1, MPI_FLOAT, y, 8, comm)

$SACC_j$ MPI_Recv (v_b, 1, MPI_FLOAT, x, 8, comm, status)

$SACC_i$ and $SACC_j$ can use the following function calls to realize the asynchronous communication:

$SACC_i$ MPI_Isend (v_k, 1, MPI_FLOAT, y, 8, comm, 5)

$SACC_j$ MPI_Irecv (v_b, 1, MPI_FLOAT, x, 8, comm, 7)

Collective Communication Routines

A collective communication involves all processes in the scope of a communicator, which are, by default, all members in the communicator MPI_COMM_WORLD. Collective operations within a subset of processes are accomplished by first partitioning the members into subsets and then attaching each subset to a new communicator. Collective communication routines do not take message tag arguments, and work only with the MPI-defined data types and not with derived types.

Collective communication routines are mainly used for the following three types of collective operations, which are all blocking:

- Synchronization processes wait until all members of the group have reached the synchronization point
- Data movement broadcast, scatter/gather, all to all
- Collective computation one member of the group collects data from the other members and performs an operation on that data

Some useful MPI collective communication routines in C are listed and explained as follows:

MPI_Barrier (*comm*): MPI_Barrier creates barrier synchronization in a group. Each task, upon reaching the MPI_Barrier call, blocks until all tasks in the group reach the same MPI_Barrier call. For example, this MPI_Barrier function call could be used to realize synchronization in parallel and distributed processing of power system.

MPI_Bcast (*buffer, count, datatype, root, comm*): MPI_Bcast broadcasts a message from the process with rank "root" to all other processes in the group. For example, the CCC of a power system can use this MPI_Bcast function call to send a time signal to all the ACCs in the system.

MPI_Scatter (*sendbuf, sendcnt, sendtype, recvbuf, recvcnt, recvtype, root, comm*): MPI_Scatter distributes distinct messages from a single source task to each task in the group. For example, the ISO can use this MPI_Scatter function call to inform the ACCs about the different LMPs in each subarea.

MPI_Gather (*sendbuf, sendcnt, sendtype, recvbuf, recvcount, recvtype, root, comm*): MPI_Gather directs distinct messages from each task in the group to a single destination task. This routine is the reverse operation of MPI_Scatter. For example, the ISO can use this MPI_Gather function call to collect the voltages of the pivot buses from ACCs in each subarea.

MPI_Allgather (*sendbuf, sendcount, sendtype, recvbuf, recvcount, recvtype, comm*): MPI_Allgather concatenates data to all tasks in a group. Each task in the group performs a one-to-all broadcasting operation within the group. For example, in parallel load flow computation, if the load flow equations are not well partitioned so that every processor needs to communicate with all other processor for synchronization, this MPI_Allgather function call could be used by each processor to reduce communication times. Figure 2.25 illustrates this case. In each synchronous round, every processor needs to communicate with other processors four times when using point-to-point communication. However, with the help of MPI_Allgather, each processor would only need a one-time MPI_Allgather function call to complete the communication with other processors. So the communication times are greatly reduced.

MPI_Reduce (*sendbuf, recvbuf, count, datatype, op, root, comm*): MPI_Reduce applies a reduction operation on all tasks in the group and places the result in one task. For example, a GENCO can use this

PARALLEL AND DISTRIBUTED POWER SYSTEMS

MPI_Reduce function call to control the entry productions of its subordinated power plants.

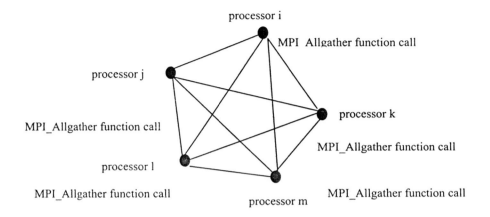

Figure 2.25 MPI_Allgather Application In Parallel Load Flow Computation

2.12.3 Attributes of MPI

In general, the procedure to execute the parallel or distributed processing using MPI is comprised of the following four steps:

i) Include MPI head files

   # include "mpi.h"

   *main(int argc, char *argv){*

   }

ii) Initialize MPI computing environment

   MPI_Init(&argc, &argv);

   MPI_Comm_size(MPI_COMM_WORLD,&numprocs);

   MPI_Comm_rank(MPI_COMM_WORLD,&pid);

iii) Parallelize computation and make message passing calls

   If (pid==i) {

   MPI_Send (buffer, count, type, dest, tag, comm);
   MPI_Recv (buffer, count, type, source, tag, comm, status);

}
iv) Terminate MPI environment

MPI_Finalize();

To further demonstrate the application of MPI to parallel and distributed computation and to compile and execute MPI programs, we consider a sample MPI program, parallel_pi.c that uses the trapezoidal rule to estimate Pi:

parallel_pi.c source code

```
#include "mpi.h"
#include <stdio.h>
#include <math.h>
#define Pi 3.14159265358979323846264

double f(double x){
        return (4.0 / (1.0 + x*x));
}
main(int argc, char *argv[]){
   int i, n, done, myid, numprocs;
   double h, y, Pi, mypi, sum;

   MPI_Init (&argc, &argv);
   MPI_Comm_size (MPI_COMM_WORLD, &numprocs);
   MPI_Comm_rank (MPI_COMM_WORLD, &myid);

   n = 0;
   done = 0;

   while (!done)
   {
      if (myid == 0) {
              if (n==0) n=100; else n=0;
      }

      MPI_Bcast (&n, 1, MPI_INT, 0, MPI_COMM_WORLD);

      if (n == 0) done = 1;
```

```
        else {
                h = 1.0 / (double) n;
                sum = 0.0;
                for (i = myid + 1; i <= n; i += numprocs) {
                        y = h * ((double)i - 0.5);
                        sum += f(y);
        }

        mypi = h * sum;
        MPI_Reduce (&mypi, &Pi, 1, MPI_DOUBLE, MPI_SUM, 0,
                                        MPI_COMM_WORLD);

        if (myid == 0)  {
        printf ("Pi is approximately %.16f, Calculation error is %.16f\n",
                                        Pi, fabs(Pi - PI));
                }
        }
    }
    MPI_Finalize ();
}
```

Suppose that this parallel program is to be executed on a COW. The MPI environment variables need to be set up properly, and a *.rhosts* file needs to be created in the home directory, which lists the names or IP addresses of the available computers on COW. In addition to the current directory, there must be created a file, called *mymachines*, which lists the names or IP addresses of the computers to be utilized for this parallel computation. The parallel_pi.c program can be compiled as follows:

$ mpicc -o parallel_pi parallel_pi.c

The compiling operation generates an executable parallel program called parallel-pi. Suppose this program is to be executed by three processors. Then it can be executed as follows:

$mpirun -machinefile mymachines -np 3 parallel_pi

Then the following information will be shown on the screen: *Pi is approximately 3.1416009869231254, and calculation error is 0.000008333333332.*

2.13 OTHER FORMS AND TECHNIQUES FOR DISTRIBUTED PROCESSING

2.13.1 Client/server Architecture

Previously most enterprise applications were self-contained monolithic programs that had difficulties in using data from other sources. These applications were cumbersome to build and expensive to maintain because for even a simple function change, the entire program had to be rewritten, re-compiled, and re-tested. A client/server architecture has the scalability and robustness required to support mission-critical applications throughout the whole enterprise. Multiple users can visit the same file server simultaneously. A typical three-tiered client/server architecture contains a service tier, a data store tier, and a MMI (man-machine interface) tier. This architecture is usually object-oriented, server-based, and database-driven and is most suitable for Web-based applications.

File Sever

A file server is employed to store documents that can be accessed by authorized or public users. In the early stages of distributed processing, efficient data communication tools were not available, so a file sever was often used to provide a mechanism for information exchange among computers on a LAN. The file server mechanism was widely adopted by many enterprises as an effective way to exchange intra-company information. Figure 2.26 depicts a file server system where an electric utility uses the file server to store common files that are accessible in different departments.

This file server architecture had wide applications in the World Wide Web. Public clients could browse Web pages provided by ISPs (Internet service providers); clients could also publish their own messages on the Internet by uploading their documents onto the Internet servers if they are given the permission. The OASIS, as shown in Figure 2.27, is an example of this application. Besides, various e-commerce platforms including online energy-trading used file servers for information exchange, which included e-mail and voice mail services.

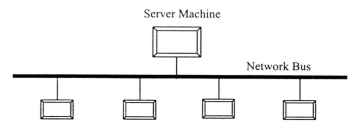

Figure 2.26 LAN with File Server

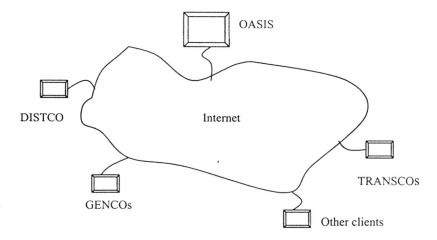

Figure 2.27 Client/Server Architecture of OASIS

When a LAN was used for distributed computation, participating computers had to write and read a common file saved on the server machine. With the increasing number of computers on the network, the efficiency of data communication of this file server architecture decreased; this is because the shared file could be read by all users at the same time but could be written by only one user at a time.

2.13.2 Network Programming

Network programming enables applications to retrieve any information stored in a local network or in the Internet. When distributed processing is programmed based on the Internet, no special hardware investment is needed for such applications. What users will need is a Web browser. Network programming is an ideal way for a user to publish information. For instance, the ISO publishes information on its OASIS Web server and thousands of customers visit the OASIS Web site at the same time and at various locations. Network programming can also be used for information exchange on the Internet. For instance, a TRANSCO can communicate with the ISO via email. When data communication time delay is no longer a problem, as it is for data communication on the private WAN of ISO, many computers on a network could be deployed in distributed processing.

2.13.3 Distributed Database

There are distributed systems other than those discussed in Section 2.2 that are designed for special purposes. For instance, the information exchange or distributed control of geographically distributed systems, with a regular or an irregular network topology, employs one of the following two communication schemes:

Distributed databases provide a mechanism for the information exchange in distributed processing. A distributed database is different from a centralized database as the latter provides data on one machine. A distributed database is composed of separate databases distributed on a computer network that are accessed by all permitted users. Information exchange may also occur among databases; in other words, a computer in the distributed database can access the data stored in other computers. Data consistency is guaranteed by intrinsic properties of databases.

One advantage of a distributed database is that local databases are used to store the local information in the system. Compared with a centralized database, a distributed database can save capital investments in communication networks with a higher security and reliability, since it can restrict local faults within the corresponding subareas.

Each SACC of a power system collects the real-time data of its subarea through its SCADA system, and stores the data in its local database. Though the information exchange between the CCC and SACCs is not actually based on a distributed database, as shown in Figure 2.28, the local databases form a distributed database from the perspective of ISO.

PARALLEL AND DISTRIBUTED POWER SYSTEMS

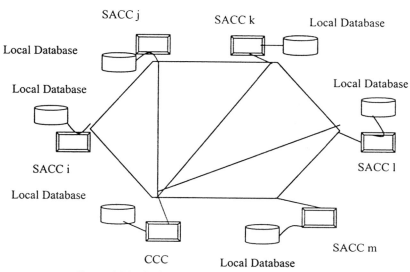

Figure 2.28 Distributed Database of Power Systems

Distributed databases can receive wide applications in power systems. For instance, a GENCO with power plants distributed throughout a power system can build a distributed database for information exchange and management. A DISTCO with several branch companies could also build its own distributed database.

2.13.4 Java Language

Java is uniquely suited for network programming. It provides solutions to many issues that are difficult to solve using other programming languages. Java applets have a higher level of safety than that of similar software packages. Java makes network programming much easier. Java applications and applet can communicate across the Internet; this feature is especially suitable for large-scale systems that are deployed over a large geographical area. Java is portable and platform-independent; Web applications are designed on various hardware and operating environments. Java executes applications in a run-time environment called a virtual machine. The virtual machine executes the platform-independent bytecode that a Java compiler generates and is easily incorporated into Web browsers or the kernel of the operating system. Java virtual machines and Java API's (application program interfaces) insulate Java programs from hardware dependencies; this is why Java's bytecode can run on a wide range of platforms. Java applications can run on a variety of platforms without modification.

Java is also used for database programming. SQLJ is a way to embed the SQL (structured query language) in Java programs and to reduce the development and maintenance costs of Java programs that require database connectivity. SQLJ provides a simple model for Java code containing SQL 'statements. SQLJ offers a much simpler and more productive programming API's than JDBC (Java database connectivity) to develop applications that access relational data, and it can communicate with multi-vendor databases using standard JDBC drivers.

Chapter 3

Information System for Control Centers

3.1 INTRODUCTION

As we saw in the preceding chapters, the ISO acts as a security coordinator for the entire control area, which is made up many subareas. Individual subarea control centers perform AGC (automatic generation control) and switching operations under the direction of the ISO. The transmission system is usually owned by a group of transmission owners, and the transmission capacity is limited by the physical characteristics of the grid. With its primary mission of ensuring the reliability of the transmission system, the ISO performs a wide range of functions. The ISO ensures that all transmission customers receive non-discriminatory and equal access to the transmission system under its jurisdiction and functional control. The ISO is responsible for the coordination of energy transfers to maximize the grid utilization in a manner that maintains the integrity of the grid. Besides administering transmission tariffs, the ISO facilitates planning studies for transmission expansion and generation siting.

To fulfill all these missions, the ISO should have a custodial trust relationship with transmission owners when performing the following functions:

- Maximizing transmission service revenues
- Distributing transmission service revenues to transmission owners
- Preventing damages to the transmission grid

3.2 ICCS IN POWER SYSTEMS

Because the ISO has to perform many complicated and interrelated operations, the ISO's control center infrastructure has formed an integrated control center system (ICCS), which is essentially an integrated information system [Web01, Web07, Web10-11] and has the following fundamental requirements:

- Flexible user interfaces to the ISO to view the system information, re-dispatch resources, and monitor and coordinate the security of transmission system.

- A reliable backup system to cover the possible loss of ICCS functionalities, all the real-time functions of ICCS should remain operational in any case of significant interruption;

- A distributed computing environment with LAN and WAN such that there would be no restrictions on the geographical dispersal of applications among ICCS computers.

- An application environment with adequate flexibility and what is economical and easy to maintain and upgrade.

All ICCS applications must conform to mainstream computing standards, and data communications of ICCS must conform to OSI standards, in particular IEC 870-6 and TCP/IP (Transmission Control Protocol/Internet Protocol) protocols.

ICCS consists of many interconnected elements and processes that must function continuously and reliably. To meet the requirement for reliability, ICCS is configured as a fully redundant distributed system so that there can be no single point of failure among critical ICCS processing units (PUs). Normally the ISO carries out its functions from a primary control center that is equipped with an exactly identical backup control center; these two control centers are equipped with the same ICCS hardware/software functionality and configuration [Nie89].

The ICCS communications system is comprised of communication interfaces and communication networks. Included are ICCS LANs, the ISO's private network, the NERC's ISN (Interregional Security Network), and the public Internet. Specially designed firewalls are installed to interface between these communications networks and to provide protection from possible security threats occurring from external

INFORMATION SYSTEM FOR CONTROL CENTERS 103

communication networks. Any data that are to be made public will be obtained from the information storage and retrieval (IS&R) system of ICCS via such a firewall. The IS&R system may consist of a commercial database management system for accommodating the long-term archival and retrieval of information produced by ICCS [Web01, Web10-14].

In this chapter we will discuss the configuration and services of ICCS including electronic tagging services, and the utilization of UCA (utility communications architecture) and ICCP (inter-control-center communication protocol) by ICCS [Rob95].

3.3 ICCS CONFIGURATION

3.3.1 ICCS LAN

Recall that the ICCS is a distributed system based on LAN. Various sophisticated ICCS applications are logically integrated on this LAN, which consists of a number of PUs dedicated to particular functions. Industry standards are used for the design and implementation of ICCS hardware, software, and user interfaces to ensure interoperability and software portability. This open architecture allows the addition of future functionality and the replacement of hardware without disruption to the initial ICCS. The LAN for this application comprises two parts: LAN for ICCS applications, which is called the ICCS LAN, and LAN for ICCS administration, which is called the ICCS administration LAN. The general architecture of ICCS LANs is depicted in Figure 3.1 [Dyl94, Bla90, Bri91].

As shown in Figure 3.1, an ICCS LAN adopts a redundant structure that consists of a primary LAN and a backup secondary LAN. Each PU is connected to these two LANs at the same time. Obviously this configuration provides a capability for meeting the specified availability and reliability in a cost-effective manner. The ICCS LAN is connected to the ISO's administrative LAN via a firewall to isolate ICCS from the administration systems LAN. The firewall provides the protection from possible security threats occurring at the administration system LAN. Various suitable user interfaces are provided for ISO operators to use ICCS for viewing the power system information, to re-dispatch resources, and to monitor and coordinate the security of power systems.

The ICCS function will be transferred automatically from a failed PU to an alternative PU as a failure occurs. Certain failure logic is applied to all PUs in ICCS. Each PU is at least monitored by another PU, configuration management software, or dedicated hardware so that appropriate actions are activated as necessary. Dedicated detection mechanisms are employed to detect software and hardware failures throughout the entire ICCS and are used to supervise the whole process of switching functions from a failed PU to a backup PU [Sko02].

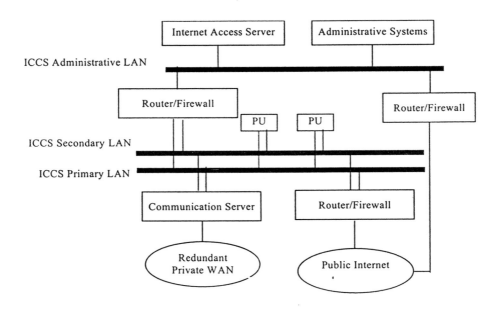

Figure 3.1 Architecture of ICCS LANs

In the case of a LAN failure, the LAN functions will be automatically transferred to an alternative LAN. The probability of two simultaneous LAN failures is very small but when it happens, ICCS functions at the ISO's primary control site can be entirely transferred to the ICCS at the ISO's backup site. Functions of any failed communication server would automatically be transferred to another available communication server. Likewise functions of any failed peripheral will be automatically transferred to another available peripheral. Any transfer will set off an alarm that announces a function transfer.

3.3.2 Availability and Redundancy of ICCS

The ability of each ICCS component to perform its specific tasks under normal conditions and during hardware and software failures is of paramount importance to the ISO. Hence, sufficient redundancy is

INFORMATION SYSTEM FOR CONTROL CENTERS

introduced in ICCS to guarantee that any single failure would only cause a brief interruption in the availability of that function. A functional processing interrupted by a failure would automatically be taken over by an alternative processor. Functional transfers are completed automatically and without any loss of data. Functions that were previously scheduled to be executed during a functional transfer would automatically be executed right after the completion of transfer.

A suitable redundancy should be provided at ICCS to prevent the loss of any critical ISO functionality. One design option indicates that the ISO could carry out its mission using a primary control center and a backup control center that are located remotely from each other. These two control centers have the same ICCS hardware and software functionality and configuration. The backup ICCS is provided to cover those situations where the loss of ICCS functionality could occur and last for an extended period of time. ICCS is configured as a single fully redundant distributed system so that there is no single point of failure among the critical ICCS PUs. Normally the ISO operates from the primary control center, which is continuously connected with an exactly identical backup control center. The primary ISO periodically checks the condition of the backup ICCS and keeps it initialized.

The backup ICCS is required to be in a ready condition without the need for on-site personnel while in the standby mode. In contingencies the backup ICCS is required to take over the entire functionality of the primary ICCS within a fraction of a minute. Once the primary ISO is ready to return to its normal service, it is initialized from the backup ICCS in order to reflect the current state of the power system. For a short period of time, both the primary and the backup ICCS need to operate in parallel as functionality is transferred back to the primary ICCS.

Redundancy for reliability is achieved both locally at each site and between the sites, as shown in Figure 3.2. ICCS LANs at each site are connected by a wide-band WAN communication link for coordinating the primary and backup sites. Although this link could be thousands of miles long and routers or bridges are installed at both terminals of the link, the LANs appear to ICCS as a single local redundant LAN for all ICCS processors.

For the sake of software reliability, ICCS applications data have a backup version that can be automatically brought up as part of a restart or transfer procedure. If a hot start-up is required, the start-up procedure will detect whether any data entry was lost and then notify users of the need for re-entry of the data. The hot start-up procedure will also detect whether any program-generated transaction was lost and when necessary will automatically initiate the recovery procedure to maintain application integrity [Web04-14].

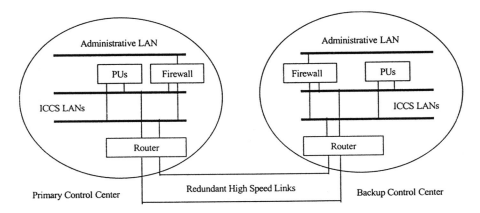

Figure 3.2 Redundant Structure of ICCS

3.4 INFORMATION SYSTEM FOR ICCS

The functional configuration of ICCS makes the ICCS to be a centralized information system that has the following features:

- Communication interfaces with a suitable firewall protection: these connect to external computer systems and services.
- Processor/server interfaces via a LAN: this interconnects various application functions.
- User interfaces: these consist of workstations, PCs and a rear projection video display matrix.
- Peripheral support facilities: these include telephones, printers, copiers, and fax machines.

As demonstrated in Figure 3.3, the ISO's functions are part of ICCS functions. The ICCS has the following major functions:

- Interfaces to communication networks connected to the ISO
- Communication services to collect and send information

INFORMATION SYSTEM FOR CONTROL CENTERS

- Data exchange and processing of information between the ISO and its participants
- Communications between the primary and the backup control centers
- Data models and databases for ICCS application functions.
- Information storage and retrieval applications used by other applications to create records of important information and to find and retrieve that information for use by ICCS applications, displays, and reports [Web10-11].

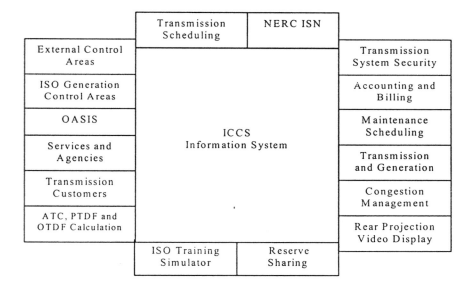

Figure 3.3 ICCS Functional Configuration

3.5 CCAPI FOR ICCS

ICCS users interact with ICCS through applications of executables, local datasets, and public datasets. Local data are private to a particular application and not a concern in plug-in interfaces. For the integration of ICCS applications, what matters is the way an executable in one application accesses another application's public data, or the way an executable exchanges messages with an executable in another application. In general, as defined in CCAPI (Control Center Application Program Interface [EPR96]), a runtime application space has a reference model as depicted in Figure 3.4.

There are two kinds of interfaces for applications to exchange information: message bus interface and data access interface. The message bus interface is a general message-passing interface that is used for program-to-program exchanges. The data access interface provides shared access to public data entities and is mainly used for program to dataset exchanges. Based on the model for a run-time application space, a CCAPI reference model for ICCS is depicted in Figure 3.5 [EPR96, Web04, Web07, Web10].

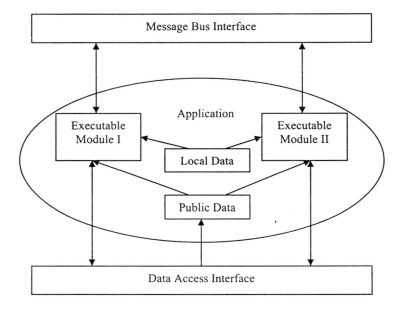

Figure 3.4 Model for Run-time Application Space

3.6 INTERFACES FOR ICCS USERS

Because ICCS is an integrated and centralized information system, ICCS users almost comprise all entities of a restructured power system, which include the ISO operators at the ISO's primary and backup control centers, transmission owners, transmission customers, and other users as shown in Figure 3.6. There are four departments of ISO, as shown in Figure 3.7, that are the major ICCS users. The respective functions of these departments are discussed next [Web10-14].

INFORMATION SYSTEM FOR CONTROL CENTERS

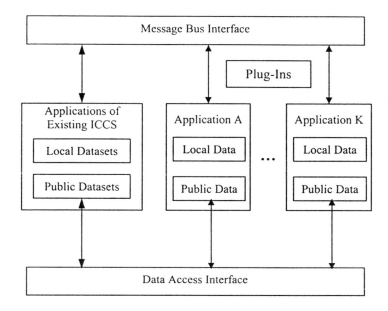

Figure 3.5 CCAPI Reference Model for ICCS

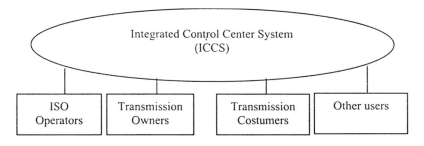

Figure 3.6 ICCS Users

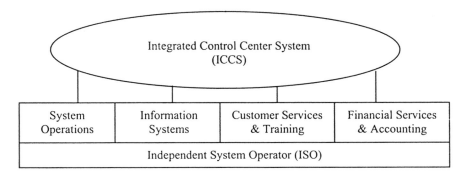

Figure 3.7 ISO Departments Related to ICCS Activities

- **Interfaces to System Operations.** The system operations department is responsible for the real-time administration of the ISO tariff, security coordination, scheduling and coordination of control area generation, operations support, transmission operational planning, and other operations engineering activities. Tasks to be fulfilled by this department mainly include real-time database maintenance and applications, technical operations, accounting and billing, communications, and hardware and software platform support.

- **Interfaces to Information System.** The information system department is responsible for all computer infrastructure activities, including application programming, data models, user interface, communications programming, and computer and communications hardware, as shown in Figure 3.8. Basic functions that information systems would provide include interfaces to communication networks connected to the ISO control center, communication services to collect information from and send information to various sources, data exchange and processing of information between the ISO and its participants, communications between the primary control center and the backup control center, and data models and databases to be used by the ICCS application functions.

- **Interfaces to Customer Services and Training.** The customer services and training department deals with customers for scheduling coordination, billing, and settlement questions, and providing information to loads and suppliers. It also provides training function for the ISO staff and for customers. The main functions are as follows: administer and register transmission customer applications for the ISO services, prepare procedure manuals for transmission customers, conduct transmission customer training, conduct the ISO staff training, coordinate meetings, coordinate transmission customer dispute resolution, and monitor the power market.

- **Interfaces to Financial Services and Accounting.** The financial services and accounting department handles the bookkeeping, billing, settlements, and accounting functions of the ISO. It mainly has the following functions: define the processes and procedures for transmission service settlement, define algorithms and formulas for accounting and billing, prepare transmission customers' invoices,

INFORMATION SYSTEM FOR CONTROL CENTERS

administer electronic transfers of funds, and administer the ISO's business accounting and billing.

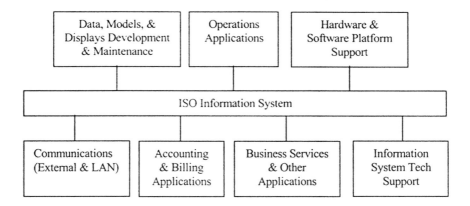

Figure 3.8 ISO Information System Functions

The latest graphics display standards are adopted for user interfaces in all ICCS functions. User interfaces and activities are normally accomplished through window, tool bar, menu, and icon operations using a mouse and keyboard. For convenience, consistent graphics standards are applied to all displays. To let the data with similar appearances have a consistent interpretation throughout ICCS, each function should be consistent in its use of graphics, commands, menus, colors, point and click procedures, and data entry.

In addition to the user interfaces discussed above, appropriate hardware and software are provided for ICCS to interface with other facilities including OASIS, time synchronization, and satellite communications backup, among other telecommunications services. The OASIS node of ICCS would meet FERC Standards and Communications Protocols (S&CP), and utilize both the public Internet and the ISO private intranet to interface with transmission customers. In other words, transmission customers can choose to use the public Internet or the ISO's private intranet to conduct business on OASIS, which requires the ICCS to support identical applications running over both the public Internet and the private intranet. Besides, interfaces to other communications service applications such as public telephone, NERC hotline, fax, weather services, voice recorder for system operations, and accounting and billing with the capability to e-mail conversations are also provided.

3.7 ICCS COMMUNICATION NETWORKS

The ISO provides the necessary components for ICCS communications infrastructure. In addition to ICCS LANs discussed in Section 3.1, the ISO would mainly use the following three types of computer networks to support various communication functions and services of ICCS:

- ISO's private WAN
- The public Internet
- NERC's interregional security network

These three networks are WAN computer networks. This section provides a general overview of the relationship between these three supporting communication networks and the corresponding ICCS functions [Web10-12, Web14].

3.7.1 Private WAN of the ISO

The ISO's infrastructure includes a private WAN that is dedicated to the ISO's participants. This network is primarily used for exchanging the data between ICCS and transmission owner control centers and serving the ISO's transmission customers. This private network is particularly built to interconnect the ISO's control center and all its member control centers which are irregularly distributed in the ISO's control territory. As part of this private network, a wideband link is utilized to connect the ISO's primary control center and the backup site. From the viewpoint of the ISO, this private network is a kind of intranet. All the ISO members' control centers would have access to this private intranet. Though it is also available to transmission customers, power exchanges, and regional reliability councils, this private intranet is principally used for data exchange between ICCS and the ISO members' control centers.

The structure of the private intranet can be of any form determined by the ISO according to its specific relation with its member control centers. Because the ISO members' control centers are usually irregularly distributed in power systems, and interconnected by different communication links, the structure of the ISO's private intranet can be quite irregular. Figure 3.9 shows a possible star type of structure for the ISO's private intranet. The reliability of the computer system in Figure 3.9 is often enhanced by a mirrored architecture across the two centers. This is analogous to the reliability of the ISO's private intranet, which consists of two separate networks: the primary WAN and the secondary WAN. This redundant configuration is illustrated in Figure 3.10.

INFORMATION SYSTEM FOR CONTROL CENTERS

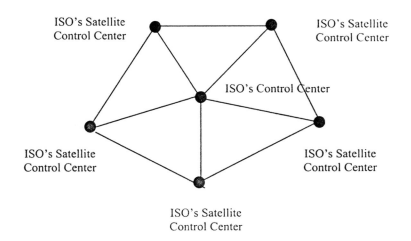

Figure 3.9 Architecture of the ISO's Private WAN

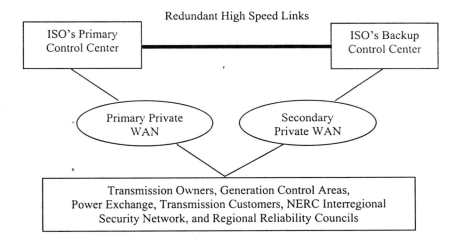

Figure 3.10 Redundant ISO Intranet Configurations

Normally the primary WAN is connected to the primary ISO's site and the secondary WAN is connected to the backup ISO's site. ICCS supports an access to redundant WANs from either the primary or the backup ISO's sites, regardless of the location of system operators or active PUs. A PU located at the primary ISO's site would have access to the communication server connected to the secondary WAN if the primary communication server fails or the primary WAN becomes unavailable, and vice versa. The two ISO's control centers coordinate their operations to ensure that at any time one site could operate in the primary mode, and the other in the backup mode. The same criterion applies to any operating PU and its complementary PU at each control center [Raj94].

Extensive data exchange would take place between the ICCS and the ISO members' control centers. An ICCP is used for ICCS to support the system monitoring and transaction processing [Eri97, Rob95, Vaa01]. Traditional real-time data are exchanged between the ISO and its members' control centers; these data are also accumulated for the historical storage, planning, and analysis. General messaging applications such as messaging communications, curtailment instructions, and dispatch requests are supported by this private intranet. Functions for file transfers are also supported, since common Internet applications such as Web servers, FTP servers, and e-mail servers are implemented over this private intranet. Likewise the dissemination of accounting information and time-related schedules such as planned generation schedules, transaction schedules, and ancillary service commitment schedules are supported.

3.7.2 NERC's Interregional Security Network

The NERC's interregional security network (ISN) is used for the ISO's communications with other NERC security coordinators, external control areas, and ISOs. As required by NERC orders, an ISO must be able to quickly and reliably communicate with neighboring security coordinators. The ISO has access to the NERC ISN for the purpose of meeting the NERC security coordinator requirements. The protocol used for ISN access is ICCP, which will not be used for peer-to-peer communication among transmission owners or between transmission owners and the ISO. The data exchange between the ISO and other security coordinators would support system security applications. When necessary, the ISO can also communicate with external control areas via the NERC ISN [Web11].

The ISO accesses the NERC ISN to collect real-time data from external control areas and to provide real-time data to other NERC security coordinators and external control areas. In many cases System Data Exchange (SDX) is also used to provide data exchange between the ISO and other security coordinators.

Figure 3.11 provides general diagram that shows how the three communications networks discussed above are interrelated and how they interface with the network users. The dotted communication lines between

INFORMATION SYSTEM FOR CONTROL CENTERS

the ISO's private WAN and the power exchange, regional reliability councils (RRCs) and transmission customers point out that these communication links are optional, because they can normally access the ISO through the public Internet.

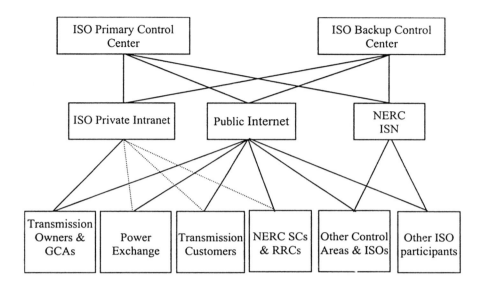

Figure 3.11 ICCS Communications Networks

3.7.3 Public Internet

The public Internet is the most common tool for the ISO to communicate with power market participants and OASIS. As a kind of public utility, the public Internet can be accessed by anyone who has been given an appropriate access authorization. For security reasons, suitable firewalls are built to prevent any unauthorized access to the ISO's control center via the Internet.

The ISO provides access for its member using both a private intranet as well as the public Internet. All services that are implemented on the ISO private intranet, including ICCP communications, are also made available over the Internet using the same communication and data processing applications. In addition to supporting access for its members to ICCS core services via the Internet connections, the ISO uses the Internet for more traditional administrative applications such as accessing Web sites, exchanging e-mails with non-member organizations, and downloading data files using FTP. However, consoles on the ICCS LAN and workstations on the administrative LAN will require the Internet applications for the completion of their daily functions and services.

Generally, transmission customers can access the ISO either via the ISO's private network or via the public Internet. Transmission owners and generation control areas normally access the ISO via the ISO's private intranet, but in the event that private intranet is not available, the communication will be via the public Internet. This switching process is quite automatic as identical software tools and services are provided by the ICCS for users to operate on these two networks.

3.8 ICCS TIME SYNCHRONIZATION

Since the concept of time is most critical to the ISO, satellite control centers, and power market participants, the system time at either the ISO control center or its satellite control centers is maintained locally using global position system (GPS) timeservers. ICCS is equipped with a time facility to determine the universal coordinated time (UCT), which is obtained from the global positioning system (GPS) satellite constellation. The ICCS is able to correctly interpret and distribute time stamps on time-sensitive data regardless of the local time zones to which the data applies.

Specifically, the ICCS time synchronization includes computer time synchronization, network time synchronization and OASIS time synchronization as discussed next [Web07, Web10-14].

- **Computer time synchronization.** Each of ICCS computers maintains a common internal calendar and 24-hour clock time. All applications on any of these computers are required to take into account any time-related issues such as local date and time that are shown on all displays and reports. When necessary, a single point adjustment is performed to keep the time synchronized.

- **Network time synchronization.** The ICCS network time is maintained for all component elements of ICCS. Distributed time services are used for synchronization among computers in the ICCS network. Every computer on the network periodically synchronizes to the time server, and all network computer clocks are automatically synchronized to within one microsecond of the time standard. When the time synchronization service is unavailable, the users can manually update the computer clock through the user interface. If a computer's internal clock and the time standard differ by more than an adjustable amount, an alarm message will be set off to inform the operator.

- **OASIS time synchronization.** The OASIS time synchronization is implemented by using the network time protocol (NTP) or by using other standard time signals such as GPS. Through time synchronization, the time stamps of all transactions on the same OASIS node can be accurate to within ± 0.5 seconds of the standard time.

3.9 UTILITY COMMUNICATIONS ARCHITECTURE

The integration of various power system applications was extremely complicated and costly because individual power system commercial software vendors had designed proprietary communications systems for their own products. To solve the problem of by providing a standard communications architecture for both electric utilities and vendors, in the late 1980s, EPRI initiated a project called Integrated Utility Communications (IUC) as a first step toward creating the necessary industry standards. The Utility Communications Architecture (UCA) is one of the IUC activities.

The objective behind the UCA effort was to build an information architecture that could meet the communication needs of electric utilities. At the beginning, UCA was used to clarify the types of information that electric utilities would need to communicate and the ways they would communicate; as these issues were resolved, UCA could be used to identify the types of protocol that utilities would use to perform these functions.

The long-term goal of UCA is to build an architecture that has vendor-independent communication tools, easily expandable services, and enterprise-wide access to the real-time information. To avoid the development of a utility-specific communication protocol, UCA has incorporated several existing international standards and technologies. Since no single transport protocol is perfect for all applications, UCA provides several protocols within communication layers from which vendors and integrators can choose to suit their different applications. A manufacturing message specifications (MMS) protocol serves as the application protocol for all applications [IEEE01].

UCA represents several advantages for utility operations. First, UCA facilitates integrations and allows utilities to choose the best-in-class equipment for their specific tasks. Second, UCA facilitates the use of distributed measurement, control, and communication schemes, and thus provides an alternative to all-in-one-box solution that usually provides more functionality than necessary. For instance, when a traditional EMS is replaced by an OFDEMS (open functional distributed EMS) [Wang97], which has several computers on a LAN, feasible EMS applications would increase. Moreover, if the LAN is fast enough, the distributed processing of OFDEMS applications would be practical. Third, depending on the communication specifications on a WAN, UCA can allow users' access to real-time data for certain functions. Thus system operators, planners, power brokers, and finance personnel can all benefit from the availability of the real-time information [EPR96].

3.9.1 Communication Scheme

The data communication in UCA is based on the seven-layer model of the OSI, which is a general model for network communication developed by the International Standards Organization. This model shown in Figure 3.12 is composed of seven separate layers with distinct functionalities.

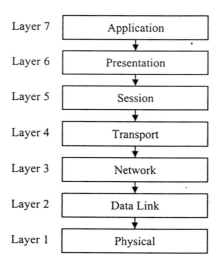

Figure 3.12 OSI Seven-Layer Model

More specifically, the OSI stack of seven layers is grouped into the following three sets of profiles:

- **A-Profiles.** Including application, presentation and session layers, which decide on the information packaging and format. In the presentation and session layers some necessary tags are added to the data that will be recognized by the communication protocol at the receiving end.

- **T-Profiles.** Including transport and network layers. The transport layer ensures a reliable message transfers across the network, and the network layer provides addressing and routing functions to allow the connection of multiple network segments to larger networks.

- **L-Profiles.** Including the last two layers of the stack, namely data-link and physical layers. The data-link layer controls the access to the network media, and the physical layer specifies wires, voltages, connectors, and other physical attributes of the system.

In the communication process, this seven-layer model will function as follows: At the sending end, the data generated from certain processes enter the application layer first, proceed through all layers to layer one, and then traverse across communication links to the receiving end. At the receiving end, the data make its way up through all the seven layers.

This seven-layer OSI model has been used for various networking schemes, such as TCP/IP, Microsoft, Novell, DNP, and UCA. Because of the prevalence of the Internet, most routers and gateways on networks are usually set up to deal with TCP/IP rather than OSI. TCP/IP was originally designed to transport data streams from one point or one application to another. Because TCP/IP streams use timers to detect the end of the data stream, TCP/IP may sit idle for a short while in awaiting the arrival of additional data. These time lags may be unacceptable for many time-critical functions of electric utilities. The OSI stacks are oriented toward packets rather than streams. These packets have a clear beginning and end signals. If a client receives an incomplete packet, it will be simply discarded and a new one may be requested at the same time. Hence, a client will not sit idle while awaiting the rest of the information. For this reason, OSI suits time-critical communications better than the TCP/IP.

The UCA component communication is illustrated in Figure 3.13, where the WAN could be the ISO's private WAN, the NERC ISN, or the public Internet. Every component could have a LAN as shown in Figure 3.14 [EPR96, Vaa01].

120 CHAPTER 3

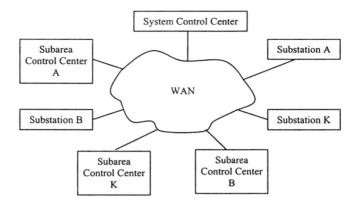

Figure 3.13 UCA Component Communications

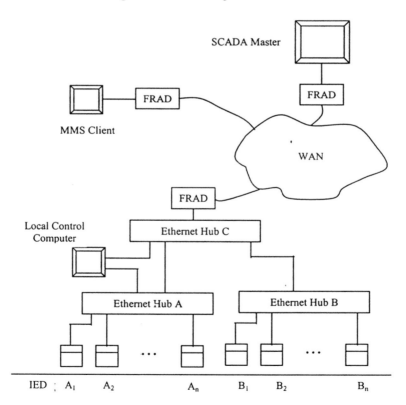

FRAD: Frame Relay Access Device
IED: intelligent electronic device

Figure 3.14 Conceptual LAN of UCA

INFORMATION SYSTEM FOR CONTROL CENTERS 121

3.9.2 Fundamental Components

UCA has four major components: manufacturing message specifications (MMS), common application service model (CASM), generic object models for substation and feeder equipment (GOMSFE) and ICCP [EPR96, EPR01a]. All are described in this section.

MMS. is an internationally standardized messaging system as defined in the International Organization for Standardization 9506. It is the application-level messaging protocol which is used for real-time data exchange applications; it is also applied to a real-time networked data exchange and supervisory control. MMS is very robust and has the capacity to perform functions that are requested by utilities. As a top-level application in the seven-layer data communication model, MMS is independent of transport layers that are located below it. No matter what networking protocols are used, one UCA compliant device always provides MMS at the application level to other devices on the network, which makes the integration of network components much simpler. Therefore, MMS has emerged as a protocol for implementing the UCA functionality. Whereas UCA was concerned with identifying functions that utilities would like to perform, the UCA 2.0 version is more concerned with methods and languages that allow devices from different vendors to work together in an electric utility substation.

CASM. gives the details of the processes that a communication service must follow within UCA, and defines these within the step-by-step logic flow of UCA operations. CASM does not specify protocols for executing services and thus is independent of any protocol. UCA maps services to MMS when it needs to perform actions specified in CASM.

GOMSFE. defines information categories, hierarchy, and naming conventions for the electric utility substation, and it therefore serves as a dictionary of names for equipment and functions within a substation. An IED (intelligent electronic device) in a substation will have all of its data and functions available to respond to these names. For data to be viewed, the location of the data within the organizational hierarchy of GOMSFE must be provided. The data in UCA is set up in a series of groupings like file folders. The names of these folders and their information are listed as follows:

- Remote terminal unit
- Measurement unit

- Load tap changer controller
- Capacitor bank controller
- Switch controller
- Automated switch controller
- Circuit breaker controller
- Re-closer controller
- Bay controller
- Power monitor
- Distribution feeder protection and control
- 138 kV transmission line protection and control
- 345 kV transmission line protection and control

The hierarchy of data proceeds from the top application level of the seven-layer data communication model. Any protocol functioning in this layer must be able to handle this organizational structure and services. The basic organizational scheme can be described as follows: Say the voltage VBX of bus X is required for a commercial application. A device with a unique alphanumeric code should exist on the network with a domain like Power_server5 within this device. There would be a set of information like MVM within this domain, within MVM there would be another level of classification such as MVMX, and within MVMX there could be another level such as MVMY. Suppose that VBX is within MVMY, then VBA can be accessed at the following address:

Domain = Power_server5
Object Name = MVM$MVMX$MVMY$VBX

GOMSFE also defines the way information will be provided. For a voltage measurement, a device can provide it as an integer or a floating-point value in the appropriate unit and format. The domain name is used here instead of the physical address of the device. There could be multiple domains at one physical address, as this allows a user to create several logical devices within one physical device on the network.

ICCP. specifies database-oriented communication methods within UCA. It is particularly designed for data communication between power system control centers. ICCP is known as IEC60870-6 TASE.2 (Telecontrol Application Service Element Number 2) [Eri97]. ICCP provides methods for the data transfer between utility control centers and defines UCA services in terms of MMS in the application layer. ICCP uses neither CASM nor GOMSFE. Compared with CASM and GOMSFE, which deal

INFORMATION SYSTEM FOR CONTROL CENTERS

with data in a device-centric view, ICCP defines the data similar to those in a SCADA system. ICCP is defined in terms of the client-server model of ISO/IEC 9506, and is modeled as one or more processes operating as a logical entity that perform certain communications [Gre92]. For example, for a model of a control center that includes several different classes of applications such as SCADA/EMS, DSM/load management, distributed application, and man/machine interface, the logical relationships of ICCP to the control center applications is that depicted by Figure 3.15.

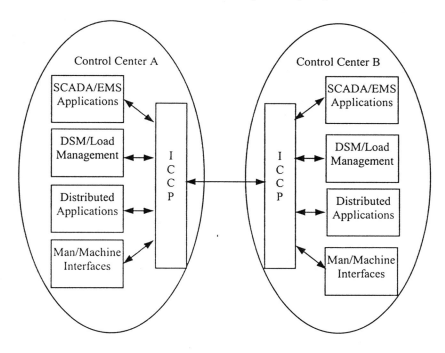

Figure 3.15 Relationship between ICCP and Control Centers

3.9.3 Interoperability

Previously one vendor could place bus X's voltage VBX in a register named XXXX when implementing DNP; other devices on the network such as RTU and PLC would need to be programmed with the location of this information. When a new vendor's product that uses the register YYYY to place voltage VBX of bus X is added to the network, all PLC, RTU, and the like, must be reprogrammed. There is no interoperability in this situation, since extensive changes have to be made to the existing system for each new device. Now with UCA 2.0 the integration of various products from different vendors becomes much easier. UCA solves this interoperability problem in the product design phase by applying many new integration-friendly concepts [EPR96, Web10-14].

Instead of specifying a register name for the VBX voltage of bus X, a variable named MVM$MVMX$MVMY$VBX is used for all devices on the network. Since every device refers to the data in the same way, reprogramming becomes unnecessary after a new vendor's product has been added. A UCA client can extract a list of self-describing objects from the server that it is trying to access; these objects detail the information and services the server can provide or perform, and thus reduce the need to specify how the clients should interact with servers. On the other hand, because of the separation of T-profiles from A-Profiles, a UCA client can be developed independently of the precise T-profile used. This implies that the client can access a device in the exact same manner because the difference in the T-profile is isolated from the application. UCA 2.0, CASM, and GOMSFE allow for more functions to be implemented in a standardized manner rather than just registering the exchanged value. Because MMS is the common language for all applications, new instructions for each involved device do not need to be translated, although instructions on how to use the new services would need to be provided. Moreover, the protocols used within UCA are international standards. They are well established and defined, and they benefit from economies of scale and a common knowledge base. Moreover the self-describing services of UCA greatly simplify the communication among devices.

EPRI and electric utilities sponsored a number of meetings where vendors of UCA compliant devices interoperated on an Ethernet LAN. For instance, the AEP LAN Substation Demonstration Initiative was a UCA proof-of-concept program centered on the substation. AEP installed a unified power flow controller (UFPC) at its Inez 765 kV station and utilized the UFPC to handle communication between the Inez station, six remote stations, area dispatching, and the corporate office [Edri98]. The communication medium was 10 Mbps Ethernet on a switched hub or 100 Mbps on a shared hub. To ensure timely response to control commands, a priority mechanism was implemented to give Inez LAN traffic the highest priority at each routing node on the AEP system WAN.

INFORMATION SYSTEM FOR CONTROL CENTERS

3.10 ICCS COMMUNICATIONS SERVICES

ICCS provides various communications services to support data exchange applications. Figure 3.16 shows the services used by participants to communicate with ICCS [EPR96, EPR01b, Cau96, Web10-14].

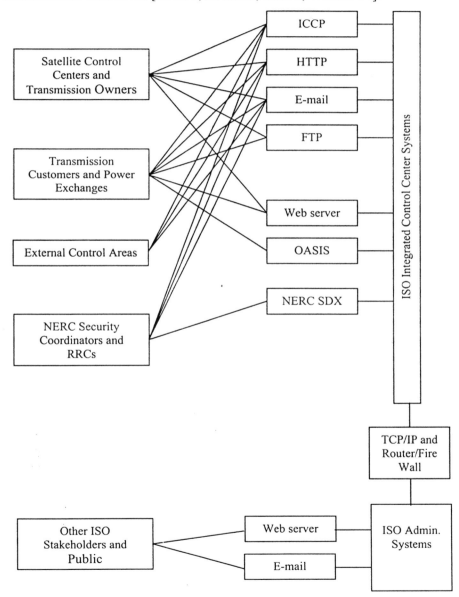

Figure 3.16 ICCS Communications Services

One of the main objectives of these communications services is to ensure the interoperability across computing platforms as ICCS, transmission owners and transmission customers usually operate on different computing platforms. Moreover, a group of application interfaces that are necessary to automate the data exchange process are provided. The applications of each service are discussed next.

ICCP. As a special communication protocol, ICCP is mainly used to support data exchanges composed of values associated with fixed time increments. Practically, ICCP is used for a real-time data exchange between ICCS and transmission owners and satellite control centers, external control areas and other entities, as illustrated in Figure 3.8. ICCP can also support certain messaging types of applications.

TCP/IP. As a popular industry standard protocol for data transmission among computer sites, TCP/IP is also used as the communications service standard for ICCS and information systems LANs. A dedicated addressing domain is used for the communication services including, FTP, Web services, e-mail, and ICCP among other special communication services.

HTTP. As a support for information distribution, HTTP based on WWW servers is installed on ICCS LANs. HTTP functionalities are included in Web server tools. All activities occurring on Web servers, including data snapshots, are date and time stamped. The Web support tools also interpret HTML, and thus ensure the interoperability among heterogeneous systems in a network.

E-mail. As a convenient tool for users to send and receive messages at different sites of the public Internet, e-mail services are provided by ICCS for all ICCS users. Usually the SMTP and POP3 are implemented to provide mail-handling services for all e-mail clients. Auditable e-mail is provided as a means by which a process is initiated, tracked and archived. Both functions belong to non-interactive messaging and thus can take advantage of many existing standard software packages.

FTP. As a standard Internet protocol for data exchange among computer sites, FTP is used for the exchange of data both into and out of ICCS. ICCS can receive data files from user sites by downloading or being uploaded using FTP. ICCS users can receive data files similarly from ICCS. The privacy of FTP data exchanges is ensured by providing the necessary security measures like the user authorization password.

NERC SDX. The NERC SDX is developed for data exchange between ICCS FTP and NERC FTP sites via the Internet. NERC SDX also provides the mechanism for retrieving data files from NERC for the calculation of ATCs and PTDFs. NERC SDX provides the means for security coordinators to enter data on the peak load information; net exports and imports; operating reserves; and generation and transmission outages.

OASIS. OASIS is the communications system used by the power industry for the coordination of transmission services over the public Internet. All businesses associated with transmission services are conducted on OASIS nodes. ICCS LAN has a special node for OASIS. The OASIS input system provides pre-validated re-dispatch and ancillary service bids, as well as generator parameters and other information to ICCS. ICCS also provides some information to OASIS which includes ancillary service commitments and other data needed to support transmission service functions and participant information requirements.

3.10.1 Other Communication Services

Certain ICCS messages, which must be acknowledged and audited by the recipient, use ICCP to transfer messages other control centers. The messaging applications include: control functions, generation re-dispatch, voltage schedules, transmission loading relief, and the ISO's reserve sharing. The messaging process consists of the following steps:

- An instructive message is sent from the ISO to a designated satellite control center.

- The designated satellite control center confirms the receipt of the message from the ISO by sending confirmation back to the ISO.

- Messages involved in the first two steps are structured to allow intelligent processing and analysis. For instance, any error in the transaction process will generate alarms at the sending site.

3.11 ICCS DATA EXCHANGE AND PROCESSING

ICCS provides a group of functions for data exchange and processing. This section will discuss how these functions are realized by utilizing the data communication services discussed above.

3.11.1 Real-Time Data Processing

Through ICCP, power system real-time data can be transmitted from transmission owners' computers to ICCS. These real-time data include but not limited to the following [Cau96, Web01, Web10-14]:

- 2-second status data
- 5-minute status data
- 10-second analog data
- 60-second analog data
- 15 minute, 30 minute, and hourly digital data
- On-demand digital data.

In the SCADA system, the 2-second status data are reported by exception, and 5-minute status data and all digital data are transmitted within a predetermined time window. Usually every five minutes, the ISO prepares and deposits a file containing the results of the state estimation for each transmission owner and other NERC security coordinators. The file appropriate to each transmission owner is placed in the transmission owner's designated directory following the execution of state estimation. ICCS sends the data via ICCP to transmission owners and other entities that are connected to the ISO's private network. Data formats conversion applications will be invoked if the format of the incoming data is different from that of the local database. A network status processing application detects abnormal circuit conditions by automatically analyzing the network model database. Usually every five minutes, the ISO prepares and deposits a file containing the results of the state estimation for each transmission owner and other NERC security coordinators.

3.11.2 Transaction Scheduling and Processing

Transmission customers post their service requests to the OASIS Web server, and within a predetermined time limit, service confirmations will be published on the same OASIS Web server by the ISO. All of these requests and confirmations are accessible by transmission customers, whether they were issued over the public Internet or the ISO private network. ICCS uses a directory structure and a file naming convention to identify each schedule and its time association. Schedule files are sent to the ISO's FTP site according to the directory structure and file-naming convention, and at the same time an e-mail message will be sent to notify ICCS that the schedule has been deposited at the ISO's FTP site. Transmission owners must also send their maintenance schedules to the ICCS for approval. These transactions use the standard FTP with auditable e-mail techniques. Requests for ancillary services are also posted to the OASIS Web server by transmission customers requiring ancillary services. Ancillary services schedules are approved and posted on the OASIS Web server by the ISO. Whether they are issued over the public Internet or the ISO private network, all bids and schedules are accessible by transmission customers. In addition, billing data are transmitted to transmission customers by ICCS on a daily basis. On a monthly basis, ICCS transmits an invoice to each transmission customer by attaching the invoice to an auditable e-mail message. Appropriate user authorization security measures are provided by ICCS to ensure the privacy of the billing and invoice data.

3.11.3 Generation Unit Scheduling and Dispatch

Unit commitment schedules are usually prepared to cover the entire ISO's operating time horizon, which is up to two weeks in the future (i.e., the current day and the next 13 days). Unit commitment schedules consisting of availability status and desired MW output are prepared daily or even more frequently when necessary by each generation's control area. On a day-ahead and hourly basis, generation owners use ICCP to submit generation bids to the ISO. At a designated deadline within the hourly time window, ICCS uses the ICCP to retrieve these bids from each generation owner. All bids will be posted on the OASIS. Depending on the outcome of the generation re-dispatch processing of bids by ICCS, it may be necessary for generation owners to send a re-dispatch schedule to the generation control area. Generation unit owners must also send their maintenance schedules to the ICCS for approval. After the ICCS has processed these maintenance data, the approval or denial of the schedules is returned to the individual transmission and generation owners [Web01, Web05-07, Web10-14].

3.12 ELECTRONIC TAGGING

Electronic tagging applies to all transactions whether or not they cross control area boundaries [Tom01]. Both the ISO and its satellite control centers are responsible for staging, or contracting to stage, electronic tagging authority and approval services which receive and parse tags. As defined by NERC, the electronic tagging process consists of the following triple-A services [Web11]:

- Agent service
- Authority service
- Approval service.

The ISO and its satellite control centers are responsible for providing these services. Necessary functions used to accomplish tagging services are integrated into ICCS. The electronic tagging system of ICCS is totally compliant with NERC standards and specifications. Hence, although it can use any proprietary mechanism to convey the tag information, ICCS has to comply with technical standards and protocols for the exchange of transaction information with tagging related services. The detailed electronic tagging process is depicted in Figure 3.17.

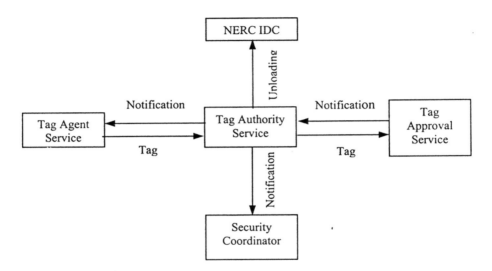

Figure 3.17 Electronic Tagging

During the process of information exchange, the electronic tagging system initially creates electronic tags that represent their respective transactions, and then disseminates these tags to all parties that are directly involved with these transactions. To achieve a good performance, the electronic tagging system must be able to identify parties in a transaction that have responsibility for the data exchange at every step and to ensure data integrity without duplicating the data entry or replicating errors. The time or the number of data transfers among parties should be minimized, and the electronic tagging system should be able to upload the approved tags to the NERC's interchange distribution calculator (IDC).

3.12.1 Tag Agent Service

The tag agent service provides the initial creation of an electronic tag representing an interchange transaction and the transfer of that information to the appropriate tag authority service. Purchasing-selling entities (PSEs) are responsible for providing this service directly or by arranging with a third party to provide this service as their agent. The tag agent service first validates the input tag data from the PSEs and then prepares all required tables and data elements as defined in the tag data model based on the PSE input data. Tags created by the tag agent service are forwarded to the tag authority service associated with the sink control area. The tag agent service assigns a tag ID and a tag key to each transaction and electronically communicates the tag ID, tag key, and tag data to the corresponding tag authority. A mechanism is provided by the tag agent service for PSEs to query tag authorities for the current status of their transactions either by simple polling or via an optional unsolicited notification mechanism. The tag agent service also provides the means for PSEs to withdraw, cancel, or terminate early any of their pending or active tags.

3.12.2 Tag Authority Service

The tag authority service provides the focal point for all interactions with a tag and maintains the single authoritative copy of record for each tag received from any tag agent service. Every control center is responsible for providing this service directly or by arranging with a third party to provide this service as its agent. The tag authority service manages each transaction's individual approval and overall composite status based on communications with tag agent and tag approval services. The tag authority service accepts input tag data from any tag agent service and identifies entities with approval rights over the transaction. All tags associated with entities identified as having approval rights over that transaction are transferred to the tag approval service for evaluation. Based on the approvals/denials received from these tag approval services, the tag authority arbitrates and sends the final disposition of the tag to the originating tag agent and all tag approval services associated with the transaction, and to that control area's security coordinator as well. The tag authority service verifies the identity of each approval entity attempting to approve or deny a tag based on the tag ID and the tag key, and updates the transaction's approval and composite status as appropriate. The tag authority service also provides the capability for both tag agent and tag approval services to review the current approval status of any transaction tag on demand. The tag authority service provides a mechanism for partial curtailment of transactions. All tags that are canceled or terminated will be forwarded to a designated location as identified by the information defined in the master registry associated with the sink control area.

3.12.3 Tag Approval Service

The tag approval service receives all tags submitted by tag agent services via the appropriate tag authority service, and communicates approval or denial information to the tag authority managing the transaction electronically in compliance with the protocol description. A mechanism is provided for the approval entity to send an approval or denial feedback to the tag authority service. The tag approval service can receive notification messages from the tag authority on each change in the composite status of the Tag. The current status of each transaction submitted for approval can be queried from the appropriate tag authority. Though initially the tag approval service is the responsibility of the control areas identified along the transaction's scheduling path, any entity that has the right to verify the contents of, and approve or deny, a tag is responsible for providing this service directly or for arranging with a third party to provide this service as their agent.

3.13 INFORMATION STORAGE AND RETRIEVAL

ICCS information storage and retrieval (IS&R) system consists of a commercial database management system that accommodates long-term archival storage and retrieval of information produced by ICCS. All ICCS data are available for collection, calculation, retention, and archiving by IS&R. The information to be stored and retrieved includes historical information required to meet regulatory archiving requirements, historical information required for audit purposes and for market dispute resolution, and some selected operational information required to support the ISO business process and decision support functions outside ICCS. Some types of information such as user log entries, ISO operator entries, functional control instructions, and alarms and events can even be automatically captured by IS&R, and the data to be captured and the periodicity can be defined through the IS&R database generation and maintenance function.

The IS&R provides services for a large number of information users. The IS&R database is capable of communicating with users through TCP/IP. The users' data can be exchanged with IS&R on a cyclic basis and on demand. All ICCS users with the appropriate authorization are able to access IS&R functions, review transaction scheduling and historical information, and even edit information [Web01-14].

3.14 ICCS SECURITY

ICCS manages all information system resources, including protocols, bandwidth, information, and address assignment. The ICCS has also become the server of communications functions, and information system users act as a client of ICCS services. In addition, some clients authorized by the ISO could have both reading and writing permission to ICCS computer systems. Hence, to protect ICCS from being violated of any security requirements, ICCS must be provided with necessary security measures. Most often the potential security risks to ICCS come from the following two areas: unauthorized access to ICCS by outside individuals and inadvertent destruction of data by individuals. In this section we will discuss these concerns and corresponding preventive measures.

3.14.1 Unauthorized Access

Because the ISO is actually connected with the public Internet, unauthorized access by outside individuals is a significant concern. The ICCS must provide security measures to prevent any unauthorized access. The generally desired security measures for this purpose mainly include user authentication, IP address authentication, system authentication, and Internet access via proxy servers. We discuss these measures below [Web 01-14]:

- **User authentication.** The ISO assigns a distinct identification and a password to each authorized user. Every user must supply its own identification and password in order to gain access to ICCS services. ICCS identifies the requestor's identification and password in the user login process; if the user identification and password are correct and matched, the login request will be granted, otherwise, the login request is refused

- **IP address authentication.** The ISO maintains an IP address assignment list for both its static and dynamic IP client address. In the user's login process, the IP address authentication function will check the user name against the IP address associated with the user on the list. If the user's name and IP address are matched, the access request will be granted. Because an IP address is globally unique to one user, this authentication function can help prevent illegal access when some unauthorized users masquerade as authorized users.

- **System authentication.** The system serial number of the user's machine is unique and is usually taken from the system hard drive. Like the authentication of an IP address, if the ISO maintains a system serial number list of all members, the system serial number of the user's machine can be used to validate the user identification. Because a system serial number is generally more difficult to forge than the IP address, the system authentication is more powerful to prevent an unauthorized user from masquerading as an authorized user.

- **Internet access via proxy servers.** Proxy servers can be used to limit the number of IP addresses required by the ISO for Internet access. All the ISO private network and LAN IP addresses can be hidden from the Internet. All access to the ISO's private network and LAN IP addresses further will be via particular proxy servers. This way not only the network security can be increased but also the entire IP address space can open up for use on the ISO's private network and LANs.

3.14.2 Inadvertent Destruction of Data

Inadvertent destruction of data can be minimized or even avoided through user access control and frequent system backup. The user access control can be realized by restricting the user's ability to read, write, delete, and execute files on ICCS and will be applied to all ICCS directories and files. In particular, to prevent users from accidentally deleting the output results of ICCS functions, users are given read-only access to some files. Meanwhile the system backup can be used as another effective approach to recover accidentally destroyed information. As a complementary tool, multi-generation backups are reserved to recover the lost data that are not detected soon after the loss [Web11].

Chapter 4

Common Information Model and Middleware for Integration

4.1 INTRODUCTION

4.1.1 CCAPI/CIM

In this era of information technology, the availability of pertinent and easily accessible information is becoming the most valuable resource in every energy company. In a competitive marketplace, real-time energy information can provide the strategic advantage that companies would require to achieve their critical goals, such as optimizing the power systems operation, enhancing asset and personnel productivity, reducing operational costs, and maintaining a vital flexibility for future change. To keep pace with rapid changes, which affect the energy industry, energy companies are seeking ways to improve accessibility to critical resources.

Improvements to the way a power company uses information could start from the EMS located in control centers. Many cutting-edge EMS applications that control room managers would utilize to keep pace with rapid industry changes are incompatible due to specialized or proprietary engineering designs or software applications in EMS. The exclusiveness of these applications precludes exchanging EMS data with other databases within an energy company.

EPRI launched a project in 1990, called Control Center Application Program Interface (CCAPI), for reducing the cost and time needed to add new applications to EMS, and protecting resources invested in existing EMS applications that are working effectively. The main goal of CCAPI was to develop a set of guidelines and specifications to enable the creation of plug-in applications in a control center environment. Essentially, CCAPI is a standardized interface that enables users to integrate applications from various sources by specifying the data that applications could share and the procedure for sharing them [CAP01, Kha01, Mau02]. To realize this project, the essential structure of a power system model is defined as the Common Information Model (CIM), and this comprises the central part of CCAPI.

CIM provides a common language for information sharing among power system applications that can be converted into extensible markup language (XML) [Bec00, Ber00, IEC01, IEC02, IEC03, EPR96, Goo99, Hir99, Lee99, Neu01, Pod99a, Pod99b, Wan00]. XML ensures that standard format is used in exchanging power system model information that any EMS can understand. The North American Electric Reliability Council (NERC) mandates security coordination centers (SCCs) to use CIM for their model information exchange.

CIM compliance enables control center personnel to combine, on one or more integrated platforms, software products for managing the power system economy and reliability. As a result personnel can upgrade or migrate their EMS systems incrementally and quickly, while preserving prior utility investments in customized software packages. Migration is perceived to reduce upgrade costs by at least 40 percent and enable energy companies to gain strategic advantages by using new applications as they become available.

4.1.2 What Is IEC?

The International Electrotechnical Commission (IEC) is a worldwide organization comprised of all national electrotechnical committees whose objective is to promote international cooperation on questions concerning the standardization in electrical and electronic fields. IEC publishes the international standards prepared by the appropriate technical committees of IEC. IEC collaborates closely with the International Organization for Standardization in accordance with conditions determined by agreement between the two organizations. In promoting international standards, IEC national committees undertake to apply IEC standards transparently in their national and regional standards. Any divergences from IEC standards are clearly indicated in the corresponding national or regional standards.

COMMON INFORMATION MODEL AND MIDDLEWARE 137

Largely based upon the results of the EPRI CCAPI research project, the IEC technical committee on power system control and associated communications has prepared an international standard series IEC 61970 on Energy Management System Application Program Interface (EMS-API). The IEC 61970 series defines the standard for EMS-API, which consists of the following parts:

- EMS-API Part 1: Guidelines and General Requirements

- EMS-API Part 2: Glossary

- EMS-API Part 301: CIM base (which specifies the basic and core CIM from the logical view of the physical aspects of EMS information)

- EMS-API Part 302: CIM Financial, Energy Scheduling, and Reservation (which specifies the CIM from the financial and energy scheduling logical view)

- EMS-API Part 303: CIM SCADA (which specifies the CIM from the SCADA logical view;

- EMS-API Part 501: CIM Resource Description Framework (RDF) Schema, which specifies a Component Interface Specification or (CIS) for EMS-API). Included are the format and rules for producing a machine readable form of the CIM as specified in EMS-API Part 301, 302, 303 standards as well as a CIM vocabulary to support the data access facility and associated CIM semantics.

RDF is the language used for expressing the metadata that machines can process simply. RDF is expressed as a special kind of XML document. The RDF schema is a schema specification language that describes resources and their properties, including how one resource is related to other resources, as this information is used in an application-specific schema. EMS-API Part 501 takes advantage of current Web standards and provides a mechanism for independent suppliers to access CIM metadata in a common format and with standard services that enables subsequent CIM data access. It also provides CIM versioning capabilities and a mechanism that is easily extensible to support site-specific needs.

4.1.3 What Is CIM?

CIM is an abstract model that represents all the objects in an electric utility enterprise that are typically contained in an EMS information model. It defines public classes and attributes for these objects as well as their relationships using object-oriented modeling techniques. The CIM specification uses the unified modeling language (UML) notation, which defines CIM as a group of packages. By CIM, the information of common interest can be defined in a common model to which individual applications translate their local designs.

By providing a standard representation of power system models such as combining multiple proprietary power system models into a single merged internal model for an RTO, CIM facilitates the integration of EMS applications developed independently by different vendors, or between EMS and other systems concerned with different aspects of power system operations, such as generation or distribution management. This is accomplished by defining standard application program interfaces to enable these applications to access public data and exchange information independent of how such information is represented internally. CIM specifies semantics for EMS-API.

However, the use of CIM goes far beyond EMS applications. This standard should be understood as a tool for integration in any domain where a common power system model is needed to facilitate interoperability and plug compatibility between applications and systems independent of any particular implementation. For instance, CIM can be used for transferring power system model data between security coordinators, and the exchange of power system models among different vendor products.

Some of the advantages of using CIM as a single unified data model for data exchange and applications integration are as follows:

- Applications written to a standard environment require less environmental customization.

- Each data supplier or customer would only need to know how to translate to a single model rather than make multiple point-to-point translations.

- Although data may be distributed, a virtual database built upon a common schema helps reduce data duplicity.

- System-wide data changes are available to allow numerous applications simultaneously.

- The simplified system architecture makes subsequent advancements in capabilities less difficult.

4.2 CIM PACKAGES

A package in CIM is used for grouping generally related model elements. Each CIM package contains a number of defined classes and one or more class diagrams showing these classes and their relationships graphically. All CIM packages describe any entity that could appear in an electric power system. There is no specific semantic for package definitions, and the chief purpose of using the package concept is to make CIM easy to design, understand, and review. An entity may have associations that cross many package boundaries, and an application may use CIM entities from more than one package. Additional packages may be defined as needed to support new views of the power system model.

A comprehensive CIM consists of a complete set of packages. As depicted in Figure 4.1 a comprehensive CIM could be partitioned into sixteen packages. In the figure, the dashed line indicates a dependency relationship, with the arrowhead pointing from the dependent package to the package on which it has a dependency.

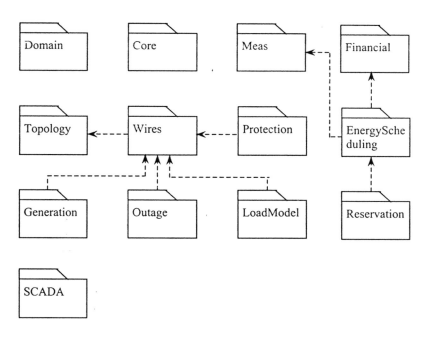

Figure 4.1 CIM Top Level Packages

The *Core, Topology, Wires, Outage, Protection, Asset, Meas, LoadModel, Generation, Production, GenerationDynamics,* and *Domain* packages are defined and described in IEC 61970-301. The *Energy Scheduling, Reservation, and Financial* packages are defined and described in IEC 61970-303. The *SCADA* package is described in IEC 61970-303. The package boundaries do not necessarily imply application boundaries.

In the following, we will give a brief introduction to the main CIM packages, and discuss attributes and relationships of classes in each package.

- **Core.** This package contains basic entities that are shared by all applications. Not all applications require the entire Core entities.

- **Topology.** This package is an extension of the Core package; in association with the Terminal class, it models Connectivity. Connectivity is the physical definition of how equipment is connected together.

- **Wires.** This package is an extension of Core and Topology packages; it models information on electrical characteristics of transmission and distribution networks. This package is used by network applications such as state estimation, load flow, and optimal power flow.

- **Outage.** This package is an extension of the Core and Wires packages; it models information on the current and planned network configuration.

- **Protection.** This package is an extension of the Core and Wires; it models information for protection equipment such as relays.

- **Meas.** This package contains entities that describe dynamic measurement data exchanged between applications.

- **Load Model.** This package provides models for energy consumers and the system load. Special circumstances that may affect the load, such as seasons and day types, are also included here. This information could be used for load forecasting and management.

- **Generation.** This package is divided into two subpackages: Production and GenerationDynamics.

- o **The Production** subpackage provides models for various kinds of generators. It also models the production cost information that is used to economically allocate the demand among committed units and calculate reserve quantities. This information is used by unit commitment and economic dispatch of hydro and thermal generating units, load forecasting, and automatic generation control applications.

- o **The GenerationDynamics** subpackage provides models for prime movers, such as turbines and boilers, which are needed for simulation and educational purposes. This information could be used for a unit modeling in dynamic training simulator applications.

- **Domain.** This package is a data dictionary of quantities and units that define data types for attributes that may be used by any class in any other package. Each data type contains a value attribute and an optional unit of measure, which is specified as a static variable initialized to the textual description of the unit of measure. Permissible values for enumerations are listed in the documentation for the attribute using UML constraint syntax inside curly braces. String lengths are listed in the documentation and specified as a length property.

- **Financial.** This package is defined for energy transaction settlement and billing. These classes represent the legal entities that participate in formal or informal agreements.

- **Energy Scheduling.** This package provides the capability to schedule and account for transactions for the exchange of electric power among companies. It includes transactions for megawatts that are generated, consumed, lost, passed through, sold, and purchased. These classes are used by accounting and billing transactions for energy, generation capacity, transmission, and ancillary services.

- **SCADA.** This package defines the SCADA's logical view in CIM. SCADA contains classes that model data points located in remote units like RTUs, substation control systems and remote control centers. The Meas classes are basic to SCADA, since they gather telemetered data from various sources and support operator control of equipment, such as opening or closing a breaker.

4.3 CIM CLASSES

CIM classes will model objects of an electric power system that need to be represented in a common way for various purposes. A class is a description of an object found in the real world, such as a power transformer, generator, or load that needs to be represented as part of the overall power system model in an EMS. Other types of objects include issues such as schedules and measurements that EMS applications need to process, analyze, and store. These objects need a common representation to achieve the goal of the EMS-API standard for the plug compatibility and interoperability. A particular object in a power system with a unique identity is modeled as an instance of the class to which it belongs.

4.3.1 Class Attributes

As in other object-oriented techniques, CIM classes have attributes that describe the characteristics of their objects and can be used to identify an instance of such a class. Each attribute has a type, which identifies the kind of attribute it is. Typical attributes types are integer, float, Boolean, string, and enumeration, which are called *primitive* types. Many additional types can also be defined; for instance, a capacitor bank may have a maximum *kv* attribute of type *Voltage*. The definitions of data type are contained in the *Domain* package of CIM. The classes in the *Domain* package also have an optional unit of measure for their attribute type.

In order to make CIM as generic as possible and easy to configure for specific implementations, CIM entities have no specific behavior other than defaults for create, delete, update, and read. Generally, it is easier to change the value or domain of an attribute than to change a class definition; therefore, CIM tries to avoid defining too many specific sub-types of classes; instead, CIM defines generic classes with attributes giving the type name so that applications may then use this information to initiate specific object types as required.

4.3.2 Class Relationships

A specific class of CIM can have various relationships with other classes. The class relationships in a CIM package can be shown with a class diagram. Class relationships reveal how individual classes are structured in terms of others. In a class diagram, if a class has relationships with classes in other packages, these classes will be shown with a note identifying the

COMMON INFORMATION MODEL AND MIDDLEWARE 143

respective packages. CIM classes are mostly related based on the following three options:

- Generalization
- Association
- Aggregation

4.3.2.1 Generalization

A generalization relationship exists between a general class and a more specific class that contain additional attributes. For instance, a *Breaker* is a more specific type of *Switch*, which in turn is a more specific type of *ConductingEquipment*; a *ConductingEquipment* is a more specific type of *PowerSystemResource*, which is a primitive class in the *Core* package. Class generalization is a powerful technique for simplifying class diagrams as it allows a specific class to inherit attributes and relationships from a more general class. In the class diagrams of CIM, a generalization relationship is represented with an arrow pointing from the specific class to the general class. The primary use of generalization in CIM is depicted in Figure 4.2.

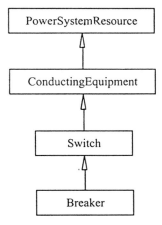

Figure 4.2 Class Generalization

The *PowerSystemResource* class is used to describe any physical object or grouping of physical object that needs to be modeled, monitored, or measured. By defining a *PowerSystemResource* class, the attributes and relationships of this class can be inherited by its subclasses such as

ConductingEquipmen, Switch, and *Breaker*. All of these subclasses can at least inherit the following relationships from *PowerSystemResource*:

- **MeasuredBy.** An object of *PowerSystemResource* may be "*measured by*" certain measurements

- **OwnedBy.** An object of *PowerSystemResource* may be "*owned by*" certain entity

- **MemberOf.** An object of *PowerSystemResource* may be a "*member of*" another *PowerSystemResource* object.

4.3.2.2 Association

An association is a kind of connection between two classes with an assigned role. The roles of two classes are generally different, so we need to assign one role to each class separately. Figure 4.3 shows a general way of representing a class association in CIM class diagrams, where (x..y) and (w..z) are the cardinalities for roles A and B respectively, and x, y, w, and z are integers within the range of (0, n).

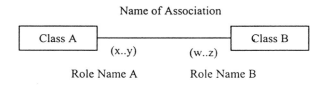

Figure 4.3 Representation of Association

For instance, there is an association named *HasA* between a *Measurement* class and a *MeasurementUnit* class, which means a *Measurement HasA* a *MeasurementUnit*. Figure 4.4, which is taken from the *Meas* class diagram, shows the *HasA* association between these two classes. As illustrated, the *Measurement* class is assigned a role named *HasUnit*, and the *MeasurementUnit* has a role named *UnitFor*.

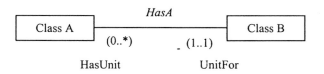

Figure 4.4 Class Association

The cardinalities are shown at both ends of the association. The notation (0..*) by the *Measurement* class means that there can be any number of *Measurement* associated with the *MeasurementUnit*, while the notation (1..1) by the *MeasurementUnit* class means that in this association there is only one *MeasurementUnit* for the associated *Measurement*. A value of zero indicates an optional association and the asterisk represents any default number.

4.3.2.3 Aggregation

Class aggregation is a special type of association. It indicates that the relationship between two associated classes is a sort of whole-part relationship, where the whole class "consists of" or "contains" the part class, and the part class is only "part of" the whole class. Unlike the class generalization, in a class aggregation the part class does not inherit any attributes or relationships from the whole class. In CIM class diagrams, the class aggregation is represented with a diamond symbol pointing from the part class to the whole class. In CIM, there are the following two types of class aggregation.

- **Composite aggregation.** The composite aggregation models the whole-part relationship of two classes where the composite multiplicity is 1, that is a part belongs to one and only one whole. Taken from the *Topology* class diagram, Figure 4.5 illustrates a composite aggregation relationship between the *TopologicalIsland* class and the *TopologicalNode* class. This aggregation is named *MemberOf*, and is represented with a diamond symbol pointing from the class *TopologicalNode* to the whole class *TopologicalIsland*. The *TopologicalIsland* class is assigned a role named *Contains*, and the *TopologicalNode* has a role named *MemberOf*. As illustrated, one or more *TopologicalNode* can be a *MemberOf* a *TopologicalIsland*, and a *TopologicalIsland* can comprise any number of *TopologicalNode*.

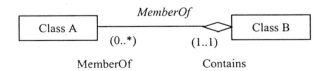

Figure 4.5 Composite Aggregation

Because some equipment is grouped into more than one container, CIM allows one class to have more than one type of composite aggregation relationship. For instance, a *Switch* can have the following three kinds of *MemberOf* relationship at the same time:

- *MemberOf*: A *Switch* maybe a *"MemberOf"* a substation
- *MemberOf*: A *Switch* maybe a *"MemberOf"* a transmission line
- *MemberOf*: A *Switch* maybe a *"MemberOf"* a company.

Besides *MemberOf*, CIM has a number of other composite aggregation patterns that can be used to model different relationships of classes in the real world.

- **Shared aggregation.** A shared aggregation is a special kind of aggregation where the part may be shared by several aggregations. In other words, shared aggregation is used to model whole-part relationships where the multiplicity of the composite is greater than one, that is, a part belongs to more than one whole. Figure 4.6 illustrates a shared aggregation between the *Measurement* class and the *MeasurementSet* class. As shown, a *Measurement* class may be a *MemberOf* any number of other *MeasurementSet* classes. We refer the reader to Appendix III for all class associations of CIM.

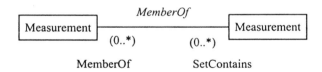

Figure 4.6 Shared Aggregation

4.4 CIM SPECIFICATIONS

Each class of a CIM package is defined in terms of its attributes and relationships to other classes, and each package contains one or more class diagrams showing its classes and their relationships graphically. The classes within a package are listed alphabetically. Native class attributes are listed first, followed by inherited attributes. The associations are described according to the role of each class participating in these associations. Similar to class attributes, native associations are listed first for each class, followed by inherited associations. Aggregations are listed only for the role that contains the aggregation. For a complete description of CIM classes, attributes, types, and relationships, refer to Annex A of IEC 61970-301. In order to help readers understand CIM more comprehensively, a power transformer is illustrated here.

The transformer model is a portion of the *Wires* package class diagram. As shown in Figure 4.7, a *PowerTransformer* is a specialized class of *PowerSystemResource*, as are *ConductingEquipment* and *TapChanger*. This generalization relationship, which is demonstrated by an arrow pointing to *PowerSystemResource*, permits *PowerTransformer* to inherit attributes and associations from *PowerSystemResource*. A *PowerTransformer* should have a *TransformerWinding*, which is modeled with an aggregation relationship using a diamond symbol pointing from *TransformerWinding* to *PowerTransformer*. As shown in the figure, a power transformer may contain one or more transformer windings, but a transformer winding belongs to only one power transformer. In the figure we also observe that a *TransformerWinding* has the following relationships with other classes:

- A generalization relationship with *ConductingEquipment*

- An association relationship with *WindingTest*, such that a *TransformerWinding* may be *TestedFrom* a *WindingTest*

- An aggregation relationship with *TapChanger*, such that a *TransformerWinding* may have a *TapChanger* associated with it.

As we noted earlier, class attributes and associations can be inherited from a general class by a more specific class. No matter whether it is a native or an inherited attribute, the attribute information includes the attribute name, type, and documentation. The class role information includes the role cardinality, the name of role, the name of the class to which the role applies, and the role documentation.

4.5 CIM APPLICATIONS

Power system applications based on CIM require a CIM context to be executed. CIM provides a run-time environment for applications. To create a CIM run-time environment, CIM needs to be implemented with a relational database. The CIM schema described in either UML or RDF will be first exported to a specific relational database, and then the power system data described with CIM formats will be loaded into the database. This process is depicted in Figure 4.8. The export of the CIM schema and the access to data can be implemented by calling certain functions in commercial software packages.

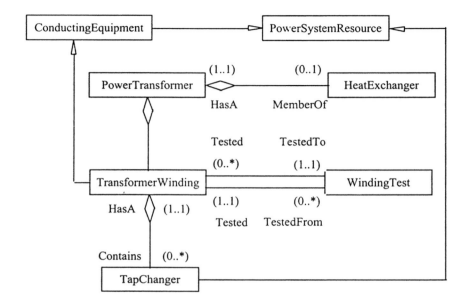

Figure 4.7 Transformer Model

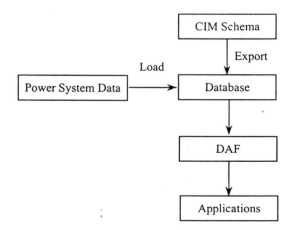

Figure 4.8 Application of CIM

COMMON INFORMATION MODEL AND MIDDLEWARE 149

For most beginners, the difficulty in applying CIM should lie in modeling various real objects of the power system with CIM specifications. In the following discussion, we use a simple example to illustrate the power system load flow computation based on CIM.

4.5.1 Example System

Let us consider a three-bus system as shown in Figure 4.9. The actual objects of this system are as follows:

- 2 generators: GT1, GT2
- 1 load: ELD
- 3 bus bars: BUS1, BUS2, BUS3
- 2 power transformer: TR1, TR2
- 3 lines: LN1, LN2, LN3
- 7 breakers: B1, B2, B3, B4, B5, B6, B7
- 18 switches: S1, S2, S3, S4, S5, S6, S7, S8, S9, S10, S11, S12, S13, S14, S15, S16, S17, S18

For a topology analysis, CIM needs to define a class of *ConnectivityNode*. A connectivity node is a physical point where terminals of conducting equipment are tied together with zero impedance. In the example system we assume that there is a connectivity node at each terminal of a real object. For instance, there is a connectivity node, denoted as CN10, between BUS1 and S1, and there is a connectivity node CN15 between S1 and B1. There are 36 connectivity nodes in the example system.

To build a CIM model for this example system, we must model all physical objects in the CIM format; in other words, all attributes defined in CIM for each physical object will be listed in a table according to the specific characteristics of this object in this concrete system. The CIM for this sample system is discussed next.

4.5.1.1 Attributes of Generators

In CIM the generator model is specified in the *GeneratingUnit* class, which is derived from *PowerSystemResource*. For simplicity we assume that GT1 and GT2 are exactly of the same type. The attributes information for GT1 and GT2 models is listed in Tables 4.1, 4.2, 4.3.

In CIM a generating unit contains an operator-approved operating schedule, which is defined in *GenUnitOpSchedule*. This operating schedule is a curve of generation over time (X-axis) showing the values of MW (Y1-axis) and MVAr (Y2-axis) for each unit of the period covered. This curve could be practically produced through unit commitment. For this example system, GT1 and GT2 *GenUnitOpSchedule* attributes inherited from *CurveSchedule* are listed in Table 4.4.

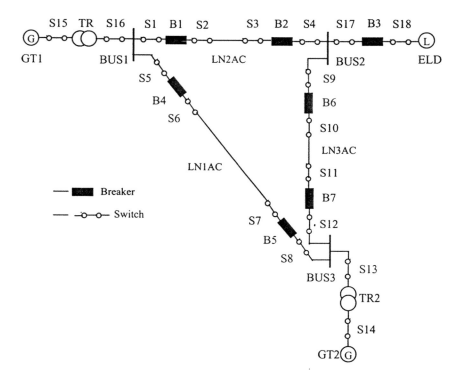

Figure 4.9 Three-Bus Example System

Table 4.1 Native Attributes of GT1 and GT2

Attribute	Value	Description
ControlDeadBand	0.05	Unit control error dead band. When a unit's desired MW change is less than the dead band, no control pulses are sent to the unit
Efficiency	0.35	Efficiency of the unit in converting mechanical energy, from the prime mover, into electrical energy
InitialMW	260	Default initial MW for storing power flows based on the unit and network configuration
LowControlLimit	100	Low limit for secondary (AGC) control
MaximumAllowableSpinningReserve	85	Maximum allowable spinning reserve, regardless of the operating point
MaximumEconomicMW	270	Maximum economic MW (will not exceed the maximum operating MW)
MaximumOperatingMW	280	Maximum operating MW limit of unit
MinimumEconomicMW	10	Minimum economic MW limit that must be greater than or equal to the minimum operating MW limit
MinimumOperatingMW	10	Minimum operating MW limit of unit
RatedGrossMaxMW	320	Unit's gross rated maximum capacity (Book Value).
RatedGrossMinMW	1	Gross rated minimum generation level at which the unit can operate safely while delivering power
RatedNetMaxMW	260	Net rated maximum capacity determined by subtracting the auxiliary power for operating the internal plant from the rated gross maximum capacity
StartupTime	10	Required time to get the unit on-line, once the prime mover mechanical power is applied

Table 4.2 GT1 Attributes Inherited from *PowerSystemResource*

Attribute	Value	Description
PowerSystemResourceName	GT1	Name or identification of an instance of a power system resource
PowerSystemResourceDescription	Generating unit 1	Description information
TypeName	CTL	Different types of power system resources that otherwise have identical attributes

Table 4.3 GT2 Attributes Inherited from PowerSystemResource

Attribute	Value	Description
PowerSystemResourceName	GT2	Name or identification of an instance of a power system resource
PowerSystemResourceDescription	Generating unit 2	Description information
TypeName	GST	Different types of power system resources that otherwise have identical attributes

Table 4.4 GT1 and GT2 GenUnitOpSchedule Attributes Inherited from CurveSchedule

Attribute	Value	Description
CurveScheduleDescription	GTX schedule	Description information
CurveScheduleName	GTXSCH	Uniquely identifies a curve instance among a set of curves or schedules
CurveStyle	X-Y	Style or shape of curve.
RampMethod	20-X	DeltaY versus deltaX units of measure. Applies to all ramps.
RampStartMethod	0 at start point	Method of applying ramp: 0 at start point, 50% at start point, 100% at start point. For methods 2 and 3, the ramp begins ahead of the start point on the X-axis. NOTE: For storage, all ramps are to be normalized to Method 1 (i.e., 0 at start point).
RampUnits	MW/HOUR	DeltaY versus deltaX units of measure. Same for "two" Y values.
xAxisType	Integer	Independent variable.
xAxisUnits	HOUR	X-axis units of measure
y1AxisUnits	MW	Y1-axis units of measure
y2AxisUnits	MVAR	Y2-axis units of measure
yAxisType	Float	Dependent variable

4.5.1.2 Attributes of Transformers

For simplicity we assume that TR1 and TR2 are of the same type. The attributes information of TR1 and TR2 are listed in Tables 4.5, 4.6, and 4.7.

COMMON INFORMATION MODEL AND MIDDLEWARE

Table 4.5 Native Attributes of TR1 and TR2

Attribute	Value	Description
BmagSat	5	Core shunt magnetizing susceptance in the saturation region, in percent
MagBaseKV	110k	Reference voltage at which the magnetizing saturation measurements were made
MagSatFlux	30	Core magnetizing saturation curve knee flux level
Phases	A	Phases carried by a power transformer. Possible values {ABCN, ABC, ABN, ACN, BCN, AB, AC, BC, AN, BN, CN, A, B, C, N}
TransfCoolingType	Wind cooling	Type of transformer cooling

Table 4.6 TR1 Attributes Inherited from *PowerSystemResource*

Attribute	Value	Description
PowerSystemResourceName	TR1	Name or identification of an instance of a power system resource
PowerSystemResourceDescription	Power Transformer 1	Description information
TypeName	STAR	Different types of PowerSystemResources that otherwise have identical attributes

Table 4.7 TR2 Attributes Inherited from *PowerSystemResource*

Attribute	Value	Description
PowerSystemResourceName	TR2	Name or identification of an instance of a power system resource
PowerSystemResourceDescription	Power Transformer 2	Description information
TypeName	DELTA	Different types of PowerSystemResources that otherwise have identical attributes

4.5.1.3 Attributes of ELD

Suppose the ELD (Equivalent Load) at BUS2 is a generic ELD for an energy consumer on a transmission voltage level. Then the native attributes of ELD are listed in Table 4.8.

Table 4.8 Native Attributes of ELD

Attribute	Value	Description
FeederLoadMgtFactor	100	Feeder's contribution to load management, in percent
MVArColdPickUpFactor	50	Nominal feeder MVAr that is picked up cold, in percent
MWColdPickUpFactor	60	Nominal feeder MW that is picked up cold, in percent
PhaseAmpRating	100k	The rated individual phase amperes
LoadAllocationFactor	30	Permit assignment of loads on a participation factor basis. Given three equivalent loads with factors of 10, 25 and 15, a feeder load of 100 amps could be allocated on the feeder as 20, 50 and 30 amps.

An equivalent load is an entity of *PowerSystemResource*. Hence, ELD inherits the attributes from *PowerSystemResource* listed in Table 4.9.

Table 4.9 ELD Attributes Inherited from *PowerSystemResource*

Attribute	Value	Description
PowerSystemResourceName	ELD	Name or identification of an instance of a power system resource
PowerSystemResourceDescription	Equivalent load	Description information
TypeName	ELDT	Different types of PowerSystemResources that otherwise have identical attributes.

An equivalent load is also an entity of *ConductingEquipment*. Hence, ELD can inherit the attributes from *Conducting Equipment* listed in Table 4.10.

Table 4.10 ELD Attributes Inherited from *ConductingEquipment*

Attribute	Value	Description
Terminals	2	Maximum number of terminals of an equipment
Phases	AN	Phases carried by a conducting equipment. Possible values { ABCN, ABC, ABN, ACN, BCN, AB, AC, BC, AN, BN, CN, A, B, C, N }.

An equivalent load is also an entity of *EnergyConsumer*. Hence, ELD can inherit the attributes from *EnergyConsumer* listed in Table 4.11.

Table 4.11 ELD Attributes Inherited from *EnergyConsumer*

Attribute	Value	Description
ConformingLoadFlag	Yes	Flag is set to YES if the load is conforming; i.e., tracks the area load to which the energy consumer belongs
CustomerCount	30	Number of customers represented by this demand
pFexp	3	Exponent of per unit frequency effecting real power.
pfixed	150	Real component of the load that is a fixed quantity, MW.
pfixedPct	80	Fixed MW as percent of load group fixed MW
pnom	200	Nominal value for real power, MW. Nominal real power is adjusted according to the load profile selected for a consumer. It equates to one per unit in the load profile.
pnomPct	80	Nominal MW as percent of load group nominal MW
powerFactor	30/50	Power factor for nominal portion of the load. Defined as MW/MVA
pVexp	2	Exponent of per unit voltage effecting real power
qFexp	2	Exponent of per unit frequency affecting reactive power
qfixed	70	A fixed reactive component of load, MVAr
qfixedPct	70	Fixed MVAr as per cent of load group fixed MVAr
qnom	260	Nominal value for reactive power, MVAr
qnomPct	80	Nominal MVAr as a percentage of load group nominal MVAr
qVexp	2	Exponent of per unit voltage effecting reactive power

In CIM a load contains a curve of load versus time (X-axis) showing the values of MW (Y1-axis) and MVAr (Y2-axis) for each unit of the period covered, which is defined in *LoadDemandModel*. This curve represents a typical pattern of load over the time period for a given day type and season. The attributes of *LoadDemandModel* of ELD are listed in Table 4.12.

4.5.1. 4 Attributes of Buses

A bus in CIM is an entity of *BusbarSection*. The attributes information needed to build the models of the three buses in the example system is listed in Tables 4.13 through 4.16.

Table 4.12 ELD LoadDemandModel Attributes Inherited from CurveSchedule

Attribute	Value	Description
CurveScheduleDescription	ELD Load Demand Model	Description information
CurveScheduleName	ELDLDM	Uniquely identifies a curve instance among a set of curves or schedules
CurveStyle	X –Y	Style or shape of the curve
RampMethod	20-x	DeltaY versus deltaX units of measure. Applies to all ramps.
RampStartMethod	0 at start point	Ramping method: 0 at start point, 50% at start point, 100% at start point. For methods 2 and 3, the ramp begins ahead of the start point on the X-axis. NOTE: For storage, all ramps are to be normalized to method "1" (0 at start point).
RampUnits	MW/HOUR	DeltaY versus deltaX units of measure. Same for "two" Y values.
xAxisType	Integer	Type of independent variable.
xAxisUnits	Hour	X-axis units of measure
Y1AxisUnits	MW	Y1-axis units of measure
Y2AxisUnits	MVAr	Y2-axis units of measure
yAxisType	Float	Type of dependent variable

Table 4.13 Native Attributes of BUS1 and BUS3

Attribute	Value	Description
BaseVoltage	220	Bus bar's base voltage in kV, expressed in terms of line to line for ac networks and line to ground for HVDC networks
HighVoltageLimit	235	Bus bar's high voltage limit in kV
LowVoltageLimit	185	Bus bar's low voltage limit in kV

Table 4.14 Native Attributes of BUS2

Attribute	Value	Description
BaseVoltage	110	Bus bar's base voltage in kV, expressed in terms of line to line for ac networks and line to ground for HVDC networks
HighVoltageLimit	125	Bus bar's high voltage limit in kV
LowVoltageLimit	85	Bus bar's low voltage limit in kV

As a kind of power system resource and conducting equipment, a bus inherits the following attributes from *PowerSystemResource* and *ConductingEquipment*.

Table 4.15 BUS1, BUS2, and BUS3 Attributes Inherited from *PowerSystemResource*

Attribute	Value	Description
PowerSystemResourceName	BUSX	Name or identification of an instance of a power system resource
PowerSystemResourceDescription	xth bus	Description information
TypeName	Main	Different types of PowerSystemResources that otherwise have identical attributes.

Table 4.16 BUS1, BUS2 and BUS3 Attributes Inherited From *ConductingEquipment*

Attribute	Value	Description
Terminals	1	Maximum number of terminals of the equipment
Phases	A	Phases carried by conducting equipment. Possible values {ABCN, ABC, ABN, ACN, BCN, AB, AC, BC, AN, BN, CN, A, B, C, N}.

4.5.1.5 Attributes of Lines

In CIM a line is supposed to be made up of AC line segments and/or DC line segments. In particular, an entity of Line "hasA" an entity of ACLineSegment and/or DCLineSegment. Suppose there are only AC line segments in the exampling system, that is, LN1, LN2, and LN3 have LN1AC, LN2AC, and LN3AC as their AC line segments, respectively. The attributes information needed to model LN1, LN2, and LN3 is listed in the following tables.

A line is an entity of *PowerSystemResource*; hence, LN1, LN2, and LN3 can inherit the attributes from *PowerSystemResource* listed in tables 4.17, 4.18, and 4.19.

Table 4.17 LN1 Attributes Inherited from *PowerSystemResource*

Attribute	Value	Description
PowerSystemResourceName	LN1	Name or identification of an instance of a power system resource
PowerSystemResourceDescription	Line 1	Description information
TypeName	AC	Different types of PowerSystemResources that otherwise have identical attributes

Table 4.18 LN2 Attributes Inherited from *PowerSystemResource*

Attribute	Value	Description
PowerSystemResourceName	LN2	Name or identification of an instance of a power system resource
PowerSystemResourceDescription	Line 2	Description information
TypeName	AC	Different types of PowerSystemResources that otherwise have identical attributes

Table 4.19 LN3 Attributes Inherited from *PowerSystemResource*

Attribute	Value	Description
PowerSystemResourceName	LN3	Name or identification of an instance of a power system resource
PowerSystemResourceDescription	Line 3	Description information
TypeName	AC	Different types of PowerSystemResources that otherwise have identical attributes

ACLineSegment is derived from *PowerSystemResource*, *ConductingEquipment*, and *Conductor*. Therefore, LN1AC, LN2AC, and LN3AC inherit the attributes from *PowerSystemResource* listed in Tables 4.20, 4.21, and 4.22.

Table 4.20 LN1AC Attributes Inherited from *PowerSystemResource*

Attribute	Value	Description
PowerSystemResourceName	LN1AC	Name or identification of an instance of a power system resource
PowerSystemResourceDescription	LN1 AC line segment	Description information
TypeName	ACS	Different types of PowerSystemResources that otherwise have identical attributes

Table 4.21 LN2AC Attributes Inherited from *PowerSystemResource*

Attribute	Value	Description
PowerSystemResourceName	LN2AC	Name or identification of an instance of a power system resource
PowerSystemResourceDescription	LN2 AC line segment	Description information
TypeName	ACS	Different types of PowerSystemResources that otherwise have identical attributes

Table 4.22 LN3AC Attributes Inherited from PowerSystemResource

Attribute	Value	Description
PowerSystemResourceName	LN3AC	Name or identification of an instance of a power system resource
PowerSystemResourceDescription	LN3 AC line segment	Description information
TypeName	ACS	Different types of PowerSystemResources that otherwise have identical attributes

LN1AC, LN2AC and LN3AC also inherit the following attributes from *ConductingEquipment* listed in Tables 4.23, 4.24, and 4.25.

Table 4.23 LN1AC Attributes Inherited from ConductingEquipment

Attribute	Value	Description
Terminals	2	Maximum number of terminals the equipment may have
Phases	A	Phases carried by conducting equipment. Possible values are {ABCN, ABC, ABN, ACN, BCN, AB, AC, BC, AN, BN, CN, A, B, C, N}.

Table 4.24 LN2AC Attributes Inherited from ConductingEquipment

Attribute	Value	Description
Terminals	2	Maximum number of terminals the equipment may have
Phases	A	Phases carried by conducting equipment. Possible values are {ABCN, ABC, ABN, ACN, BCN, AB, AC, BC, AN, BN, CN, A, B, C, N}.

Table 4.25 LN3AC Attributes Inherited from ConductingEquipment

Attribute	Value	Description
Terminals	2	Maximum number of terminals the equipment may have
Phases	A	Phases carried by conducting equipment. Possible values are {ABCN, ABC, ABN, ACN, BCN, AB, AC, BC, AN, BN, CN, A, B, C, N}.

LN1AC, LN2AC and LN3AC also inherit the following attributes listed in Tables 4.26, 4.27, and 4.28 from *Conductor*.

Table 4.26 LN1AC Attributes Inherited from *Conductor*

Attribute	Value	Description
B0ch	0.023	Zero sequence shunt susceptance, uniformly distributed, of the entire line section
bch	0.348	Positive sequence shunt susceptance, uniformly distributed, of the entire line section
G0ch	0.512	Zero sequence shunt conductance, uniformly distributed, of the entire line section
gch	0.489	Positive sequence shunt conductance, uniformly distributed, of the entire line section
length	30	Segment length for calculating line section capabilities
r	0.231	Positive sequence series resistance of the line section
r0	0.386	Zero sequence series resistance of the line section
x	0.325	Positive sequence series reactance of the line section
X0	0.392	Zero sequence series reactance of the line section

Table 4.27 LN2AC Attributes Inherited from *Conductor*

Attribute	Value	Description
B0ch	0.03	Zero sequence shunt susceptance, uniformly distributed, of the entire line section
bch	0.480	Positive sequence shunt susceptance, uniformly distributed, of the entire line section
G0ch	0.412	Zero sequence shunt conductance, uniformly distributed, of the entire line section
gch	0.498	Positive sequence shunt conductance, uniformly distributed, of the entire line section
length	40	Segment length for calculating line section capabilities
r	0.215	Positive sequence series resistance of the line section
R0	0.436	Zero sequence series resistance of the line section
X	0.245	Positive sequence series reactance of the line section
X0	0.502	Zero sequence series reactance of the line section

Table 4.28 LN3AC Attributes Inherited from *Conductor*

Attribute	Value	Description
B0ch	0.045	Zero sequence shunt susceptance, uniformly distributed, of the entire line section
bch	0.356	Positive sequence shunt susceptance, uniformly distributed, of the entire line section
G0ch	0.544	Zero sequence shunt conductance, uniformly distributed, of the entire line section
gch	0.543	Positive sequence shunt conductance, uniformly distributed, of the entire line section
length	60	Segment length for calculating line section capabilities
r	0.656	Positive sequence series resistance of the line section
r0	0.434	Zero sequence series resistance of the line section
x	0.233	Positive sequence series reactance of the line section
x0	0.544	Zero sequence series reactance of the line section

4.5.1.6 Attributes of Breakers

There are in total seven breakers in the example system. For simplicity we assume that all these seven breakers are of the same type, and the only difference between them being in their names. Then, the attributes information needed to model these breakers is listed in Table 4.29.

Table 4.29 Native Attributes of Breakers

Attribute	Value	Description
ampRating	300KA	Fault interrupting rating in amperes
inTransitTime	0.02	Transition time from open to close, in seconds

A breaker is also an entity of *PowerSystemResource* and *ConductingEquipment*, and *Breaker* is derived from *Switch*; therefore, a breaker can inherit certain attributes from these classes, which are listed in Tables 4.30, 4.31, and 4.32.

Table 4.30 Breaker Attributes Inherited from *PowerSystemResource*

Attribute	Value	Description
PowerSystemResourceName	BX	Name or identification of an instance of a power system resource
PowerSystemResourceDescription	xth breaker	Description information
TypeName	SF6	Different types of PowerSystemResources that otherwise have identical attributes

Table 4.31 Breaker Attributes Inherited from *ConductingEquipment*

Attribute	Value	Description
Terminals	2	Maximum number of terminals that the equipment may have
Phases	A	Phases carried by conducting equipment. Possible values are {ABCN, ABC, ABN, ACN, BCN, AB, AC, BC, AN, BN, CN, A, B, C, N}

Table 4.32 Breaker Attributes Inherited from *Switch*

Attribute	Value	Description
NormalOpen	1	Set if the switching device is normally open
SwitchOnCount	10	Switch on count since the switch was last reset or initialized
SwitchOnDate	9:30am on 10/01/2001	Date and time when the switch was last switched on

4.5.1.7 Attributes of Switches

There are eighteen switches in the example system. They are S1 through S18. For simplicity we assume that all of these switches are of the same type, and the only difference among them being in their names. The attributes information needed to model these switches is listed in Table 4.33.

Table 4.33 Native Attributes of Switches

Attribute	Value	Description
NormalOpen	1	Set if the switching device is normally open
SwitchOnCount	10	Switch on count since the switch was last reset or initialized
SwitchOnDate	9:30am on 10/01/2001	Date and time when the switch was last switched on

In CIM, *Switch* is derived from *PowerSystemResource* and *ConductingEquipment*. The attributes of a switch inherited from these classes are listed in Tables 4.34 and 4.35. The type name attribute in Table 4.34 is used to optionally indicate that the database switch does not represent a corresponding real device and has been introduced for modeling purposes only. "Yes" means dummy and "No" means real.

Table 4.34 Switch Attributes Inherited from *PowerSystemResource*

Attribute	Value	Description
PowerSystemResourceName	SX	Name or identification of an instance of a power system resource
PowerSystemResourceDescription	xth switch	Description information
TypeName	NO	Different types of PowerSystemResources that otherwise have identical attributes

Table 4.35 Switch Attributes Inherited from *ConductingEquipment*

Attribute	Value	Description
Terminals	2	Maximum number of terminals the equipment may have
Phases	A	Phases carried by conducting equipment. Possible values {ABCN, ABC, ABN, ACN, BCN, AB, AC, BC, AN, BN, CN, A, B, C, N}

4.5.1.8 Attributes of Connectivity Nodes

Take CN10 as an example, a connectivity node has the following attributes.

COMMON INFORMATION MODEL AND MIDDLEWARE

Table 4.36 CN10 Attributes

Attribute	Value	Description
ConnectivityNodeDescription	BUS1TOS1	Description information
ConnectivityNodeName	CN10	Name of connectivity node
ConnectivityNodeNumber	10	Numerical identification, indicating electrical location

4.6 ILLUSTRATION OF CIM APPLICATIONS

All the attributes information of power system objects will be loaded in the database that is built based on the CIM schema. The management system of this database will manage the relationships of these objects and provide data for applications.

4.6.1 Load Flow Computation

The load flow calculation of the example system can be performed from the CIM database. The following is a procedure for the load flow computation.

- **Topology analysis.** The topology analysis program will scan the terminals of conducting equipment in the system and then figure out the final topology of the system by analyzing the connectivity nodes of the system, since a connectivity node has a native "*ConnectedTo*" association with classes. Suppose that all switches and breakers are normally closed. For instance, the topology program will find that CN10, CN15, CN20, and CN25 can be merged into one connectivity node since CN10 connects BUS1 and S1, CN15 connects S1 and B1, CN20 connects B1 and S2, CN25 connects S2 and LN2, and B2 is supposed to be closed. Finally, several similarly merged connectivity nodes are merged again to form a topological node, which is the BUS1.

- **Data preparation.** The load flow program can get its input data automatically from the database and these include line and transformer parameters. In most cases the load flow program needs to convert the data to a format that fits its requirements. The load flow program can obtain the MW and the MVAR data of the generation and load from the operator-approved operating schedules of GT1 and GT2 and the load demand curve of ELD by specifying a specific time.

- **Load flow solution.** After obtaining the required data, the load flow program will be executed to obtain the load flow solution of the example system.

4.6.2 Other Applications

The preceding example demonstrated the CIM application to the load flow computation. When CIM is applied to the online state estimation, the SCADA measurement attributes will be loaded into the CIM database so that the measurement data, including the status information of breakers and switches, the generation and load, and the line flow measurements, can be obtained from the database. When the system's dynamics must be studied, the dynamics of each generating unit should be loaded into the CIM database. Likewise, when the power trading is to be implemented on CIM, the financial information of the system should be loaded into the database.

4.7 APPLICATIONS INTEGRATION

The components of a large complex system are usually separate and independent of one another. Integration refers to the consolidation of system components for some purpose through information exchange among related components. Integration is never used to change the functions of components in the integration process.

4.7.1 Previous Schemes

The power system information technology (IT) infrastructure traditionally had a point-to-point connection structure like a spider web where information exchange is custom designed and implemented. For example, in Figure 4.10, for N entities of a DPS (distributed processing system) we need $N-1$ individual proprietary application program interfaces (APIs) for each entity [EPR01a]. In this scheme the communication protocols are designed for point-to-point and connection-oriented communication to provide guaranteed data delivery. In cases where multiple applications are required to receive the same data simultaneously, multiple connections to each information source would be required.

Another limitation of this approach is that it does not allow CIM to convey the identity of the data being exchanged. Standard data communication protocols address the syntax issue quite effectively but do not address the semantics. That is, each interconnected entity is responsible for assigning an identity to the data received by associating a received data object with an internally maintained power system model before the entity can use the data. Since the application program interface is typically left to each end user as a local implementation issue, the communication protocols do not fall into the category of middleware, which solves the application integration problem in a more efficient manner.

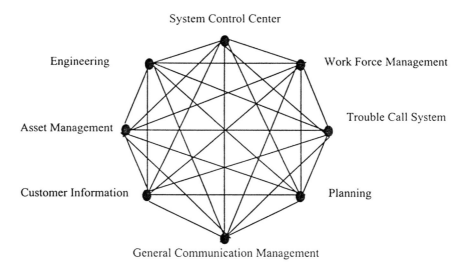

Figure 4.10 DPS Integration through Individual Connections

This kind of integration produces a tangle of inconsistent, individually customized software with a very low extendibility and varying reliability, integrity, and maintainability. A more efficient and cost-effective application integration scheme might be based on a single internationally recognized standard protocol for all interconnections. Such a scheme could change the one to N-1 interconnection problem to a one to one interconnection problem.

4.7.2 Middleware

Middleware is a family of technologies for connecting distributed software entities into an integrated logical entity. Middleware mainly facilitates the integration of applications and end users which may be physically distributed across an enterprise, across a country, or even across a continent. The transaction management system (TMS) is such an example.

Middleware can be utilized for two types of integration. One is an application system that is distributed across a network, and the other is a single autonomous application system that needs to be coordinated or integrated. The primary distinction between these two is the autonomy of system components.

Middleware can be divided into two categories. One is the intra-application problem in which the elements are closely related under the control of one managing agency; for instance, it is reasonable to use locking to synchronize transactions which affect multiple information exchanges. The other is the inter-application problem in which the elements are under separate controls and must maintain at arm's-length. The middleware technology is primarily concerned with the autonomous and heterogeneous application problem because intra-application integration is more of a local implementation problem rather than an integration for a distributed processing system entity across a large geographical area.

In the following we provide an overview of the different types of middleware technologies that are available today for inter-application integration. We start with the most fundamental approach of creating connections among individual application systems and culminate with the latest approaches that build on message-brokering systems with event channels.

4.8 MIDDLEWARE TECHNIQUES

Middleware technologies provide application interfaces and mechanisms for information exchange among applications. Suppose that the information system of a power company comprises the information components shown in Figure 4.10. The functions of middleware in this power system operation are depicted in Figure 4.11 [EPR01a].

Middleware has a critical function in the distributed processing of power systems, as shown in Figure 4.12. For example, in the EMS arena, a CIM-based software provides the key link for the aggregation of vital EMS model data from different vendors into a single data model in yet another vendor's data schema. Leveraging CIM can reduce the cost of creating a complicated metadata repository as the master source.

In using CIM as the model for all data exchanges, utilities can avoid the extensive costs and effort involved in creating their own model. Since CIM is nonproprietary, utilities are more likely to use it, and they upgrade CIM-based applications with little effort. This is particularly beneficial to ISOs, and RTOs, as CIM greatly facilitates the implementation of hierarchical and distributed computation and control of ISO/RTO.

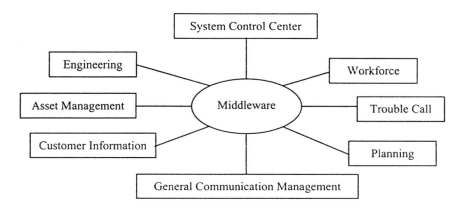

Figure 4.11 Roles of Middleware

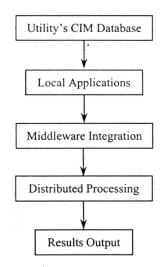

Figure 4.12 Middleware for Distributed Processing

As we discussed in Chapter 1, a geographically distributed DEMS is the basis of distributed computation and control of an RTO/ISO. If all EMS components of a DEMS adopted CIM, the information exchange problem would become much easier and efficient. CIM provides substantial support for distributed computation and control with respect to methodological and technological issues; although its initial intention was not directed toward distributed processing applications.

4.8.1 Central Database

A central database can be used for the integration of individual components as shown in Figure 4.13. Such a database can be implemented using commercial database technologies. Because the database is a separate entity outside the specific components that are being connected, the component that is providing the information will send the information to the database, and the component that needs this information will obtain this information by querying the database.

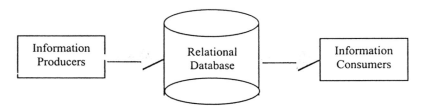

Figure 4.13 Integration with a Central Database

The obvious advantage of this database scheme is that it disconnects information producers and users. For instance, the load forecasting application of an EMS will get its input data from the CIM database and then write its forecasting results back into the CIM database. The database scheme makes the system design flexible as the information structure may be extended and new components may be added without causing wholesale changes to the codes of participating components.

The central database scheme is widely deemed as a fragile and non-scalable solution. One major limitation of this scheme is the problem with its notification. Users will not be informed of new information on which they should act unless a notification system is added to the scheme. Where there is no notification, users will be forced to check periodically to see if anything new has popped up. For some applications such as daily updates, this method works reasonably well, but in many situations, periodical query poses impractical communication burdens on the network and database.

4.8.2 Messaging

Messaging middleware provides a solution to the notification problem. Usually, notification requires a message delivery from the information producer to the user. Traditional two-party communication methods bind the information producer to the user, and thus lack flexibility when several notifications are in the system.

Messaging middleware is developed based on the publish-subscribe method, where the information producer registers the types of the information that it produces, and the user subscribes the information that it needs by type. By the messaging bus, the messaging middleware disconnects the information producers and users, and obtains many advantages over the central database scheme, mainly anonymity and flexibility.

Messaging middleware can be combined with a central database. A major difference between the messaging-plus-database and a central database is that the former can be queried as shown in Figure 4.14. For instance, a power system application that normally runs by obtaining the incremental changes in breaker status would also need to be initialized by the current breaker status when it starts.

4.8.3 Event Channels

An event channel is essentially a sort of advanced messaging, a publish-subscribe pipe with storage. It can be configured to retain event-driven messages for arbitrary lengths of time. Not only can query (in the sense of selecting) which messages are delivered as they occur, it can also be queried for past messages that have been stored. Hence, the initialization will no longer need to get information from a relational database; instead, specialized event channels that directly subscribe to the main event channel will preserve the necessary information they need for initialization as shown in Figure 4.15. The event channel is a great idea in advancing middleware architectures for integration, because most applications work more naturally and efficiently with a stream of messages rather than with just a few.

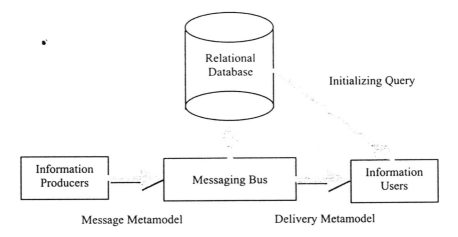

Figure 4.14 Components Integration via Messaging

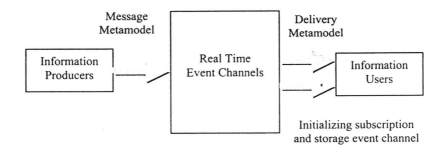

Figure 4.15 Integration via Event Channels

The application of event channels can be substantially extended when event channels are enhanced with a storage facility. Event channels with a storage facility can be used very effectively for information exchange between components even when an immediate delivery is desired. In these situations, event channels are used as brief-but-indefinite parking spaces for information that needs to be moved between locations under a regulated control.

Often a relational database is the right vehicle for the middleware architecture. However, an event channel may replace a relational database if relational databases are to be used within applications rather than between applications. When there is more than one event channel in use, such as with distributed event channels, relational databases may even be used for initialization.

4.8.3.1 Event Channel Applications

Distributed processing of power systems can be roughly classified into two categories: distributed transaction, and distributed computation and control. Event channels can be applied to both categories. Distributed transactions include transactions among customers and scheduling coordinators and transactions among participating entities and the Power Exchange. Notification is important to TMS as information exchanges of TMS are time-critical. Under time-critical action-and-response situations, requests or events at one component must affect a prompt subsequent processing at another component. This feature pushes the TMS into an environment where messaging plays a main role. Accordingly, the user may need to determine the architecture of TMS, namely whether to use the simple messaging combined with a central database or to adopt the concept of event channels. Sometimes event channels are more desirable than necessary.

Now we review the application of even channels to the transaction information system (TIS). As part of TMS, TIS is a good example of event channels with storage. In TIS, a marketer may need to deal with trades, while a security coordinator may need to look at all TS and AS reservations. As depicted in Figure 4.16, for event channels a tag is formatted as a message, and each tag is assigned an event channel.

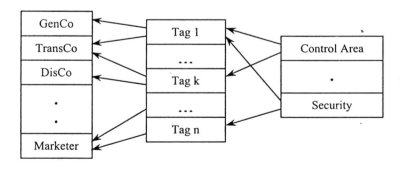

Figure 4.16 Event Channels with Storage Facility

Here a tag is a file that relates a set of individual agreements to a complete transaction deal. In such a scheme, each tag subscriber, (e.g., a marketer or a security coordinator) sets up a persistent query for the type of data required by one event channel; each tag possesses one event channel. The information sources send their messages to their destination tags via message metamodel, and messages stored in tags are delivered to the receiving application via the delivery metamodel. The publish-subscribe nature of event channels can be used to maintain all the essentials of the tag in one physical place while giving every participant the information it needs.

The optimal power flow (OPF) application is another illustration of event channels interacting with a storage facility. OPF usually produces a set of changes for optimizing the operation of a power system. Normally, these changes are reviewed before they are implemented, since OPF can directly publish these changes to a local event channel with storage. These events can be viewed by system operators by subscribing to the channel and can be released to local channels for the implementation in real time.

Usually the real-time computation and control for middleware must meet more stringent requirements than those of the energy transaction management and monitoring when data communication speed, guaranteed delivery, and the like, are critical considerations. Event-based communication, where events are used to both trigger data integration actions and automate the business process without human intervention, are another application for the distributed computing and control.

4.9 CIM FOR INTEGRATION

CIM provides standard objects for applications such as energy production, transmission, distribution, marketing, and retailing functions. As a standard data model, CIM together with a set of standard APIs can greatly facilitates the interchange and interoperability of power system applications.

4.9.1 Integration Based on Wrappers and Messaging

When the utility does not want the integration to disturb existing applications, an intermediate wrapper can be embedded for this purpose. The main function of such a wrapper, as depicted in Figure 4.17, is to expose the functionality of the wrapped application in the CIM format and translate received messages in the CIM format into the local data model. A significant advantage of this scheme for integration is that legacy applications do not need to be rewritten. However, the problem is that the wrappers are proprietary. When electric power system applications are equipped with their own wrappers, the integration of these applications can be shown in Figure 4.18 [Bec00]. The wrapper is only for temporary use in a transitional process as all applications will eventually be implemented with CIM.

Utilities need a way for automating data communication in addition to managing the common information. To do this, as we mentioned earlier, one could leverage CIM to create a central database to which all applications are linked. However, this approach has proved to be non-scalable. A widely accepted practice is to encode messages using XML to move structured data into a text file. Just like HTML, XML allows users to create custom tags and describe how message data are used, and this way provides facilities for self-describing messages.

XML has been adopted by the World Wide Web Consortium (W3C) and has become the favorite approach for exchanging complex data between applications. XML and associated APIs facilitate the interchange and interoperability of various applications. NERC has mandated the use of the Resource Description Framework (RDF) as the XML schema/syntax for CIM, which is defined in the draft IEC 61970-501 CIM RDF Schema standard. Many electric utility software vendors have utilized the XML version of the CIM to test the capability of their software products to exchange and correctly interpret the power system CIM model.

4.10 SUMMARY

The CIM's flexible and open architecture can help RTOs meet the FERC Order 2000 requirements and enable them to respond quickly at low costs to changes in organizational, regulatory, market, and technical requirements. Integration will enable electric power companies to achieve primary objectives, such as improving staff productivity and maximizing transmission and distribution system assets, as well as providing flexibility for the future.

In the past, power systems met with considerable difficulties in information exchange when they tried to integrate their major systems used for T&D real-time operations, including their transmission and substations SCADA system, outage management system (OMS), stand-alone radio control system for distribution switches. The same difficulties would occur when they tried to link their transmission system EMS with their distribution facility management system, and their local EMS with the EMS of their neighboring power systems.

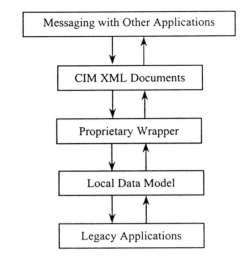

Figure 4.17 Wrappers Functions

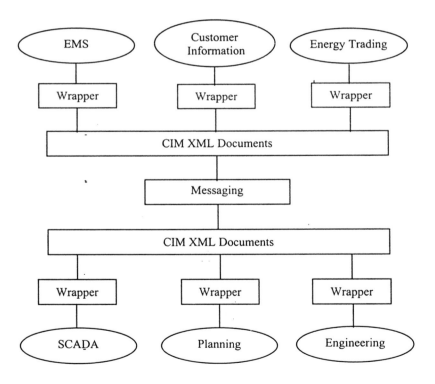

Figure 4.18 Integration Via Wrappers

COMMON INFORMATION MODEL AND MIDDLEWARE

Using CIM as the foundation, integration to various systems can be implemented by employing middleware. With middleware for integration, bus data synchronization and logging can be initiated automatically, several T&D operation systems can talk to each other in real time, and regulatory material can be developed automatically with improved accuracy and completeness.

While the control center remains the nerve center for electric power companies, the information it generates and receives from EMS and other applications is used throughout the energy enterprise. Executives, distribution engineers, and transmission planners also have certain applications for real-time system data. The CIM-enhanced EMS will foster an inter-disciplinary approach to conducting business by enabling interdepartmental teams to access all necessary data via the open systems. Hence, in innovative applications, energy companies are implementing the CIM outside the control center to achieve these goals.

As to the real-time distributed computation and control of RTO/ISO, based on DEMS, the distributed computation and control application for a special purpose such as the distributed load flow algorithm is generally provided by one vendor. Hence, the integration of this kind of distributed computation and control can be realized via intra-application middleware. On the other hand, the real-time distributed computation and control are in most cases based on a dedicated computer network, though the Internet can be used to accomplish similar tasks. The event-driven communication is used for the distributed LF and SE computation and interim results will be exchanged by special communication tools like MPI.

Chapter 5

Parallel and Distributed Load Flow Computation

5.1 INTRODUCTION

The load flow program is one of the most basic and important tools for power system operation, optimization, and control at control centers. To a large extent, load flow is the core of EMS applications, which may include unit commitment, economic dispatch, security analysis, transmission congestion management, and energy trading. The system operator will use the load flow results to monitor and control a power system in real-time. Many load flow computation approaches such as Newton-Raphson load flow, fast decoupled load flow, and dc load flow have received wide applications in power systems [Amb01, Ama96, Bar95, Che93, Che02, Exp02, FuY01, Osa9, Sch02, Vla01, Yan01, Zur01].

In a restructured power system, the load balance condition will be satisfied when the PX clears the power market. The ISO executes load flow program for reliability analysis and transmission congestion management. The ISO will examine the feasibility of the generation schedules submitted by market participants based on the load flow computation results. If no transmission congestion is discovered, the ISO will approve the proposed schedules; otherwise, the ISO will return the congestion information to the participants for the rescheduling of their proposed generation and demand.

As was described in Chapter 2, DEMS (distributed energy management system) is based on a WAN in the system. DEMS is a large and complex computing and control system, which rightly suits the monitoring and control requirements of an electric power system that is naturally distributed on a vast geographical area. A DEMS is composed of several EMS clusters which can be several hundreds, sometimes even thousands, of kilometers apart. The computers of a DEMS are usually heterogeneous and interconnected via various types of communication links including digital microwave and optical fiber links. Compared with a traditional centralized energy management system (CEMS), a DEMS has many advantages in reliability, flexibility, and economy. A DEMS can provide many new functions that are hard to realize in CEMS; for instance, individual EMS clusters in a DEMS can provide on-line external equivalent data to their counterparts for the real-time security analysis. The rapid developments of modern computer science and communication technologies have provided substantial technology supports for many on-line DEMS applications.

In this chapter, we discuss the design methodology and applications of parallel and distributed load flow algorithms for transmission system. We also discuss the methodology for the design of parallel and distributed load flow algorithms and analyze the convergence properties of these algorithms. We begin with the design of a parallel load flow algorithm, and then consider the unpredictability of communication time delay in a distributed load flow algorithm design. Theoretically, a parallel load flow algorithm should have the same convergence property as its serial counterpart, so the convergence analysis and mathematical proofs of the distributed load flow algorithm is discussed here. The simulation methods for a distributed computation of an ISO/RTO and its satellite control centers will also be discussed.

5.1.1 Parallel and Distributed Load Flows

Parallel load flow and distributed load flow represent two distinct approaches for large-scale power system load flow computation. Parallel load flow is implemented on a parallel machine with multiple processors at the system control center; however, distributed load flow is based on DEMS. Since the design of parallel and distributed algorithms is closely related to their associated computer systems, it is essential to design parallel and distributed load flow algorithms according to the specific topologies and communication characteristics of their computing system.

The synchronous distributed load flow algorithm is difficult to use in DEMS because of the serious synchronous penalty caused by tasks imbalance, computer heterogeneousness and the unpredictability of data communication delays. Therefore, it is necessary to seek a novel load flow algorithm, which is suitable for DEMS applications. The asynchronous distributed computation can adapt to the unpredictable time delay of the data communication between subarea control centers. Asynchronous algorithm designs can also be used for other applications in DEMS such as state estimation, which will be discussed in Chapter 7.

5.2 MATHEMATICAL MODEL OF LOAD FLOW

Let us assume for a specific system that the grid structure, network parameters, nodal generations, and nodal loads are given. The system load flow computation is based on the solution of the following equation

$$F(x, S, P, L, G) = 0 \tag{5.1}$$

where the vector x represents system state variables, S represents the network topology, P represents the network parameter, L represents nodal loads and G represents nodal generation respectively. Here we further assume that the system is modeled using CIM, and S, P, L, and G are stored in a rational database. Equation (5.1) is rewritten in a simpler form as

$$g(x) = 0 \tag{5.2}$$

which is further written in a fixed-point form as

$$x = f(x) \tag{5.3}$$

where $x \in X \subset R^n$, $f(.)$ is a nonlinear mapping function $f:X\ R^n$. Several solution methods such as Newton-Raphson, fast P-Q decoupled and DC load flow are employed to solve this load flow equation.

5.3 COMPONENT SOLUTION METHOD

Suppose that $X_1,...,X_i, ... , X_N$ are subsets of the Euclidean space, $n = n_1 + ... + n_N$, and $1 \leq n_i \leq n$. An $x \subset R^n$ can be decomposed as $x = (x_1,---,x_N)$, where $x_i \in R^{n_i}$ ($i = 1,\ ,N$) is the ith component of x. Suppose that f is a mapping function defined by $f(x) = (f_1(x),..., f_N(x))$, $\forall x \in X$, $f_i:X\ X_i$.

Accordingly, the load flow problem formulated by (5.3) is solved concurrently by multiple processors as follows:

$$x_i = f_i(x_i, \cdots, x_N), \forall i \in \{1, \cdots, N\} \tag{5.4}$$

Equation (5.4) is called the component solution of (5.3); when processor i is updating x_i, all other components of x remain unchanged. This component solution method is the basis of parallel and distributed computation, which we discuss next.

5.4 PARALLEL LOAD FLOW COMPUTATION

5.4.1 System Partitions

As discussed earlier, the only purpose of parallel computation is to obtain a faster solution speed. There are many factors that can affect the parallel solution speed, but the most important are as follows:

- Load balance among participating processors
- Performance of processors
- Speed of data communication between processors.

Therefore, to achieve a high speed computation, it is necessary to divide the system load flow equations into several balanced subsets for parallel computation, or to divide the entire system into a number of balanced zones for distributed computation.

Theoretically, we may split (5.3) in an arbitrary manner. However, the division may not represent the physical characteristics of the system, and the nodes in the subvector may not be located in one geographical area of the power system. Furthermore, this partitioning method may give rise to a complicated communication requirement in both parallel and distributed algorithms. As shown in Figure 5.1, each processor would have to communicate with many other processors in exchanging information at every iteration step. The algorithm's execution will be greatly compromised if a large amount of data has to be transmitted among processors. To keep data communication at a minimum, we assume that the data to be exchanged among processors are represented by boundary state variables.

PARALLEL AND DISTRIBUTED LOAD FLOW

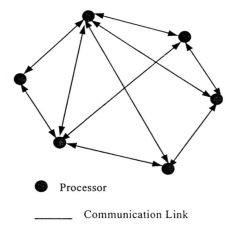

Figure 5.1 Communications among Processors

Let us divide a large-scale power system into N subareas, the first $(N - 1)$ subareas have approximately an equal number of nodes, and the Nth subarea which, is composed of all boundary nodes and lines, is called the boundary subarea. More specifically, let S_i represent the node set of subarea i, then

$$S_i \cap S_j = \phi, \quad i, j = 1,...,N-1 \tag{5.5}$$

$$S_i \cap S_N = S_{bi}, \quad i = 1,...,N-1 \tag{5.6}$$

$$\bigcup_{i=1}^{N-1} S_{bi} = S_N \tag{5.7}$$

where φ is an empty set, S_{bi} is the set of boundary nodes of the ith subarea and its boundary subarea. The system partitioning based on this method is shown in Figure 5.2.

5.4.2 Parallel Algorithm

Suppose that each subarea is assigned to a processor of a parallel machine, and that x_k ($k = 1, ..., N$) is the state vector of the kth subarea.

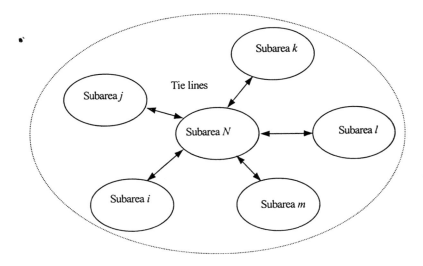

Figure 5.2 System Partition for Parallel Computation

The load flow can then be formulated with the following bordered block diagonal matrix (BBDM) form:

$$\begin{bmatrix} J_{1,1} & & & J_{1,N} \\ & J_{k,k} & & J_{k,N} \\ J_{N,1} & J_{N,k} & & J_{N,N} \end{bmatrix} \begin{bmatrix} \Delta x_1 \\ \Delta x_k \\ \Delta x_N \end{bmatrix} = \begin{bmatrix} \Delta s_1 \\ \Delta s_k \\ \Delta s_N \end{bmatrix} \qquad (5.8)$$

Equation (5.8) can be decomposed into (5.9) and (5.10) as follows:

$$J_{k,k}\Delta x_k + J_{k,N}\Delta x_N = \Delta s_k \qquad (5.9)$$

$$J_{N,k}\Delta x_1 + ... + J_{N,k}\Delta x_k + ... + J_{N,N}\Delta x_N = \Delta s_N \qquad (5.10)$$

where $J_{k,k}$ is the Jacobian matrix of subarea k, $k = 1,..,N$ and $J_{k,N}$ is the Jacobian matrix of subarea k with respect to the boundary subarea N, $k = 1,...,N-1$. From (5.9) we have

$$\Delta x_k = J_{k,k}^{-1}\{\Delta s_k - J_{k,N}\Delta x_N\}, \quad k = 1,..., N-1 \qquad (5.11)$$

Let

$$J_k = J_{N,k}J_{k,k}^{-1}J_{k,N} \qquad (5.12)$$

PARALLEL AND DISTRIBUTED LOAD FLOW

$$S_k = J_{N,k} J_{k,k}^{-1} S_k \qquad (5.13)$$

By substituting Δx_k in (5.11) into (5.9), we get

$$\Delta x_N = \left\{ J_{N,N} - \sum_{k=1}^{N-1} J_k \right\}^{-1} \left\{ \Delta s_N - \sum_{k=1}^{N-1} s_k \right\} \qquad (5.14)$$

The parallel load flow computation is obtained by solving (5.11) and (5.14) alternatively.

Now, choosing the processor in the boundary subarea to be the coordinator, we can design the parallel load flow algorithm as follows:

Step 1: use flat start, i.e., $x_k = x_k(0)$, $k = 1, \ldots, N$

Step 2: processor k ($k = 1, \ldots, N-1$) calculates $J_{k,k}$, $J_{k,N+1}$, $J_{k,k}^{-1}$, and ΔS_k concurrently, and processor N calculates $J_{N,N}$, $J_{N,k}$, and ΔS_N concurrently

Step 3: processor k ($k = 1, \ldots, N-1$) calculates $J_{N+1,k} J_{k,k}^{-1} J_{k,N+1}$, $J_{N,k} J_{k,k}^{-1}$ and ΔS_k concurrently and transmits the results to processor N

Step 4: processor N solves Δx_N according to (5.14) and transmits the result to processor k ($k = 1, \ldots, N-1$)

Step 5: processor k ($k = 1, \ldots, N-1$) solves Δx_k according to (5.11)

Step 6: stop the iterative process if convergence conditions are satisfied for every processor; otherwise go to Step 2.

The design of this parallel algorithm is quite explicit and easy to implement on a parallel machine or even on a conventional serial machine. Such a parallel algorithm will have the same convergence property as its serial counterpart.

5.5 DISTRIBUTED LOAD FLOW COMPUTATION

5.5.1 System Partitioning

Distributed computation and control can be applied to large, widely spread out systems to avoid costly investments in telecommunication lines and to enable systems to be monitored more effectively and efficiently.

Historically, most large-scale power systems were composed of a number of subareas, each with its local control center called subarea control center (SACC). Therefore, the distributed load flow computation can be directly based on the existing SACCs. As depicted in Figure 5.3, each SACC will be responsible for the load flow computation of its own subarea. For example, if the power system has N SACCs, the distributed load flow will be performed right on the distributed system composed of N computers at the N corresponding SACCs.

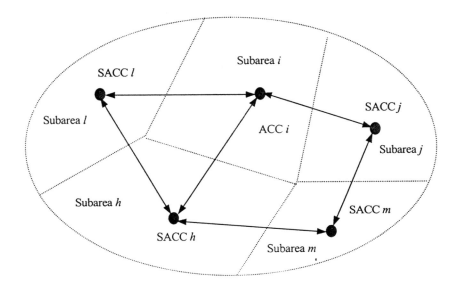

Figure 5.3 System Partitioning for Distributed Load Flow Computation

5.5.2 Distributed Load Flow Algorithm

Theoretically, distributed computation is quite similar to parallel computation, and the only difference between the two is that the distributed computation is based on a computer network rather than a parallel machine where communications among processors are assumed to be fixed and reliable. Usually the data communication process in a distributed system is more complicated than in a parallel machine.

5.5.2.1 General Iteration Equation. Based on the system partitioning paradigm, shown in Figure 5.3 for the distributed load flow algorithm, each computer at its SACC will perform the following iteration:

$$x_k(t+1) = x_k(t) - \gamma J_{kk}^{-1}(x) g_k(x), \forall\ k \in \{1, \cdots, N\} \quad (5.15)$$

where $J_{kk}^{-1}(.)$ is the Jacobian matrix of the ith subarea and γ is the relaxation factor. As in conventional relaxation iteration algorithms, this algorithm is called either low or exceedingly relaxed iteration algorithm based on $\gamma < 1$ or $\gamma > 1$ respectively. The $J_{kk}(.)$ in (5.15) is determined as follows. Suppose, as shown in Figure 5.4 (a), that there is one tie line whose terminal buses i and j belong to subarea k and h, respectively. Buses i and j will be grounded for the formulation of J_{kk} and J_{hh} as shown in Figure 5.4 (b).

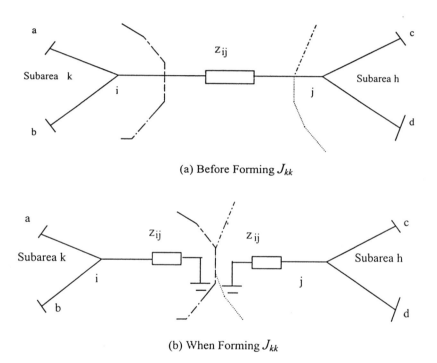

(a) Before Forming J_{kk}

(b) When Forming J_{kk}

Figure 5.4 Formation of J_{kk}

To be more specific, we suppose the kth subarea has n_k buses, which are consecutively put together in the system Jacobian matrix J from the $(m+1)$th row to the $(m+n_k)$th row and from the $(m+1)$th column to the $(m+n_k)$th column. Accordingly, J_{kk} in (5.17) is part of the matrix shown in Figure 5.5.

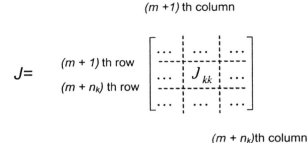

Figure 5.5 Elements of J_{kk}

5.5.2.2 Nodal Power Calculation. After system partitioning, the nodes of a subarea can be sorted either as internal nodes such as nodes a, b, c, and d in Figure 5.6, or as boundary nodes such as i and j in Figure 5.6.

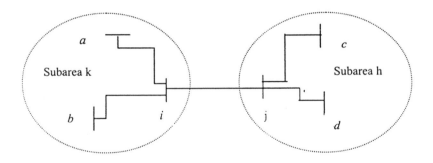

Figure 5.6 Boundary Nodes between Subareas

The calculation of nodal power for internal nodes of a subarea, either active or reactive power, is the same as that in the conventional load flow computation. The nodal power of internal nodes can be calculated based on internal state variables:

$$P_{i,I} = f_{P,i,I}(x_i), \quad \forall\, i \in \{1,\cdots,n_i\} \tag{5.16}$$

$$Q_{i,I} = f_{Q,i,I}(x_i), \quad \forall\, i \in \{1,\cdots,n_i\} \tag{5.17}$$

where $P_{i,I}$ and $Q_{i,I}$ are the functions for the active and reactive power computation of an internal node.

PARALLEL AND DISTRIBUTED LOAD FLOW

The nodal power calculation of boundary nodes will require the information on boundary state variables transmitted from the neighboring subareas. Specifically, in Figure 5.6, the power calculation of node i in subarea k needs the state variable of node j in subarea h, and the power calculation of node i can not proceed before subarea h has transmitted the state variable of node j to subarea k. In general, the power calculation of a boundary node can be formulates as follows:

$$P_{i,B} = f_{P,i,B}(x_i, x_{i,B}) \qquad (5.18)$$
$$Q_{i,B} = f_{Q,i,B}(x_i, x_{i,B}) \qquad (5.19)$$

where $P_{i,B}$ and $Q_{i,B}$ are the functions for the active and reactive power computation of a boundary node. These values are expressed as a function (f) of internal state variables x_i and the associated boundary variables $x_{i,B}$ of neighboring subareas. All boundary state variables will be communicated among distributed computers during the distributed iterative process.

5.5.3 Synchronous Algorithm of Load Flow

Since synchronization allows the participating computers to take the consistent iteration steps, the synchronous distributed computation is quite similar to the parallel computation. The synchronous distributed load flow algorithm is described as follows:

$$x_i(t+1) = x_i(t) - \gamma J_{ii}^{-1}(x_1(t), \cdots, x_N(t)) g_i(x_1(t), \cdots, x_N(t)),$$
$$\forall i \in \{1, \cdots, N\} \qquad (5.20)$$

However, the Jacobian matrix segment J_{ii} that we are using here for the local computation does not take into account the non-diagonal elements of the system Jacobian matrix. So the convergence process of the synchronous distributed algorithm could become sluggish compared with serial and parallel algorithms.

From (5.18) and (5.19), we derive the following equations for the nodal power calculation:

$$P_{i,I}(t+1) = f_{P,i,I}(x_i(t)), \qquad \forall i \in \{1, \cdots, N\} \qquad (5.21)$$
$$Q_{i,I}(t+1) = f_{Q,i,I}(x_i(t)), \qquad \forall i \in \{1, \cdots, N\} \qquad (5.22)$$
$$P_{i,B}(t+1) = f_{P,i,B}(x_i(t), x_{i,B}(t)) \qquad (5.23)$$

$$Q_{i,B}(t+1) = f_{Q,i,B}(x_i(t), x_{i,B}(t)) \tag{5.24}$$

A significant shortcoming of the synchronous distributed load flow computation is that, in most cases, synchronization will have to bear a high penalty incurred by communication time delay, load imbalance, heterogeneousness of computer performance, and so on.

5.5.4 Asynchronous Algorithm of Load Flow

In a distributed system located over a vast geographical area, time delays for data communication among computers are hard to predict. Hence, to avoid a high synchronization penalty, communication network characteristics in the distributed system must be taken into consideration for the algorithm design.

Let us suppose that T^i is the set of time (t) when x_i is updated by computer i of the distributed system and $x_j(\tau_j^i(t))$ is the value of x_j transmitted to computer i by computer j at the time $\tau_j^i(t)$. Then the asynchronous distributed algorithm of load flow is formulated as

$$x_i(t+1) = f_i(x_1(\tau_1^i(t)), \cdots, x_i(t), \cdots, x_N(\tau_N^i(t))), \forall\ t \in T^i, \forall\ i \in \{1, \cdots, N\} \tag{5.25}$$

where $\tau_j^i(t)$ obviously satisfies

$$0 \le \tau_j^i(t) \le t, \qquad \forall\ t \ge 0 \tag{5.26}$$

For all $t \notin T^i$, $x_i(t)$ remains fixed, and there exists

$$x_i(t+1) = x_i(t), \quad \forall t \in T^i, \forall\ i \in \{1, \cdots, N\} \tag{5.27}$$

After considering communication time delays and introducing a relaxation factor γ, the asynchronous distributed algorithm of load flow based on the Newton-Raphson method can be formulated as

$$x_i(t+1) = x_i(t) - \gamma J_{ii}^{-1}(x_1(\tau_1^i(t)), \cdots, x_i(t), \cdots, x_N(\tau_N^i(t)) g_i(x_1(\tau_1^i(t)), \cdots, x_i(t),$$
$$\cdots, x_N(\tau_N^i(t)), \quad \forall t \in T^i, \forall\ i \in \{1, \cdots, N\} \tag{5.28}$$

where $g_i : x \to x_i$ is the power mismatch function for the ith subarea, and $J_{ii}(.)$ is the Jacobian matrix of the load flow equation of the ith subarea, which is formulated as in section 5.5.2.1.

The main characteristic of an asynchronous algorithm is that local computers do not need to wait at predetermined points for predetermined messages to become available. Instead, some computers are allowed to compute faster and execute more iterations than others; furthermore, a few computers are allowed to communicate more frequently than others, with communication delays that can become substantial or even unpredictable. In some special cases it is assumed that communication channels can deliver messages in a different order that the one in which they were expected.

The asynchronous algorithm appears to offer several advantages. First, it can reduce the synchronous penalty and potentially gain speed over synchronous algorithms under various circumstances. Second, it has a greater implementation flexibility and tolerance for data alterations while executing the algorithm. The negative side of an asynchronous algorithm is that the conditions for its validity could become more stringent than those for its synchronous counterparts.

The following assumptions [Ber98] are essential for studying the data communication characteristics of a distributed system.

- **Totally asynchronous assumption.** Suppose that T^i is indefinite and $\{t_k\}$ is an indefinite sequence of the elements of T^i, then we have $\lim_{k \to \infty} \tau_j^i(t_k) = \infty, \ \forall j$.

- **Partially asynchronous assumption.** Suppose there is a positive integer M so that

i) For each i and $t = 0$, the set $\{t, t+1, \ldots, t+m-1\}$ has at least one element which belongs to T^i;

ii) For all i and $\forall t \in T^i$, there is

$$t - M < \tau_j^i(t) \leq t \qquad (5.29)$$

iii) For all i and $\forall t \notin T^i$, there is

$$\tau_j^i(t) = t \qquad (5.30)$$

5.5.4.1 Totally Asynchronous Algorithm. Suppose that the algorithm starts from $x(0) \in X$. As the preceding totally asynchronous assumption is satisfied, the algorithm represented by (5.27) and (5.28) is called the distributed totally asynchronous Newton-Raphson load flow method.

5.5.4.2 Partially Asynchronous Algorithm. From (5.20), we see that the data that are to be communicated in the distributed load flow computation are merely the boundary state variables. Compared with the traditional serial load flow computation, the immediate impact of asynchronization is that the computation results of the boundary power will change, which will eventually cause the difference in convergence properties of serial and asynchronous distributed load flow algorithms. In this regard the most effective method to improve the convergence property of the asynchronous distributed load flow algorithm is to reduce the boundary power mismatches caused by asynchronization. Intuitively, the boundary power mismatches could be reduced by restraining the extent of asynchronization, namely the difference between t and $\tau(t)$. Hence, we introduce the following partially asynchronous assumption.

Under the partially asynchronous assumption, the distributed asynchronous load flow algorithm described by (5.27) and (5.28) is referred to as partially asynchronous distributed load flow algorithm, and M is called the asynchronization measure. Since practically the communication time delays are within certain limits, the partially asynchronous distributed load flow algorithm will be more suitable for DEMS applications.

5.5.4.3 Boundary Power Calculation. Since we suppose that there is no communication delay for the local information, the power calculation for internal nodes in asynchronous distributed load flow algorithms is the same as that in the synchronous distributed algorithm. However, because of the communication delay of boundary state variables, the calculation of boundary node power is different from that of the synchronous distributed algorithm. The nodal power calculation of boundary nodes does not have to be delayed until it receives boundary state variables from its neighboring subareas. So the available boundary state variables will be used to compute the boundary power. For instance, the injection power calculation of node j in subarea h will not be delayed for the latest value of the state variable of node i in the subarea k. Rather, the calculation will proceed by using the available state variable of node i that was received previously. The formulation is given as

PARALLEL AND DISTRIBUTED LOAD FLOW

$$P_{i,B}(t+1) = f_{P,i,B}(x_i(t), x_{i,B}(\tau_j^i(t))) \qquad (5.31)$$

$$Q_{i,B}(t+1) = f_{Q,i,B}(x_i(t), x_{i,B}(\tau_j^i(t))) \qquad (5.32)$$

5.5.5 Boundary Power Compensation

To derive the mathematical formulation of the distributed load flow computation with boundary power compensation, we first take a look at the load flow formulation for the entire system. Suppose that the system is partitioned exactly the same way as the distributed load flow; then the block matrix for the load flow equation is formulated as

$$-\begin{bmatrix} J_{11}(x) & \cdots & J_{1i}(x) & \cdots & J_{1N}(x) \\ \cdots & \cdots & \cdots & \cdots & \cdots \\ J_{i1}(x) & \cdots & J_{ii}(x) & \cdots & J_{iN}(x) \\ \cdots & \cdots & \cdots & \cdots & \cdots \\ J_{N1}(x) & \cdots & J_{Ni}(x) & \cdots & J_{NN}(x) \end{bmatrix} \begin{bmatrix} \Delta x_1 \\ \cdots \\ \Delta x_i \\ \cdots \\ \Delta x_N \end{bmatrix} = \begin{bmatrix} g_1(x) \\ \cdots \\ g_i(x) \\ \cdots \\ g_N(x) \end{bmatrix} \qquad (5.33)$$

In comparing (5.33) with (5.15), we see that the distributed load flow algorithm has omitted non-diagonal blocks of the system Jacobian matrix. One possible way to improve the convergence of the distributed load flow computation would be to compensate the boundary power caused by such reductions. When the boundary power is compensated, the distributed load flow computation can be generally formulated as follows:

$$x_i(t+1) = x_i(t) - J_{ii}^{-1}(x)\left[\gamma g_i(x) + \beta \sum_{j \neq i} J_{ij}(x)\Delta x_j\right] \quad \forall t \in T^i,$$
$$\forall i \in \{1, \cdots, N\} \qquad (5.34)$$

where β is a scale factor for compensation.

Specifically, for the synchronous distributed computation, the boundary power compensation algorithm can be formulated as

$$x_i(t+1) = x_i(t) - J_{ii}^{-1}(x(t))\left[\gamma g_i(x(t)) + \beta \sum_{j \neq i} J_{ij}(x(t))\Delta x_j(t)\right] \quad \forall t \in T^i,$$
$$\forall i \in \{1,\cdots,N\} \quad (5.35)$$

While for the asynchronous distributed computation, the boundary power compensation algorithm can be formulated as

$$x_i(t+1) = x_i(t) - J_{ii}^{-1}\left[x_1(\tau_1^i(t)),\cdots,x_N(\tau_N^i(t))\right]\left[\gamma g_i(x_1(\tau_1^i(t)),\cdots,\right.$$
$$\left. x_N(t_N^i(t))) + \beta \sum_{j \neq i} J_{ij}(x(\tau_j^i(t)))\Delta x_j(\tau_j^i(t))\right], \forall t \in T^i, \forall i \in \{1,\cdots,N\}$$
$$(5.36)$$

It is obvious in the compensation algorithm that the boundary state vector $x_{i,B}$ and its increment $\Delta x_{i,B}$ need to be transmitted to neighboring subareas.

5.6 CONVERGENCE ANALYSIS

The parallel algorithm is just another computation method of the load flow problem and its convergence property will be the same as that of its serial counterpart. Therefore, in this section, we only analyze the convergence of distributed computation.

5.6.1 Convergence of Partially Asynchronous Distributed Algorithm

According to the partially asynchronous assumption, the next iteration solution $x(t + 1)$ depends on the previous solutions $x(t), x(t - 1),…, x(t - M + 1)$. Hence, we introduce a vector $z(t) = (x(t), x(t - 1),…, x(t - M + 1))$ which summaries all the information to be used for the next iteration. Since $x(t + 1)$ is decided by $z(t)$, it indicates that $z(t + 1)$ will be decided by $z(t)$. Suppose that l is a positive integer, then $z(t + l)$ is determined by $z(t)$. Further, if $f(.)$ is continuous, $z(t + l)$ is also a continuous function of $z(t)$.

Let Z be the Cartesian product of M copies of X, that is, $Z = (x^1,…,x^M)$ and $Z^* = \{(x^*,…,x^*) \mid x^* \in X^*\}$. It is conceivable that if $z(t) = z^* \in Z^*$ then $z(t+l) = z^*$. So, we offer the following proposition.

Proposition 5.1. Say there exist some positive integer t^* and a continuous Lyapunov function $d: Z \to (0,+\infty)$. If the partially asynchronous assumption holds, for every scenario of the partially asynchronous load flow computation we have:

i) For every $z(0) \notin Z^*$, there exists $d(z(t^*)) < d(z(0))$;
ii) For every $z(0) \in Z$ and for every $t \geq 0$, there exists $d(z(t+1)) < d(z(t))$;

Then there is $z^* \in Z^*$ for every limiting point $z^* \in Z$ of the sequence $\{z(t)\}$ generated by the partially asynchronous load flow algorithm. Proposition 5.1 is too general to provide a practical convergence implication. Hence, in the following, we continue to seek more specific convergence conditions for the partially asynchronous load flow computation.

Before we go on to analyze the convergence condition of asynchronous distributed load flow computation, we present the following fact. Suppose that the set of the fixed points of the load flow problem are expressed as $X^* = \{x \in R^n \mid x = f(x)\}$, then the following fact holds.

Fact 5.1. 1) X^* is non-empty; 2) $f(x)$ is continuous;

Proof. 1) there is always a load flow solution for a steady state power system.

2) $f(x)$ is continuous and derivable when the system is stable under small disturbances[Ara81].

Let us define the following function on R^n

$$h(x) = \inf_{x^* \in X^*} \left\| x - x^* \right\|_\infty \tag{5.37}$$

Then the following proposition holds.

Proposition 5.2. Suppose that the iterative load flow function $f : X \to R^n$ is non-expansive, which satisfies

$$\left\| f(x) - x^* \right\|_\infty \leq \left\| x - x^* \right\|_\infty, \forall x \in R^n, \forall x \in X^* \tag{5.38}$$

Then its properties are as follows:
1) The set X^* is closed
2) For every $x \in X$, there exists some $x^* \in X^*$ which satisfies $h(x) = \|x - x^*\|_\infty$

3) The function $h: X \to R^n$ is continuous
4) For every $x \in X$, there is $h(f(x)) \le h(x)$.

For the purpose of simplification, we introduce a few notations. First we let $S(x, x^*)$ represent the set of indices of coordinates of x that are the farthest from x^*, that is,

$$S(x, x^*) = \{i \mid |x_{(i)} - x^*_{(i)}| = \|x - x^*\|_\infty\} \tag{5.39}$$

Then, we define

$$E(x, x^*) = \{y \in R^n \mid y_{(i)} = x_{(i)}, \forall i \in I(x, x^*)\} \cup \{y \in R^n \mid \|y_{(i)} - x_{(i)}\| \le \|x - x^*\|_\infty,$$
$$i \notin I(x, x^*)\} \tag{5.40}$$

To simplify the mathematical proof, we assume that $n_i = 1 (i = 1,..., N)$. Let us consider the following assumption:

Assumption 5.1. The iterative load flow function $f: X \to R^n$ has the following features:
i) X^* is convex
ii) For every $x \in R^n$ and $x^* \in X^*$, there is $\|x - x^*\|_\infty = h(x) > 0$, and there exists $i \in S(x; x^*)$ such that $f_i(y) \ne y_{(i)}$ for all $y \in E(x; x^*)$
iii) If $x \in R^n$, $f_i(x) \ne x_i$, and $x^* \in X^*$, then $|f_i(x) - x^*_i| < \|x - x^*\|_\infty$.

Based on the assumption above, we offer the following proposition for the partially asynchronous load flow computation:

Proposition 5.3. If the iterative load flow function $f: X \to R^n$ is non-expansive and the Assumption 5.1 holds, then the solution sequence

PARALLEL AND DISTRIBUTED LOAD FLOW

$\{x(t)\}$ generated by the partially asynchronous load flow iteration converges to a fixed point $x^* \in X^*$.

Proof. For any $z = (x^1,...,x^M) \in Z$, we define the Lyapunov function

$$D(z;x^*) = \max_{1 \leq i \leq M} \left\| x^i - x^* \right\|_\infty \tag{5.41}$$

and

$$d(z) = \max_{x^* \in X^*} D(z;x^*) \tag{5.42}$$

Before we prove Proposition 5.3 using Proposition 5.1, we prove the following four lemmas.

Lemma 5.1. Suppose that $x \notin X^*, x^* \in X^*$, and $h(x) = \inf_{x^* \in X^*} \left\| x - w^* \right\|_\infty < \left\| x - x^* \right\|_\infty$. If $z = (y^1,...,y^M) \in Z$, $y^k \in E(x;x^*)$ $k = 1,...,M$, then there is $d(z) < \left\| x - x^* \right\|_\infty$.

Assume $z(t) = (x(t), x(t-1), \cdots, x(t-M+1))$ and for $t \in T^i$, we have $x^i(t) = (x_1(\tau_1^i(t)),...,x_N(\tau_N^i(t)))$.

Lemma 5.2.

1) If $x^* \in X^*$, then for every $t \in T^i$, there is $\left\| x^i(t) - x^* \right\|_\infty \leq D(z(t); x^*)$;

2) If $x^* \in X^*$, then for all $t \geq 0$, there is $D(z(t+1); x^*) \leq D(z(t); x^*)$;

3) For all $t \geq 0$, there is $d(z(t+1)) \leq d(z(t))$.

The item 3) of Lemma 5.2 guarantees the validity of condition ii) of Proposition 5.1. In the following, we will prove that if $z(0) \notin Z^*$ then $d(z(2nM + M)) < d(z(0))$, which guarantees the validity of condition i) of Proposition 5.1 when $t^* = 2nM + M$. To prove this, we consider $z(0) \notin Z^*$ and a particular iteration scenario. We denote $d^* = d(z(0))$ and use x^* to indicate a particular element of X^* which satisfies $d^* = d(z(0))$. We also denote $J(t) = \{i \mid |x_i(t) - x_i^*| = d^*\}$, and for any limited set J, we use $|J|$ to represent its cardinality.

Lemma 5.3. 1) If $x_i(t+1) \neq x_i(t)$ for some $t \geq 0$, then $i \notin J(t+1)$
2) For every $t > 0$, there is $J(t+1) \subset J(t)$.

Lemma 5.4. If $d^* = d(z(0)) > 0$, then at least one of the following statements is true:

1) $d(z(2nM + M)) < d^*$
2) For every t_0 in the range $0 \leq t_0 \leq 2(n-1)M$, if $|J(t_0)| > 0$, then $|J(t_0 + 2M)| < |J(t_0)|$.

Now we return to Proposition 5.3. By Lemma 5.4, if $d(z(0)) > 0$, there are only two cases to consider. Either $d(z(2nM + M)) > d(z(0))$ or the cardinality of $J(t)$ decreases every $2M$ time units until it is empty. In the latter case, we conclude that $J(2nM)$ is empty. By Lemma 5.3, it follows that $J(t)$ is empty and $\|x(t) - x^*\|_\infty \leq d^*$ for all t satisfying $2nM \leq t < 2(n+1)M$. This implies that $d(z(2nM + M)) > d^*$. We therefore can conclude that the inequality $d(z(2nM + M)) > d(z(0))$ holds in both cases. This establishes condition i) of Proposition 5.1. Notice that the sequence $\{z(t)\}$ is bounded and therefore has a limiting point z^*. By Proposition 5.1, we have $z^* \in Z^*$. Let x^* be such that $z^* = (x^*,....,x^*)$. Then $x^* \in X^*$ and $\lim \inf_{t \to \infty} D(z(t); x^*) = 0$. Since $D(z(t); x^*)$ is non-increasing, we conclude that it converges to zero. This establishes that $z(t)$ converges to z^* and $x(t)$ converges to x^*.

Detailed proofs of Lemmas 5.1 through 5.4 are found in [Ber89]. Although in all of these cases, we consider $n_i = 1$, the outcome can be extended to $n_i > 1$ as well [Wan94].

5.6.2 Convergence of Totally Asynchronous Distributed Algorithm

To find the exact convergence conditions for the totally asynchronous distributed load flow algorithm, we first look at the following general convergence theorem for a totally asynchronous computation.

Proposition 5.4. Suppose $X = \prod_{i=1}^{N} X_i \subset \prod_{i=1}^{N} R^{n_i}$, if for every $i \in \{1,..., N\}$, there is a subset sequence $\{X_i(k)\}$ of X_i, which satisfies the followings:

i) For all $k \geq 0$, $X_i(k+1) \subset X_i(k)$

ii) For all $x \in X(k)$, where $X(k) = \prod_{i=1}^{N} X_i(k)$, there is $f(x) \in X(k+1)$

iii) The limit of sequence $\{x(k) \mid x(k) \in X(k), \forall k\}$ is a fixed point of f.

Then every limit point of the sequence $\{x(t)\}$ generated by the totally asynchronous iteration with $x(0) \in X(0)$ is a fixed point of f.

Condition i) of Proposition 5.4 is the Box Condition. This condition implies that by combining components of vectors in $X(k)$, we can still obtain vectors in $X(k)$. Condition ii) of Proposition 5.4 is the Synchronous Convergence Condition, and it implies the limiting points of sequences generated by the synchronous iteration are fixed points of f. Let us define a weighted maximum norm on R^n as follows:

$$\|x\|_{\infty}^{w} = \max_{i} \frac{\|x_i\|_i}{w_i} \tag{5.45}$$

where $x_i \in R^{n_i}$ is the ith component of x; $\|\cdot\|_i$ is certain kind of norm defined on R^{n_i} and $w \in R^n$, w_i ($i = 1,..., N$) is a scalar. Then we have the following general theorem for totally asynchronous distributed load flow algorithm:

Proposition 5.5. If the iterative load flow function $f : R^n \to R^n$ is the weighted maximum norm contracting map, then the sequence $\{x(t)\}$

generated by the totally asynchronous load flow iteration with $x(0) \in X(0)$ is a fixed point of f.

Proof. if f is the weighted maximum norm contracting map, then there exists an $\alpha \in [0,1)$ which obtains

$$\|f(x) - x^*\| \le \alpha \|x - x^*\|, \quad \forall x \in X \quad (5.46)$$

where x^* is the fixed point of f.

Suppose that the totally asynchronous iteration starts with $x(0) \in X$, and there is

$$X_i(k) = \{x_i \in R^{n_i} \mid \|x_i - x_i^*\|_i \le \alpha^k \|x(0) - x^*\|\} \quad (5.47)$$

Then, the set generated by (5.47) will satisfy the conditions of Proposition 5.4 and converge to a fixed point of f.

To further study the convergence property of the totally asynchronous distributed load flow algorithm, let us consider the following proposition [Ber83].

Proposition 5.6. Suppose that

i) $X \in R^n$ is convex and $f : X \to R^n$ is continuous and differentiable;

ii) There exists a positive constant k which satisfies

$$\nabla_i f_i(x) \le k, \; x \in X, \forall i \quad (5.48)$$

iii) There exists $\beta > 0$ which satisfies

$$\sum_{j \ne i} |\nabla_j f_i(x)| \le \nabla_i f_i(x) - \beta, \quad \forall x \in X, \forall i \quad (5.49)$$

Then, if $0 < \gamma < 1/k$, the mapping function $T : X \to R^n$ determined by $T(x) = x - \gamma f(x)$ is a maximum norm contraction mapping.

Now we consider the following iterative equation of load flow

$$x(t+1) = x(t) - \gamma A^{-1}(x(t))g(x(t)) \quad (5.50)$$

PARALLEL AND DISTRIBUTED LOAD FLOW

where $A(\cdot)$ is the Jacobian matrix of $g(\cdot)$. The term $A^{-1}(x)g(x)$ in (5.50) is equivalent to $f(x)$ in Proposition 5.5. When a power system is at steady state, the difference in phase angles of any two terminals of transmission lines would satisfy $|\theta_i - \theta_j| \le \pi/2$. Under this condition, X is convex, and $A^{-1}(x)g(x)$ is continuous and differentiable.

From Proposition 5.6, we know that the iterative load flow equation (5.50) must be a maximum norm contraction mapping if the Jacobian matrix of the vector function $A^{-1}(x)g(x)$ is diagonal dominant. Furthermore, from Proposition 5.4, we know that the totally asynchronous algorithm of load flow converges to a fixed point of $g(x) = 0$. Let $R = \dfrac{\partial \left(A^{-1}(x)g(x)\right)}{\partial x}$ and the maximum value of the diagonal elements of R is k_R. Then, we obtain the following convergence theorem for the totally asynchronous algorithm of load flow:

Proposition 5.7. (Convergence theorem of the totally asynchronous algorithm of load flow). If R is diagonal dominant and γ is within $(0, 1/R_k)$, then, the totally synchronous algorithm of load flow would converge to a fixed point of f.

Because it is very difficult to analyze the Jacobian matrix of $A^{-1}(x)g(x)$ in general, in the following we provide the proof for the P-Q fast decoupled load flow algorithm.

Proof. Suppose that C and D are the constant Jacobian matrices for the θ and v iteration of the P-Q fast-decoupled load flow algorithm respectively, and let

$$H = \frac{\partial g_p(x)}{\partial \theta} \quad (5.51)$$

$$U = \frac{\partial g_q(x)}{\partial v} \quad (5.52)$$

where

$$h_{ii} = v_i \sum_{j \ne i} v_j (G_{ij} \sin \theta_{ij} - B_{ij} \cos \theta_{ij}) \quad (5.53)$$

$$h_{ij} = -v_i v_j (G_{ij} \sin\theta_{ij} - B_{ij} \cos\theta_{ij}) \tag{5.54}$$

$$u_{ii} = -\sum_{j \neq i} v_j (G_{ij} \sin\theta_{ij} - B_{ij} \cos\theta_{ij}) + 2v_i B_{ii} \tag{5.55}$$

$$u_{ij} = -v_i (G_{ij} \sin\theta_{ij} - B_{ij} \cos\theta_{ij}) \tag{5.56}$$

If we apply P-Q fast decoupled conditions (i.e., $\cos\theta_{ij} \approx 1$, and $G_{ij} \sin\theta_{ij} \ll B_{ij}$ to (5.53) through (5.56), we find that $h_{ij} = c_{ij}$ and $u_{ij} = d_{ij}$, where h_{ij}, n_{ij}, c_{ij}, d_{ij} are the (i,j)th element of matrices H, U, C, and D, respectively. Hence, $C^{-1}H = E$ and $D^{-1}U \approx E$, where E is an identity matrix. Because $C^{-1}H$ and $D^{-1}U$ are equivalent to R, R is obviously diagonal dominant. Let us set k = 1+ξ, where ξ is a very small positive factor which depends on the system characteristics. Then, as γ varies in (0,1/(1+ξ)), the P-Q fast decoupled totally asynchronous algorithm of load flow will converge.

In a broader sense, we could prove the convergence of P-Q fast decoupled load flow algorithm as follows. Let the inductive norm for a matrix A be defined as

$$\|A\|_{ij} = \max_{x \neq 0} \frac{\|Ax\|_i}{\|x\|_j} \tag{5.57}$$

where $\|\cdot\|_i$ is a certain kind of norm on R^{n_i}, and $\|\cdot\|_{ij}$ is the block maximum norm. Suppose that Ω is a very small δ field of the load flow solution x^*, i.e., $\Omega = \{x \in R^n \mid \|x - x^*\| < \delta\}$, and define $J_{ii}(x) = \dfrac{\partial g_i(x)}{\partial x_i}$ and $J_{ij}(x) = \dfrac{\partial g_i(x)}{\partial x_j}$, $\forall i, j = 1, \cdots N$. Accordingly, we provide the following theorem.

Proposition 5.8. Suppose γ is a small relaxation factor. If the distributed load flow algorithm satisfies

PARALLEL AND DISTRIBUTED LOAD FLOW

$$\left\|I - \gamma J_{ii}^{-1} J_{ii}(x)\right\|_{ii} + \sum_{j \neq i} \left\|\gamma J_{ii}^{-1} J_{ij}(x)\right\|_{ii} < 1, \forall i, j = 1, \cdots, N \quad (5.58)$$

Then, the set generated by P-Q fast decoupled totally asynchronous load flow algorithm converges to a fixed point of f.

Proof. Ω is convex when the absolute value of the difference of the phase angles of the two terminals of a transmission line is less that $\pi/2$. Suppose that for $x, y \in \Omega$, there exists a scalar β that obtains

$$h_i(\beta) = \beta x_i + (1-\beta) y_i - \gamma J_{ii}^{-1} g_i(\beta x + (1-\beta) y)$$

where $h_i : [0,1] \to R^{n_i}$ is continuous and differentiable. Let us further define a mapping function $T_i(x) = x_i - \gamma J_{ii}^{-1} g_i(x)$. Then we have

$$\left\|T_i(x) - T_i(y)\right\|_i = \left\|h_i(1) - h_i(0)\right\|_i = \left\|\int_0^1 \frac{dh_i(\beta)}{d\beta} d\beta\right\|_i$$

$$\leq \int_0^1 \left\|\frac{dh_i(\beta)}{d\beta}\right\|_i d\beta \leq \max_{\beta \in [0,1]} \left\|\frac{dh_i(\beta)}{d\beta}\right\|_i$$

Applying the chain rule, we obtain

$$\left\|\frac{dh_i(\beta)}{d\beta}\right\|_i = \left\|x_i - y_i - \gamma J_{ii}^{-1} J_{ii}(\beta x + (1-\beta) y)(x-y)\right\|_i$$

$$= \left\|[I - \gamma J_{ii}^{-1} J_{ii}(\beta x + (1-\beta) y)](x_i - y_i)\right.$$

$$\left. - \sum_{j \neq i} \gamma J_{ii}^{-1} J_{ij}(\beta x + (1-\beta) y)](x_j - y_j)\right\|_i$$

$$\leq \left\|I - \gamma J_{ii}^{-1} J_{ii}(\beta x + (1-\beta) y)\right\|_{ii} \left\|(x_i - y_i)\right\|_i +$$

$$\sum_{j \neq i} \left\|I - \gamma J_{ii}^{-1} J_{ij}(\beta x + (1-\beta) y)\right\|_{ij} \left\|(x_j - y_j)\right\|_j$$

Let $\alpha = \left\| I - \gamma J_{ii}^{-1} J_{ii}(x) \right\|_{ii} + \sum_{j \neq i} \left\| I - \gamma J_{ii}^{-1} J_{ij}(x) \right\|_{ij}$. Then we have

$\left\| T_i(x) - T_i(y) \right\|_i \leq \alpha \max_j \left\| x_j - y_j \right\|_j = \alpha \| x - y \|$, and this implies that the

mapping function $T : \Omega \to R^n$ is a block maximum norm contracting map. Then by applying Proposition 5.6, the set generated by P-Q fast decoupled totally asynchronous load flow algorithm converges to a fixed point of f.

5.7 CASE STUDIES

5.7.1 System Partition

Consider the IEEE 118-bus system as a test system. The system generation and load data are provided in Appendix A. The system is partitioned into three subareas (Refer to Appendix A for system partitioning), and the nodes of each subarea are listed in Tables 5.1, 5.2, and 5.3, respectively. Between these three subareas, there are a total of twelve tie lines which are listed in Table 5.4.

Table 5.1 Nodal Data of Subarea 1

Internal nodes (37)	1, 2, 3, 4, 5, 6, 7, 8, 9, 10, 11, 12, 13, 14, 16, 17, 18, 20, 21, 22, 23, 24, 25, 26, 27, 28, 29, 31, 32, 71, 72, 73, 74, 113, 114, 115, 117
Boundary nodes (5)	15, 19, 30, 70, 75

Table 5.2 Nodal Data of Subarea 2

Internal nodes (25)	83, 84, 85, 86, 87, 88, 89, 90, 91, 92, 93, 94, 95, 101, 102, 103, 104, 105, 106, 107, 108, 109, 110, 111, 112
Boundary nodes(3)	82, 96, 100

Table 5.3 Nodal data of Subarea 3

Internal nodes (39)	35, 36, 37, 39, 40, 41, 42, 43, 44, 45, 46, 47, 48, 49, 50, 51, 52, 53, 54, 55, 56, 57, 58, 59, 60, 61, 62, 63, 64, 65, 66, 67, 68, 76, 78, 79, 80, 81, 116
Boundary nodes(9)	33, 34, 38, 69, 77, 97, 98, 99, 118

PARALLEL AND DISTRIBUTED LOAD FLOW

Table 5.4 Tie Line Number

| Tie line number (12) | 44, 45, 54, 108, 116, 120, 128, 148, 157, 158, 159, 185 |

There is a subarea control center in each subarea of the sample system; the structure of this wide computer network which is composed of the computers at these three SCCs is shown in Figure 5.8.

In order to simulate the distributed load flow computation, a COW is employed as shown in Figure 5.9, and MPI is used for the data communication among the computers.

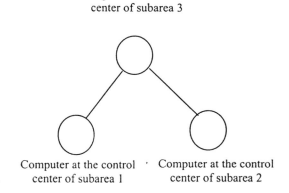

Figure 5.8 Computer Network Structure of the DCCS of the Example System

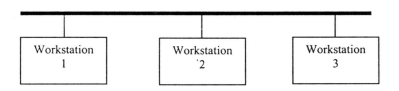

Figure 5.9 COW for Parallel and Distributed Load Flow Computation

5.7.2 Simulation Results

5.7.2.1 Synchronous Computation. In this simulation, synchronous communication is implemented using the MPI communication function MPI_Send and MPI_Recv.

Case 1: Distributed computation without boundary power compensation. In this case, non-diagonal blocks of the system Jacobian matrix are neglected. The relationship between the iterations of the distributed synchronous load flow algorithm versus γ is shown in Figure 5.10. Although at the beginning the number of iterations is high, the iteration number of the distributed load flow computation decreases gradually with γ increasing. When $\gamma = 1.0$, the iteration numbers for computers 1, 2, and 3 are 30, 34, 32, respectively.

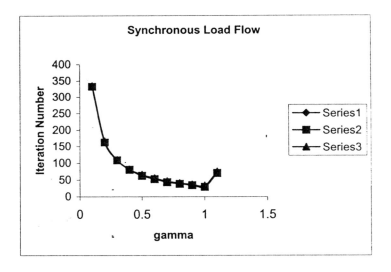

Figure 5.10 Synchronous Load Flow Computation

Note: series 1, 2, and 3 represent the iterations for computers 1, 2, and 3 respectively.

Case 2: Distributed computation with boundary power compensation. In this simulation we keep the value of γ constant and allow β to change. We set $\gamma = 1.0$ for we learned from Figure 5.10 that the synchronous distributed load flow algorithm converges most quickly when $\gamma = 1.0$. The most difficult part of this simulation is to choose an appropriate value for β. The elements of the Jacobian matrix are usually large, so the right hand

PARALLEL AND DISTRIBUTED LOAD FLOW

side vector of the load flow equation decreases and becomes very small when the iterative process approaches the final solution. For this reason, β should be a very small value. The relationship between the iteration number of the distributed synchronous load flow computation with boundary power compensation and β is shown in Figure 5.11 for γ = 1.0.

In Figure 5.11, the boundary power compensation can not always guarantee a fast convergence of the algorithm. The iteration number of distributed synchronous load flow computation with boundary power compensation varies non-homogenously with β, and when β exceeds a certain value (e.g., 0.04 in our case), the algorithm will no longer converge.

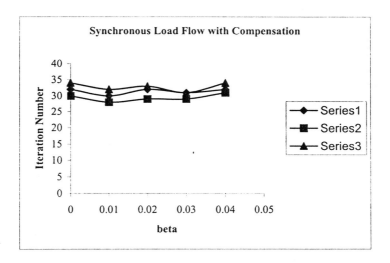

Figure 5.11 Synchronous Load Flow Computations with Boundary Power Compensation

Note: series 1, 2, and 3 represent the iteration numbers of computers 1, 2, and 3, respectively.

5.7.2.2 Partially Asynchronous Computation. To simulate the partially asynchronous distributed load flow computation, we first need to choose an asynchronous measure M. As we stated in the partially asynchronous assumption, boundary state variables (i.e. voltage magnitudes and phase angles) are transmitted to neighboring SCCs at least once in every M iterations. When the nodal phase angles of subarea 1 is delayed M iterations to be transmitted to its neighboring subarea 3, the iteration number of the partially asynchronous load flow computation versus M, is shown in Figure 5.12 for $\gamma = 0.1$.

In Figure 5.12 the iteration number of the partially asynchronous distributed load flow increases with increasing M. When $M > 14$, the load flow algorithm will not converge.

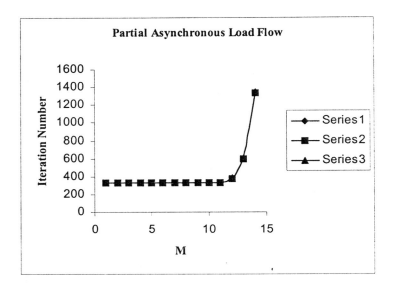

Figure 5.12 Partially Synchronous Load Flow Computation

Note: series 1, 2, and 3 represent the iterations of computers 1, 2, and 3, respectively.

5.7.2.3 Totally Asynchronous Computation. Totally asynchronization can be realized using MPI_Isend and MPI_Irecv. In this case the data communication between computers is completely random, so it is very difficult for the totally asynchronous distributed load flow algorithm to converge. However, theoretically, convergence should be possible when γ becomes small enough to converge.

5.8 CONCLUSIONS

From the case studies of this chapter, we reach the following conclusions:

- The distributed asynchronous algorithm of load flow can be established based on the unpredictability of data communication time delay. The partially asynchronous algorithm can simulate the normal operating case of a power system, while the totally asynchronous algorithm can simulate the abnormal case when the data communication time delay tends to be infinite or a certain communication link is at fault.

- The proposed distributed load flow algorithms converge to the right solution when γ is small enough. However, it converges most quickly when γ is near 1.0.

- The convergence property of the asynchronous distributed load flow algorithms can be improved by limiting data communication time delay. In partial asynchronization, the smaller the asynchronous measure is, the faster the asynchronous algorithm will converge.

- Boundary power compensation is a useful improvement to the convergence property of the distributed load flow algorithm. Boundary power compensation can accelerate the iteration process, but the compensation factor β must be properly selected.

- Compared with its synchronous counterpart, the distributed asynchronous algorithm needs additional iterations to converge. However, the asynchronous algorithm does not need to wait for data communication the way the synchronous algorithm does, and this can save computation time.

Chapter 6

Parallel and Distributed Load Flow of Distribution Systems

6.1 INTRODUCTION

The extension of computer monitoring and control to the customer level has become a new trend in modern power systems. This issue is of particular interest in a restructured environment. The new development has greatly motivated the application of parallel and distributed processing to distribution management systems. Distribution management systems (DMSs) collect and process a large amount of real-time data on the operation of distribution systems and its customers. However, it may be unnecessary and impractical to transmit all these data to the system control center for processing. In particular because the localized data processing has proved to be the most appropriate approach for distributed and restructured power systems.

To facilitate the monitoring and control of the distribution system operation, and to minimize capital investment in communication networks, a distribution system is usually divided into zones. Each zone has a dispatching center that is responsible for the real-time monitoring and control of energy distribution in that zone. Coordination is rare among these local dispatching centers because of the radial network topology of the distribution system. The energy distribution in this distribution system, shown in Figure 6.1, is through the coordination of individual dispatching centers with the substation, where the distribution system is usually modeled as an equivalent load in transmission system analyses.

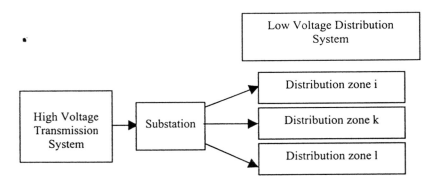

Figure 6.1 Transmission and Distribution Systems

In a restructured distribution system, DISTCOs are responsible for the energy supply and the operation of distribution zones. The monitoring and control of the distribution system reliability is the responsibility of the DNO (distribution network operator), which plays a role similar to that of the ISO in transmission systems. The DNO utilizes various monitoring and control facilities which are inherited from the traditional integrated distribution utilities. On the other hand, a DISTCO may establish its own computer network for telemetering and operation purposes. For instance, a DISTCO may impose distribution network reconfigurations based on its own computer network

A new phenomenon that has occurred with the restructuring of power industry is that distributed generation units (or distributed energy resources) have been widely deployed, especially in distribution systems. The installation of various distributed generation units greatly improves the existing monitoring and control schemes in the operation of a distribution system. Likewise, distributed generation will largely change the traditional characteristics of a distribution system. The utilization of distributed generation, however, could pose stability problems in the system operation. In this new circumstance, the proper monitoring and control of distribution systems is more important than ever.

The reason that power systems are purposely divided into a transmission system and a distribution system is the large differences in voltage levels as well as network topologies of the two systems. There also exists a significant difference in the parameters of these two systems; hence, convergence problems may occur if these two systems were directly combined for load flow computation.

LOAD FLOW OF DISTRIBUTION SYSTEMS

In general, the transmission system has a mesh network structure, while the distribution system has a radial network structure. In most computations for a distribution system, it is usually assumed that distribution networks have a tree-like radial structure. With rapid developments in modern power systems, however, distribution networks have become more complicated with a meshed structure. Although the switches that connect two radial distribution networks are supposedly to be open in the normal state, which indicates that there is a very weak current flowing through the connections if the switches are closed, in some systems the coupling effect ought to be taken into account.

The rapid development and wide application of computer network technology will provide a substantial support for the distributed processing of distribution systems. Because the geographical area of a distribution system is much smaller than that of a transmission system, its data communication will not need so many stages as in the transmission system, and computers at local control centers could be interconnected directly. As a result, to some extent, distributed processing in distribution systems will become much easier, more reliable, and less expensive than that in transmission systems.

Similar to that of the transmission system, load flow is the most useful tool for the real-time monitoring and control of a distribution system. In this chapter we discuss the parallel and distributed computation of load flow for the distribution system with respect to specific characteristics of the distribution system and based on a distributed computing system. Our discussion will focus on design methods for parallel and distributed load flow algorithms and their convergence analyses. The designed parallel and distributed load flow computation are simulated on a COW, and the unpredictability of communication delay is taken into account by designing asynchronous distributed algorithms.

6.2 MATHEMATICAL MODELS OF LOAD FLOW

Compared with transmission systems, distribution systems have a relatively low voltage level. This is because the distribution system is used for providing energy to customers while the transmission system is used for the long-distance transportation of bulk energy. The distance between two neighboring buses on the distribution network is rather minor, and the transmission system that has a mesh structure, while the distribution system has a radial structure. Even if there are some loops in the distribution system, corresponding switches that connect the two radial parts are usually located where a very weak electric current will pass through when the system is operated under normal conditions.

The P-Q fast decoupled load flow method which has received wide applications in transmission systems cannot be easily applied to a distribution system. This is because feeders and laterals of the distribution system have a high ratio of resistance to reactance. If one were to apply the traditional Newton-Raphson method to the distribution system by choosing voltage magnitudes and phase angles as state variables, the solution process would be very slow because the system needs to re-calculate the Jacobian matrix at every iteration. Hence, new and more efficient load flow algorithms are developed for the distribution systems [Aug01, Das94, Keb01, Los00, Mek01, Nan00, Nan98, Sri00]. However, additional improvements are deemed necessary to represent the new trend in distribution system operation. For instance, DistFlow has been approved as an effective load flow method for distribution systems where there is no distributed generation in the system. The load flow calculation gets more cumbersome as distributed generation is added to distribution systems. In the following, we discuss the design of parallel and distributed load flow algorithms for distribution systems.

6.2.1 Model for One Feeder

The simplest case of a distribution network is depicted in Figure 6.2 with one main feeder that does not have any laterals. This system has n branches and $(n+1)$ buses, We assume V_0 is the bus voltage at the substation and that V_0 is constant in the load flow computation.

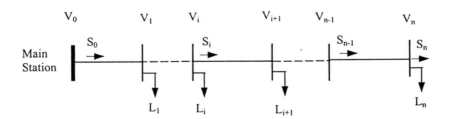

Figure 6.2 A Distribution System with Only One Feeder

We let $z_i = r_i + jx_i$ be the impedance of the ith branch, $S_{Li} = P_{Li} + jQ_{Li}$ the load extracted from bus i, and denote $S_{0,0} = P_{0,0} + jQ_{0,0}$ as the power extracted from the substation which flows

LOAD FLOW OF DISTRIBUTION SYSTEMS

into the distribution system. Then the power injected into the first branch of the main feeder is calculated according to

$$S_1 = S_{0,0} - S_{loss1} - S_{L1} = S_{0,0} - z_1 \frac{|S_{0,0}|^2}{V_0^2} - S_{L1} \qquad (6.1)$$

$$V_1 = V_0 - z_1 I_0 = V_0 - z_1 \frac{S_{0,0}^*}{V_0} \qquad (6.2)$$

By applying the above calculation to main feeder branches, we obtain the following recursive formulations:

$$P_{i+1} = P_i - r_{i+1} \frac{P_i^2 + Q_i^2}{V_i^2} - P_{Li+1} \qquad (6.3.1)$$

$$Q_{i+1} = Q_i - x_{i+1} \frac{P_i^2 + Q_i^2}{V_i^2} - Q_{Li+1} \qquad (6.3.2)$$

$$V_{i+1}^2 = V_i^2 - 2(r_{i+1} P_i + x_{i+1} Q_i) + (r_{i+1}^2 + x_{i+1}^2) \frac{P_i^2 + Q_i^2}{V_i^2} \qquad (6.3.3)$$

$$i = 0, 1, \cdots, n-1$$

where P_i and Q_i are the active and reactive power injected into the distribution branch between buses i and $i+1$, and V_i is the voltage of bus i. Formulation (6.3.1) through (6.3.3) can be further generalized as

$$x_{0,i+1} = f_{0,i+1}(x_{0,i}) \qquad i = 0, 1, \cdots, n-1 \qquad (6.4)$$

where $x_{0,i} = \begin{bmatrix} P_i & Q_i & V_i^2 \end{bmatrix}^T$. Equation (6.4) has the following boundary conditions:

$$P_n = 0 \qquad (6.5.1)$$

$$Q_n = 0 \qquad (6.5.2)$$

$$V_0 = V_0^{set} \qquad (6.5.3)$$

Equations (6.5.1) and (6.5.2) describe the condition of the last bus of the main feeder. Equation (6.4) and the boundary conditions (6.5.1) through

(6.5.3) comprise the equations for distribution load flow, which can be formulated in a general form

$$G(x_0) = 0 \tag{6.6}$$

where $x_0 = \left[x_{0,0}^T, \cdots, x_{0,n}^T\right]^T$. For any given load profile, the state vector x_0 can be determined by solving for $3(n+1)$ equations in (6.6).

6.2.2 Load flow Model for One Feeder with Multiple Laterals

Now consider a distribution network with one main feeder, n branches, and m laterals, as is shown in Figure 6.3.

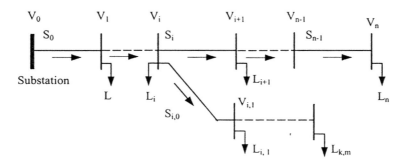

Figure 6.3 A Distribution System with One Main Feeder and Multiple Laterals

We choose the injection power $(P_{0,0}, Q_{0,0})$ at the starting bus of the main feeder and the injection power $(P_{0,k}, Q_{0,k})$, $k = 1, 2, \cdots, m$ at the starting bus of each branch, as state variables. Then the load flow problem is solved by the following $2(m+1)$ boundary equations.

$$P_{k,n_k}(z_{01}, \cdots, z_{0m}, z_{00}) = 0, \qquad k = 0,1,2,\cdots m \tag{6.7}$$

$$Q_{k,n_k}(z_{01}, \cdots, z_{0m}, z_{00}) = 0, \qquad k = 0,1,2,\cdots m \tag{6.8}$$

where P_{k,n_k}, Q_{k,n_k} $(k = 0,1,\cdots,m)$ are the active and reactive power flowing out of the last bus of each branch of the main feeder and each lateral, n_k is the number of the branches of the main feeder or the kth lateral $(k = 1, \cdots, m)$.

LOAD FLOW OF DISTRIBUTION SYSTEMS

We denote $z = (z_{01}^T, \cdots, z_{0m}^T, z_{00}^T)^T$, where $z_{00} = (P_{0,0}, Q_{0,0})^T$, $z_{0k} = (P_{0,k}, Q_{0,k})^T$, $k = 1, \cdots, m$. Then (6.7) and (6.8) will be synthesized into the following formulation

$$F(z) = 0 \tag{6.9}$$

By the chain rule, the Jacobian matrix of (6.9) will have the following form:

$$J = \begin{bmatrix} J_{11} & J_{12} & \cdots & J_{1m} & J_{10} \\ J_{21} & J_{22} & \cdots & J_{2m} & J_{20} \\ \vdots & \vdots & & \vdots & \vdots \\ J_{m1} & J_{m2} & \cdots & J_{mm} & J_{m0} \\ J_{01} & J_{02} & \cdots & J_{0m} & J_{00} \end{bmatrix} \tag{6.10}$$

where $J_{ki} = \begin{bmatrix} \dfrac{\partial P_{k,n_k}}{\partial P_{0,i}} & \dfrac{\partial P_{k,n_k}}{\partial Q_{0,i}} \\ \dfrac{\partial Q_{k,n_k}}{\partial P_{0,i}} & \dfrac{\partial Q_{k,n_k}}{\partial Q_{0,i}} \end{bmatrix}.$

Then the load flow solution is obtained by iterating the equation

$$J\Delta z(k+1) = -F(z(k)) \tag{6.11}$$

where $\Delta z(k+1)$ is the correction of the state variables in the $(k+1)$th iteration, and $F(z(k))$ is the residual vector of the injection power. Except for the last row, all non-diagonal blocks of matrix J are of the form $J_{ij} = \begin{bmatrix} \varepsilon & \varepsilon \\ \varepsilon & \varepsilon \end{bmatrix}$, where ε is a very small positive number, e.g., $0 < \varepsilon < 1$. However, on the last row of matrix J, $J_{ij} = \begin{bmatrix} 1 & \varepsilon \\ \varepsilon & 1 \end{bmatrix}$. Hence, matrix J can be rewritten as

$$\tilde{J} = \begin{bmatrix} J_{11} & 0 & \cdots & 0 & 0 \\ 0 & J_{22} & \cdots & 0 & 0 \\ \vdots & \vdots & & \vdots & \vdots \\ 0 & 0 & \cdots & J_{mm} & 0 \\ J_{01} & J_{02} & \cdots & J_{0m} & J_{00} \end{bmatrix} \quad (6.12)$$

Then the load flow problem can be resolved be iterating the equation

$$\tilde{J} \Delta z(k+1) = -F(z(k)) \quad (6.13)$$

When the distribution system is operated in a normal state, the system's Jacobian matrix J is nonsingular and there would be a solution for (6.11). This Jacobian matrix has nothing to do with network parameters of the distribution system, and thus this load flow method should have a strong numerical stability. This load flow solution method has shown a very good convergence property [108]. In practice, $\tilde{J}$ is assumed to be constant and this significantly simplifies the load flow computation.

6.3 PARALLEL LOAD FLOW COMPUTATION

Based on the load flow model discussed in the section above, parallel computation techniques can be employed for a fast load flow solution. The first step for this parallelization is to divide the coefficient matrix of (6.11). If we denote $\Delta Z(k) = (\Delta Z_1(k), \cdots, \Delta Z_m(k), \Delta Z_0(k))$ and $F(Z(k)) = (F_1(Z(k)), \cdots, F_m(Z(k)), F_0(Z(k)))$. Then we obtain the iteration equation for the parallel computation:

$$-J_{ii}\Delta Z_i(k+1) = F_i(Z(k)), \quad i=1,\cdots,m \quad (6.14)$$

$$-J_{00}\Delta Z_0(k+1) = F_0(Z(k)) + \sum_{i=1}^{m} J_{0i}\Delta Z_i(k) \quad (6.15)$$

The load flow computations of the laterals and the main feeder are given by (6.14) and (6.15), respectively. We suppose there are $(m+1)$ processors to solve the parallel load flow equation; each of the m processors will compute the load flow of a lateral except one processor which will process the load flow computation for the main feeder. In cases where the number of processors is less than $(m+1)$, we assume that some of the processors will take more than one laterals. Nevertheless, we assume that one of the processors will be responsible for the computation of the main feeder, because there is no waiting for the data exchange between different zones except for the main feeder.

6.4 DISTRIBUTED COMPUTATION OF LOAD FLOW

6.4.1 System Division

Suppose that a distribution system, which is partitioned into N zones, satisfies the following requirements:

- The main feeder is taken as one separate zone. This is because the main feeder is most important, and a fault occurring on the main feeder may affect a much larger distribution area than that on a lateral branch.

- One or a number of geographically adjacent laterals are grouped to form a zone. This grouping usually complies with the existing control zones in the distribution system.

- The performances of computers available in the system for distributed computation are the same. The load balance can then be obtained by partitioning the distribution system into zones with almost the same number of buses.

6.4.2 Synchronous Distributed Algorithm

Denote $z = (z_1, \cdots, z_N)$, $F = (F_1, \cdots, F_N)$. The synchronous distributed load flow algorithm is then formulated as

$$-J_{ii}(z(k+1))\Delta z_i = F_i(z(k)) \quad i=1,\cdots,N-1, \quad N \leq m+1 \quad (6.16)$$

$$-J_{00}\Delta z_0(k+1) = F_0(z(k)) + \sum_{i=1}^{m} J_{0i}\Delta z_i(k) \quad (6.17)$$

Equations (6.16) and (6.17) appear to be almost the same as (6.14) and (6.15), i.e., the latent difference. For parallel computation, we divide the load flow equations according to the architecture of the parallel machine while for distributed computation, we design distributed algorithms with respect to the existing system partitions. Particularly when $N = m+1$, except for the computer that computes the main feeder, each remaining computer computes one lateral. In the synchronous distributed load flow algorithm, the boundary state variables are communicated among processors at every iteration.

6.4.3 Asynchronous Distributed Computation

We denote $z^i(t) = (z_1(\tau_1^i(t)), \cdots, z_N(\tau_N^i(t)))$ where $z_i(\tau_j^i(t))$ represents the component z_i at time t which is sent from processor j to computer i. When the possible communication delay is taken into account, the asynchronous distributed algorithm of (6.13) is formulated as

$$-J_{ii}(z^i(t))\Delta z_i(t+1) = \gamma F_i(z^i(t)) \quad i = 1, \cdots, N-1, N \leq m+1 \quad (6.18)$$

$$-J_{00}(z^0(t))\Delta z_0(t+1) = \gamma F_0(z^0(t)) + \sum_{i=1}^{m} J_{0i}\Delta z_i(\tau_i^0(t)) \quad (6.19)$$

where γ is a relaxation factor. Asynchronization could be caused by a number of factors. These include

1) different sizes of zones, which usually results in load imbalances among computers
2) heterogeneous computers with different performances
3) different communication media among computers.

The asynchronous load flow algorithm proposed above can be used especially when communication delays among computers of a distributed system are hard to predict. To improve the convergence performance of asynchronous computation, a relaxation factor γ is introduced into (6.18) and (6.19).

6.4.3.1 Totally asynchronous algorithm. If the communication delays in (6.18) and (6.19) satisfy the totally asynchronous computation assumption as defined in Chapter 5, (6.18) and (6.19) are then referred to as the totally asynchronous load flow algorithm for distribution systems.

6.4.3.2 Partially asynchronous algorithm. If the communication delays in (6.18) and (6.19) satisfy the partially asynchronous assumption, defined in Chapter 5, (6.18) and (6.19) are then referred to as the partially asynchronous load flow algorithm for distribution systems.

6.5 CONVERGENCE ANALYSIS

As we saw in the last chapter, if the Jacobian matrix of the load flow equation is diagonally dominant, distributed asynchronous load flow algorithms will converge when the relaxation factor γ is small enough. In (6.10) and (6.12), matrices J and $\tilde{J}$ are apparently diagonally dominant, and $\Delta J_{kk} \gg \Delta J_{ki}$, $\forall i \neq k$; therefore, distributed asynchronous load flow algorithms for distribution systems will converge when the relaxation factor γ is small enough.

6.6 DISTRIBUTION NETWORKS WITH COUPLING LOOPS

Earlier, we assumed that the current on the tie line of two laterals is negligible. In case the current cannot be neglected as loops are formed in the distribution networks. So the load flow algorithms proposed above should be modified for this application.

The cases where the tie line current must be taken into account fall into the following two categories:

a) The power for the boundary load is supplied by one network, indicating that the power from the other network to the boundary load is zero. In this case, we can hypothetically shift the tie line to the new position and the previously proposed load flow algorithms can be applied.

b) The power for the boundary load is supplied by the two feeders in the two adjacent networks, as illustrated in Figure 6.4.

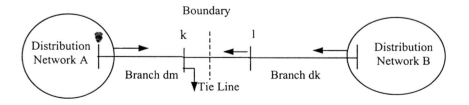

Figure 6.4 Intensely Coupled Distribution Systems

Suppose that the power flow on tie line $k - l$ is $P_{tie} + jQ_{tie}$. Then boundary conditions for the proposed load flow algorithms are modified as follows:

For distribution network A,

$$P_{jk} = -P_{tie} \tag{6.20}$$
$$Q_{jk} = -Q_{tie} \tag{6.21}$$

For distribution network B,

$$P_{gl} = P_{tie} \tag{6.22}$$
$$Q_{gl} = Q_{tie} \tag{6.23}$$

The proposed distributed load flow algorithms are used accordingly for the load flow problem of distrbution systems with different boundary conditions.

6.7 LOAD FLOW MODEL WITH DISTRIBUTED GENERATION

When there is distributed generation in the distributed system, the load flow model proposed in Section 6.4 cannot be used any more. We have to use the traditional Newton-Raphson method to solve the load flow problem of the distribution system with distributed generation [Hat93, Nak01]. Furthermore, the P-Q fast decoupled technique cannot be used due to the high ratio of r/x in distribution systems, and the Jacobian matrix of load flow equation would have to be updated at every iteration, which is time consuming. Fortunately, however, parallel and distributed processing can help speed up the entire computation process.

Similarly, we choose the bus voltage magnitude and phase angle as state variables of the distribution system. Then the mathematical model of the load flow for the distribution system with distributed generation will be formulated as

$$-J(x(k))\Delta x(k+1) = \Delta S(x(k)) \tag{6.24}$$

Suppose that the original distribution system is divided into N zones, each with one or more laterals. Also suppose that the main feeder

LOAD FLOW OF DISTRIBUTION SYSTEMS

forms the *Nth* zone. If X_k ($k = 1,...,N$) is the state vector of the *k*th zone, (6.24) can be rewritten more specifically in the format

$$-\begin{bmatrix} J_{1,1}(x) & & J_{1,N}(x) \\ & J_{i,i}(x) & J_{i,N}(x) \\ J_{N,1}(x) & J_{N,i}(x) & J_{N,N}(x) \end{bmatrix} \begin{bmatrix} \Delta x_1 \\ \Delta x_i \\ \Delta x_N \end{bmatrix} = \begin{bmatrix} \Delta S_1(x) \\ \Delta S_i(x) \\ \Delta S_N(x) \end{bmatrix} \qquad (6.25)$$

where $J_{k,k}$ is the Jacobian matrix of zone k, $k = 1,...,N$; $J_{k,N}$ is the Jacobian matrix of zone k with respect to zone N; $J_{N,k}$ is the Jacobian matrix of zone N with respect to zone k, $k = 1,..,N$.

Equation (6.25) is similar to the load flow model for the transmission system discussed in Chapter 5. However, since the P-Q fast decoupled technique cannot be used here, for every pair of bus injection P_i, Q_i, the Jacobian matrix is calculated by

$$\begin{bmatrix} \dfrac{\partial P_i}{\partial \theta_j} & \dfrac{\partial P_i}{\partial v_j} \\ \dfrac{\partial Q_i}{\partial \theta_j} & \dfrac{\partial Q_i}{\partial v_j} \end{bmatrix} \qquad (6.26)$$

The non-diagonal elements of the Jacobian matrix cannot be omitted for parallel and distributed computation load flow algorithms of distribution systems with distributed generation. This is because the r/x ratio of distribution systems is high. In other words, if we simply decompose (6.25) similar to that of parallel and distributed load flow algorithms for transmission system, we obtain the following formulation:

$$-J_{ii}(x(k))\Delta x_i(k+1) = \gamma \Delta S_i(x(k)) \quad i = 1, \cdots, N-1, \text{ for laterals} \qquad (6.27)$$

$$-J_{N,N}\Delta x_N(k+1) = \gamma \Delta S_N(x(k)), \text{ for the main feeder} \qquad (6.28)$$

where $N - 1$ is the number of zones for laterals.

Because the non-diagonal elements of the Jacobian matrix in (6.26) are not negligible, the parallel and distributed algorithm described in (6.27) and (6.28) will have a convergence problem. Even if we take the boundary information into account for the main feeder formulation and use

$$J_{N,1}(x)\Delta x_1 + \ldots + J_{N,i}(x)\Delta x_i + \ldots + J_{N,N}(x)\Delta x_N = \Delta S_N(x) \quad (6.29)$$

instead of (6.28), the load flow algorithm will have a convergence problem because each boundary bus will delete four elements from the Jacobian matrix as shown in (6.26). For instance, if there are n_b boundary buses in the distribution system, a total of $4*n_b$ elements are to be deleted from the Jacobian matrix to form the preceding parallel and distributed load flow algorithm, which could adversely affect the convergence property of the load flow algorithm.

Therefore, the non-diagonal submatrices of the Jacobian matrix must be taken into consideration for the parallel and distributed load flow algorithm of the distribution system with distributed generation. To implement this, (6.25) is decomposed into (6.30) and (6.31):

$$J_{i,i}\Delta x_i + J_{i,N}\Delta x_N = -\Delta s_i \quad i = 1, \cdots, N-1, \text{ for laterals} \quad (6.30)$$

$$J_{N,1}\Delta x_1 + \ldots + J_{N,k}\Delta x_k + \ldots + J_{N,N}\Delta x_N = -\Delta s_N, \text{ for main feeder} \quad (6.31)$$

From (6.30), we have

$$\Delta x_i = J_{i,i}^{-1}\{-\Delta s_i - J_{i,N}\Delta x_N\}, \quad i = 1, \ldots, N-1 \quad (6.32)$$

We define

$$J_i = J_{N,i}J_{i,i}^{-1}J_{i,N} \quad (6.33)$$

$$S_i = J_{N,i}J_{i,i}^{-1}S_i \quad (6.34)$$

By substituting Δx_k into (6.31) we have

$$\Delta x_N = \left\{J_{N,N} - \sum_{i=1}^{m}J_i\right\}^{-1}\left\{-\Delta s_N + \sum_{i=1}^{m}s_i\right\} \quad (6.35)$$

The parallel and distributed load flow for the distribution system with distributed generation will solve (6.35) and (6.32) alternatively. Suppose that the processor for the computation of the main feeder acts as the coordinator, then the parallel and distributed load flow algorithm is further described as follows. Take a flat start, i.e., $x_i = x_i(0)$, $i = 1, \ldots, N$.

LOAD FLOW OF DISTRIBUTION SYSTEMS

i) Processor i ($i = 1, \ldots, N$-1) calculates $J_{i,i}$, $J_{i,N}$, $J_{i,i}^{-1}$, and ΔS_i concurrently, as processor N calculates $J_{N,N}$, $J_{N,i}$, and ΔS_i at the same time;

ii) For $J_{N,i} = J_{i,N}$, processor i ($i = 1, \ldots, N$-1) calculates $J_{N,i}J_{i,i}^{-1}J_{i,N}$ and $J_{N,i}J_{i,i}^{-1}\Delta S_i$ concurrently and transmits the communication results to processor N;

iii) Processor N solves Δx_N according to (6.35) and then transmits the result to processor i ($i = 1, \ldots, N$-1);

iv) Processor i ($i = 1, \ldots, N$-1) solves Δx_i according to (6.32);

v) Check whether the convergence condition is satisfied. If not, then processors modify state variables and go back to step ii); if yes, then they stop the computation.

The algorithm above is easy to implement either on parallel machines or a computer network.

Now we consider the communication delays among processors. Denote $x^i(t) = (x_1(\tau_1^i(t)), \cdots, x_N(\tau_N^i(t)))$, and let $x_j(\tau_j^i(t))$ represent the component x_j at time t sent by processor j to processor i. The asynchronous distributed algorithm of (6.32) and (6.35) is formulated as

$$\Delta x_i(t+1) = \gamma J_{i,i}^{-1}(x^i(t))\left\{-\Delta S_i(x^i(t)) - J_{i,N}(x^i(t))\Delta x_N(\tau_N^i(t))\right\}$$
$$i = 1, \ldots, N-1 \qquad (6.36)$$

$$\Delta x_N(t+1) = \left\{J_{N,N}(x^N(t)) - \sum_{i=1}^{N-1} J_i(t)\right\}^{-1}\left\{-\Delta s_N(x^N(t)) + \sum_{i=1}^{N-1} s_i(t)\right\}$$

$$(6.37)$$

where

$$J_i(t) = J_{N,i}(x^i(t))J_{i,i}^{-1}(x^i(t))J_{i,N}(x^i(t)) \qquad (6.38)$$

$$S_i(t) = J_{N,i}(x^i(t))J_{i,i}^{-1}(x^i(t))S_i(x^i(t)) \qquad (6.39)$$

By choosing $\Delta x_N(0)$ as the initial value, we set up (6.36) and (6.37) to be computed in a totally distributed manner. If the communication delays in (6.36) and (6.37) satisfy the totally asynchronous assumption defined in Chapter 5, (6.36) and (6.37) are referred to as the totally asynchronous load flow algorithm for distribution system with distributed generation. If the communication delays in (6.36) and (6.37) satisfy the partially asynchronous assumption defined in Chapter 5, (6.36) and (6.37) are referred to as the partially asynchronous load flow algorithm for distribution system with distributed generation.

6.8 JOINT LOAD FLOW COMPUTATION OF TRANSMISSION AND DISTRIBUTION SYSTEMS

6.8.1 Problem Description

The load flow computation of the transmission system and the distribution system are performed separately because of the large differences in their network topologies and parameter values. However, because the two systems are physically interconnected and have to be solved concurrently for an accurate load flow solution, there could be some errors especially in the computation of power mismatches at the boundary buses of the two systems.

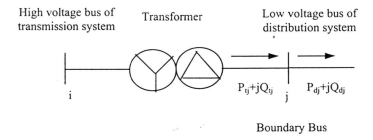

Figure 6.5 Joint Transmission and Distribution Networks

As shown in Figure 6.5, the power flow at the boundary bus of the transmission and distribution systems is represented as a load for the load flow computation of the transmission system. However, the power injected from the transmission system to the distribution network is taken as state variables in the load flow computation of the distribution network.

If the difference between the two load flow solution is substantial, the corresponding flows may introduce adverse effects on the monitoring and control of the entire power system.

LOAD FLOW OF DISTRIBUTION SYSTEMS

A joint load flow computation may help prevent this phenomenon as it can provide the ISO and DNO with a more accurate picture for the operation of their respective systems. In this regard distributed processing techniques provide the possibility for the joint load flow computation of power systems.

6.8.2 Joint Computation Based on Distributed Processing

For the implementation of distributed computation, we allow a distribution system to be a zonal power system, and choose bus voltage magnitudes and phase angles as state variables of distribution and transmission systems. Accordingly, the joint load flow can be realized quite similar to the distributed load flow for the transmission system discussed in Chapter 5. Figure 6.6 depicts this method.

This joint distributed load flow method can be implemented in either synchronous or asynchronous manner, as will be determined in the algorithm design process according to the specific characteristics of data communication networks. The only problem here is that the P-Q fast-decoupled load flow method cannot be applied to the distribution system, so the joint distributed load flow can only be implemented using the conventional Newton-Raphson method.

Theoretically, the load flow for the transmission system can use the P-Q fast-decoupled method and the load flow for distribution systems can also use the Newton-Raphson method in the joint load flow computation. However, the distributed computation may encounter a convergence problem because of the difference in convergence rates of the two computation methods.

6.8.3 Joint Computation Based on Separate Computations

We suppose that the load flow computation for transmission and distribution systems is performed separately. For simplicity, we first take a look at the case where the system has only one transmission and one distribution systems. We assume that in the transmission system and the distribution system the boundary conditions for their individual load flow computations are known. The transmission system takes the injection power P_{dj} and Q_{dj} at the substation bus as part of the input data for its load flow computation, and the load flow computation result presents the voltage magnitude V_j of the substation bus. Let us suppose that V_j is presented as part of the input data for the load flow computation of the distribution system and that the flow computation will result in the new values of P_{dj} and Q_{dj}.

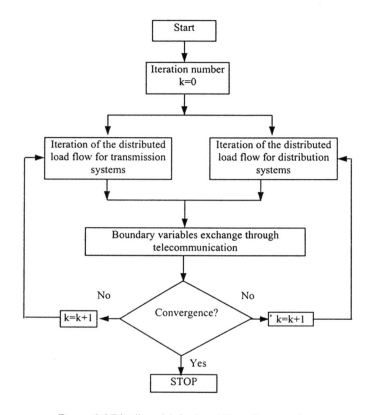

Figure 6.6 Distributed Joint Load Flow Computation

By assuming that $V_j^{(0)}$ is known, we let the distribution system perform the load flow computation and provide the boundary load $P_{dj}^{(new)}$ and $Q_{dj}^{(new)}$ after the load flow is completed. After taking P_{dj} and Q_{dj} as the boundary power injection, the transmission system performs its load flow computation and provides a new value for $V_{tj}^{(new)}$. If $|V_{tj}^{(new)} - V_j^{(0)}| < \varepsilon$, then the boundary conditions satisfy both systems, and the joint load flow converges; otherwise, the joint load flow will need more iterations. This joint load flow computation scheme is depicted in Figure 6.7.

LOAD FLOW OF DISTRIBUTION SYSTEMS

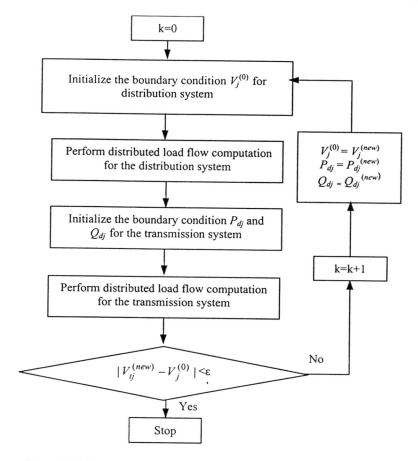

Figure 6.7 Joint Load Flow Computation Based on Separate Computations

If there are distributed generating units in the distribution system, the distribution system will take the substation bus as the slack bus. Say that its voltage magnitude is $V_j^{(0)}$, and so its load flow generates P_{dj} and Q_{dj}. Then, the transmission system takes P_{dj} and Q_{dj} as part of its input data for the load flow computation, which will generate $V_{tj}^{(new)}$ for the substation bus. This value of $V_{tj}^{(new)}$ will be used for the next iteration of the load flow computation of the distribution system. When $|V_{tj}^{(new)}(k+1) - V_{tj}^{(new)}(k)| < \varepsilon$, the convergence condition for the joint load flow computation is satisfied and the joint load flow computation is complete.

The joint load flow computation method proposed above can be easily be extended to the case where the system has more than one distribution system, as shown in Figure 6.8. In this circumstance the load flow computation for each distribution system will be performed separately in a distributed manner. The load flow for the entire system can be computed using either of the methods discussed above.

When the separate computation method is used for the joint load flow computation of the system with more than one distribution system, the scalar boundary variables used before should be substituted for a state variable vector.

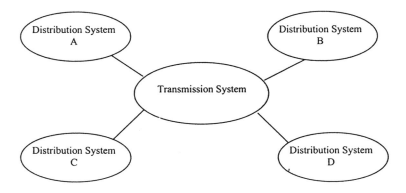

Figure 6.8 Joint Load Flow with Multiple Distribution Systems

In the joint load flow computation, the boundary information is exchanged only after the convergence of the load flow computation in transmission and distribution systems. This is because different load flow solution methods are used for transmission and distribution systems. The co-existence of multiple load flow methods is the main advantage of distributed processing.

6.9 CASE STUDIES

6.9.1 The Test System

We illustrate the distributed load flow for distribution systems with distributed generation. As shown in Figure 6.9, a distribution system can be created by adding five distributed generation units to the IEEE 37-bus distribution system. This test system is divided into three zones, and we

LOAD FLOW OF DISTRIBUTION SYSTEMS

suppose there is a local area control center (LACC) in each zone that monitors and controls the corresponding area.

The three LACCs shown in Figure 6.10 consist of distributed systems. To keep the computation load of each LACC as balanced as possible, each zone will approximately have the same number of buses. The distributed load flow for this distribution system is performed on a COW composed of three workstations, which stand for the three LACCs.

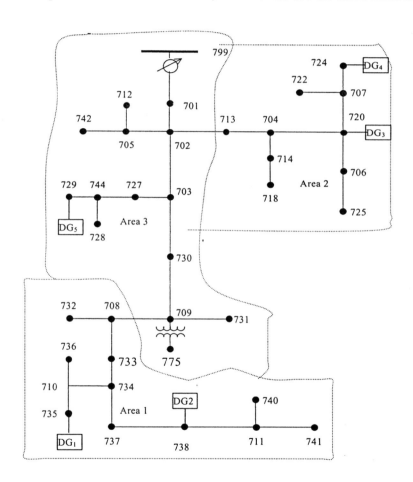

Figure 6.9 Distribution System with Distributed Generation

The network parameters, bus loads, and distributed generation injections, and the corresponding changes of the bus loads after the system is modified are listed in Table 6.1 through 6.4, respectively.

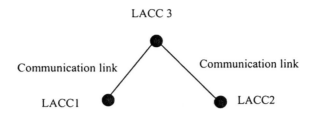

Figure 6.10 Distributed Computing System

Table 6.1 Line Parameters of Example Distribution System

i	j	r	x	k	i	j	r	x	k
701	702	0.000200	0.000136	1.0	710	736	0.000267	0.000181	1.0
702	705	0.000083	0.000057	1.0	711	741	0.000083	0.000057	1.0
702	713	0.000075	0.000051	1.0	711	740	0.000042	0.000028	1.0
702	703	0.000275	0.000187	1.0	713	704	0.000108	0.000074	1.0
703	727	0.000050	0.000034	1.0	714	718	0.000108	0.000074	1.0
703	730	0.000125	0.000085	1.0	720	707	0.000192	0.000013	1.0
704	714	0.000017	0.000011	1.0	720	706	0.000125	0.000085	1.0
704	720	0.000167	0.000113	1.0	727	744	0.000058	0.000004	1.0
705	742	0.000067	0.000045	1.0	730	709	0.000042	0.000028	1.0
705	712	0.000050	0.000034	1.0	733	734	0.000117	0.000079	1.0
706	725	0.000058	0.000040	1.0	734	737	0.000133	0.000091	1.0
707	724	0.000158	0.000108	1.0	734	710	0.000108	0.000074	1.0
707	722	0.000025	0.000017	1.0	737	738	0.000083	0.000057	1.0
708	733	0.000067	0.000045	1.0	738	711	0.000083	0.000057	1.0
708	732	0.000067	0.000045	1.0	744	728	0.000042	0.000028	1.0
709	731	0.000125	0.000085	1.0	744	729	0.000058	0.000004	1.0
709	708	0.000067	0.000045	1.0	775	709	0.000002	0.000001	1.0
710	735	0.000042	0.000028	1.0	799	701	0.000386	0.000262	1.0

LOAD FLOW OF DISTRIBUTION SYSTEMS

Table 6.2 Bus Injection Power

Bus	P	Q	Bus	P	Q
735	0.85	0.4	707	0.0	0.0
710	0.0	0.0	724	-0.42	-0.21
736	0.42	0.21	722	-1.61	-0.8
734	-0.42	-0.21	731	-0.85	-0.4
737	-1.4	-0.7	775	0.0	0.0
738	-1.26	-0.62	709	0.0	0.0
711	0.0	0.0	730	-0.85	-0.4
740	-0.85	-0.4	703	0.0	0.0
741	-0.42	-0.21	727	-0.42	-0.21
733	-0.85	-0.4	744	-0.42	-0.21
732	-0.42	-0.21	729	-0.42	-0.21
708	0	0	728	-1.26	-0.63
713	-0.85	-0.4	702	0.0	0.0
704	0.0	0.0	705	0.0	0.0
714	-0.38	-0.18	742	-0.93	-0.44
718	-0.85	-0.4	712	-0.85	-0.4
720	-0.85	-0.4	701	-6.3	-3.15
706	0.0	0.0	725	-0.42	-0.21

Table 6.3 Injection Power of Distributed Generation

DG	Bus	P	Q
DG1	735	0.4	0.2
DG2	738	0.8	0.4
DG3	720	0.35	0.17
DG4	724	0.6	0.3
DG5	729	0.8	0.4

Table 6.4 Load Increments

Bus	P	Q
707	0.18	0.09
728	0.38	0.19

6.9.2 Simulation Results

The traditional distribution load flow method cannot be used here because there is distributed generation in this distribution system. We use the parallel and distributed load flow algorithms proposed in Section 6.7 because these are more general load flow algorithms and can be used for distribution systems with distributed generation.

First, we test the parallel and synchronous distributed load flow algorithm described by (6.27) and (6.28). When it was implemented on COW, this algorithm converged to a wrong solution, even when γ was as small as 0.01. This outcome verifies the conclusion that there is a convergence problem if the non-diagonal elements of the system Jacobian matrix are omitted in the load flow computation, as discussed in Section 6.7. The corresponding asynchronous algorithms including partially and totally asynchronous algorithms will not converge to the right solution either. These simulation results have shown that we cannot build the parallel and distributed load flow algorithm for the distribution system similar to that of the transmission system in Chapter 5.

Now let us take bus 799 as the slack bus with its voltage magnitude set at 1.0 and phase angel at 0.0. When the parallel and synchronous distributed load flow algorithm described by (6.32) and (6.35) is applied, the load flow algorithm converges in two iterations, as shown in Table 6.5, which is exactly the same as that of the serial algorithm. When γ is around 1.0, the parallel and synchronous distributed load flow algorithm described by (6.32) and (6.35) will converge in two iterations.

Figure 6.11 depicts the errors in parallel and synchronous distributed load flow computation as a function of relaxation factor γ as compared with the serial load flow solution, where series one is for the voltage magnitude and series two is for the phase angle. We see from Figure 6.11 that the computation errors are sensitive to the value of γ, especially the phase angles. This shows that the convergence of the load flow computation is sensitive to the increments of state variables. This phenomenon can be explained by the high r/x ratio that results in power injections becoming more sensitive to network parameters.

LOAD FLOW OF DISTRIBUTION SYSTEMS

Table 6.5 Load Flow Solution

Bus	V	θ	Bus	V	θ
735	0.978006	-0.18361	707	0.98272	-0.14367
710	0.978031	-0.18336	724	0.982759	-0.14336
736	0.977878	-0.18455	722	0.982666	-0.14411
734	0.978159	-0.18214	731	0.979714	-0.16935
737	0.977591	-0.18703	775	0.979857	-0.16802
738	0.977395	-0.18882	709	0.979857	-0.16802
711	0.977252	-0.19013	730	0.980227	-0.16506
740	0.977204	-0.19055	703	0.981477	-0.15427
741	0.977204	-0.19052	727	0.981385	-0.15499
733	0.978861	-0.17626	744	0.981312	-0.15560
732	0.979301	-0.17257	729	0.981342	-0.15535
708	0.97934	-0.17228	728	0.981262	-0.15597
713	0.984264	-0.13076	702	0.98473	-0.12668
704	0.983714	-0.13562	705	0.984532	-0.12856
714	0.983686	-0.13584	742	0.984448	-0.12929
718	0.983563	-0.13701	712	0.984475	-0.12909
720	0.983141	-0.14034	701	0.988818	-0.09155
706	0.983069	-0.1409	725	0.983036	-0.14118

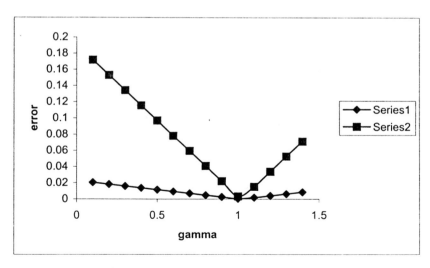

Figure 6.11 Errors in Parallel and Synchronous Distribute Load Flow Computation

For the three workstations to start the distributed load flow computation described by (6.32) and (6.35) simultaneously, we assign an initial value for $\Delta x_N(0)$ to all computers. For this test system, N = 3, a simple way is to choose $\Delta x_3(0) = 0.0$. However, the simulation results show that it may not work as expected because $\Delta x_3(0)$ will be very close to its final solution rather than at 0.0 as that the algorithm would converge in two iterations. It is also very difficult to estimate an initial value for $\Delta x_3(0)$. Therefore, the asynchronous load flow algorithms described by (6.36) and (6.37) do not easily converge even if γ takes a value as small as 0.01.

Our simulation results have shown that it is hard to implement asynchronous distributed load flow in distribution systems. Fortunately the communication links of the distributed monitoring and control system composed of LACCs in the distribution system are usually short. Therefore, the data communication among LACCs can easily be synchronized and parallel or synchronous distributed load flow computation implemented.

On the other hand, since the Jacobian matrix of the load flow for distribution systems needs to be updated at every iteration, the serial load flow algorithm will need more time for calculating the new Jacobian matrix, especially where the distribution system is large. This way the parallel and synchronous distributed computation can give a better performance. Moreover, the larger the distribution system, the better is the parallel and synchronous load flow computation performance.

Chapter 7

Parallel and Distributed State Estimation

7.1 INTRODUCTION

In the day to day operation and control of large-scale electric power systems, operators depend on an array of measured quantities, such as bus voltage magnitudes, line flows, and bus loads and injections, in order to monitor the present status of the network and to initiate control actions consistent with the system's operational goals. In real-time monitoring and control of a power system, the SCADA system at the power system control center must scan the power system periodically and acquire hundreds and thousands of measurement data, which include active and reactive power measurements, voltage measurements, current measurements, and switches and breakers status. These data are mostly acquired by remote terminal units (RTUs), which are scattered throughout the entire power system, and then transmitted through long communication links to either the system control center or an area control center of the power system.

Because the data acquisition process involves a number of complex procedures, these measured data inevitably contain errors which are usually caused by the following factors:

- Metering errors including those caused by unbalanced phases
- Transducer errors including those caused by CT and PT transformation
- A/D conversion errors
- Communication noise

In addition, errors of measured data may be caused by the following factors:

- Accidental faults of metering instruments or communication systems
- Random disturbances in metering instruments or communication systems
- Simultaneous metering of certain measurement spots
- Measurements of a process in transition

Without any external filtering or means of detection, these erroneous measurements can at least modestly distort the operator's perception of what is occurring in the network. Consequently the operator may be misled into adopting a course of action that is suboptimal, at the very least, or even highly inappropriate for the current situation. In some extreme cases, for instance, control actions based on a number of measurements that are blatantly erroneous may seriously undermine the stability of the network. Mostly, however, the errors are modest and pose no catastrophic threat to the system's security. Nevertheless, true optimization of system security and economy is critically dependent upon the availability of optimally accurate information. Since it is usually impossible to determine by mere inspection whether or not, and to what extent, a particular measurement faithfully represents a monitored system quantity, there exists a strong motivation for general methods of extracting accurate information from raw measurement data. Since all measurements are theoretically correlated because power system elements are physically interconnected, an accurate state of the power system can be obtained by applying certain optimization criteria to the measured data.

The goal of power system state estimation is to provide a reliable, accurate, and complete set of data for the real-time monitoring and control of power systems. By processing available measurements, together with the knowledge of the network topology and line model parameters, power system state estimation can obtain an accurate estimate of the state variables, which include bus voltages and phase angles. Given an accurate estimate of the complex bus voltages, exact knowledge of the network topology, and precise values for the line parameters, accurate values for all system quantities can then be computed via power flow equations and Kirchhoff's laws.

PARALLEL AND DISTRIBUTED STATE ESTIMATION   237

As shown in Figure 7.1, state estimation not only turns the "raw" measured data into "ripe and ready-to-use" data, but also detects and identifies bad data that might contaminate other data. A power system state estimator is a collection of computer programs that converts telemetered data into a reliable estimate of the system state and topology by accounting for small random metering and communication errors, bad data due to transients and telemetry failures, uncertainties in system parameter values, and errors in the network model due to faulty circuit breaker statuses. Specifically, a state estimator provides the following four basic functions for the real-time monitoring and control of a power system:

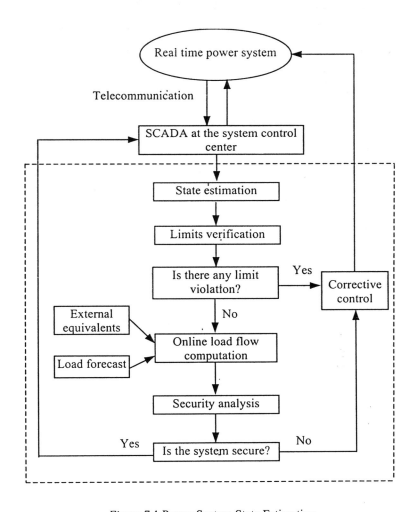

Figure 7.1 Power System State Estimation

- Eliminate measurement errors
- Detect and identify bad measurements
- Estimate power system parameters such as line impedance
- Provide data for lines and buses that are not metered

State estimation has become the very basic and most powerful tool for system operators to use in monitoring and controlling power systems. Thus, this has become the most basic and core function of an EMS at the power system control centers. The performance of the optimization and control of a modern power system relies heavily on the performance of the state estimator that the EMS utilizes. State estimation is mostly computed in a serial manner at system control centers, and thus there may be difficulty in meeting rigid time requirements for online power systems monitoring and control when a large amount of measured data must be processed within a limited time.

Previously, there were two approaches to obtain a fast computation speed for state estimation. One was to employ a higher performance computer and the other was to purposely reduce the computational burden by simply limiting the observable region to a certain range of a power system. The former approach required a large investment on computing facilities, and the latter approach provided a limited amount of information to system operators. This is especially what occurs when system operators stretch out their analyses to encompass the entire power system.

A distributed computer network is now capable of providing super computing as powerful as a high-performance parallel mainframe, and this represents recent developments in computer and telecommunication techniques. The computers at a system control center and its satellite control centers are mostly interconnected via dedicated telecommunication links with a powerful WAN. By utilizing distributed computing techniques, many complicated computation and control tasks can be performed on this WAN. Furthermore, computing system faults and shortcomings can be localized in WAN with a limited effect on the entire system, which has obviously enhanced the fault tolerance capability in the power system optimization and control. This feature can specifically be utilized in state estimation, voltage/VAR control, transmission congestion management, and so on.

In this chapter, we discuss parallel and distributed techniques for state estimation. We cover the design of parallel state estimation, which will run on a parallel mainframe; then we design an asynchronous distributed state estimation method by taking the unpredictability of communication delays into account. The mathematical proofs and analyses for the convergence of asynchronous distributed state estimation algorithms are included. The proposed distributed state estimation algorithms will run on a distributed system that consists of computers at the ISO and RTO sites, and their satellite control centers.

7.2 OVERVIEW OF STATE ESTIMATION METHODS

In all, an ideal state estimation method should exhibit an excellent performance in the following three ways:

- Numerical stability
- Computation efficiency
- Implementation complexity

The applications of estimation theory to the monitoring and control of electrical power systems was first proposed by Schweppe and Wildes in 1970 [Sch70]. After establishing the motivation for state estimation as an important tool in the control and operation of modern power systems, the authors formulated a state estimator based on the classical weighted least squares (WLS) estimation criterion. Uncertainties arising from measurement and modeling errors were also considered, and the essential logic behind error detection and identification was developed. This basic WLS state estimation method was referred to as the normal equation (NE) state estimation method (see equation 7.4).

Larson and Tinney [Lar70a, Lar70b] expanded on the proposals of [Sch70] and addressed some of the practical aspects of applying the WLS estimator to actual power systems. A great deal of attention was focused on the measurement scheme, since the location, type, and the number of measurements are important considerations in any estimator implementation. Computational aspects of WLS estimation were treated in detail, and a strategy was proposed for partitioning the measurement set into a basic subset and a redundant subset, with subsequent sequential processing of the redundant measurements.

It was demonstrated [Dop70a, Dop70b] that a measurement set consisting solely of real and reactive line flows involves a number of computational advantages and substantial enhancement of estimate robustness. In this proposed method, fictitious line element voltages are computed directly from measured line flows. Subsequently, a WLS estimate of the complex bus voltages is obtained using the computed complex line element voltages as "measurements." Despite its sizable advantages, the main drawback of this method is the large, and possibly prohibitive, number of line flow measurements that are required.

To account for the presence of bad data that severely degrades the standard WLS estimate, while circumventing the need for lengthy error analysis procedures, Merrill and Schweppe formulated an enhanced WLS estimator known as the bad data suppression (BDS) estimator [Mer71]. In BDS estimation the cost function is generally non-quadratic and it is reduced to the quadratic WLS cost function in the absence of gross errors. By successive re-weighting of large residual terms, the cost function becomes less sensitive to large errors, thereby allowing the estimator to provide a modest degree of bad data rejection.

A survey of almost all aspects of WLS estimation in power systems is given in [Sch74]. Different approaches to solving the basic WLS problem on a real-time basis, modeling of structural, parametric, and measurement errors, and a rudimentary error analysis procedure are all discussed, in addition to dynamic models which account for factors ignored in the static estimation problem.

It was quickly observed that the NE method, as applied to power systems, is particularly prone to the problem of ill-conditioned equations [Lar70a]. This becomes intolerable in the presence of round-off errors and associated limitations of finite-precision computation, especially when the system size is large or when the difference of measurements weights is large. The two-fold consequence is the degradation of estimation accuracy and near singularity of the information matrix, while the latter problem introduces sizable numerical difficulties into the estimation process. A procedure for eliminating the ill-conditioning problem of WLS estimation via successive applications of Householder orthogonal transformations, known as the Golub's method, was proposed in [Cos81]. The proposed method incorporates only perfectly conditioned matrices while avoiding the explicit calculation of the information matrix.

Motivated by the desire to avoid the destruction of information matrix sparsity, which is an unfavorable consequence of the method outlined in [Cos81], Gu and Clements presented an analysis of ill-conditioning for a decoupled WLS formulation and quantified the problem by using the Jacobian matrix's condition number. This was shown to be proportional to the degree of ill-conditioning present in the composite estimation equations [GuJ83]. By applying a series of transformations to the Jacobian matrix developed by Peters and Wilkinson, the condition number was improved (lowered). In [GuJ83] the authors attributed ill-conditioning to the size of the estimation problem and claimed that this could be avoided by decomposing the global estimation task into a number of smaller subproblems. An important distinction is drawn between critical and non--critical measurements, and the measurement set is partitioned along these lines. The Jacobian sub-matrices are decomposed based on the network topology and measurement locations. Subsequently, if the critical measurements are related to only a subset of the states, it is possible to solve two separate estimation problems in lieu of the global problem.

To overcome the ill-conditioning problem of the NE state estimation method, in the last three decades researchers have proposed the following state estimation methods:

- Orthogonal transformation (OT) method [Sim81]

- Hybrid method (HM)[GuJ83]

- Normal equation with constraints (NE/C) method [Asc77]

- Hachtel augmented matrix (HACHTEL) method [Gje85]

These state estimation methods have different computational performances, although they all make use of the maximum likelihood and weighted least square principles. They were designed with two ideas in mind. One was to avoid the calculation of the gain matrix, and the other was to take the virtual measurements as equality constraints to eliminate the adverse impact of the difference in measurement weights on the accuracy of state estimation. Virtual measurements are zero measurements at buses where there is neither injection nor load. The weight of a virtual measurement would tend toward infinity since there are no errors in virtual measurements.

In the OT method, the least square equation is solved through orthogonal transformation, and the gain matrix does not appear. In the HM method, the gain matrix is replaced by the product of the transpose matrix of the orthogonal matrix and the orthogonal matrix, and the gain matrix is not used in the iteration. In the NE/C method, virtual measurements are transferred into equality constraints rather than generic measurements. Thus the weight differences between measurements are reduced, but the gain matrix appears in the iteration. In the HACHTEL method, not only virtual measurements are taken as equality constraints, but also the incremental residual is taken as an equality constraint and augmented in the coefficient matrix of the iteration equation. There is no gain matrix in this method which seems to satisfy the objectives of new state estimation techniques. However, the dimension of the coefficient matrix is large, which means more computation time will be needed.

In the five state estimation methods mentioned above, the NE method, especially when implemented with the fast P-Q decoupling technique, has high computational efficiency, although it sometimes gives rise to ill conditioning. The OT method has the highest numerical stability, but the orthogonal transformation needs much computation. Furthermore, it is hard to use the P-Q decoupling technique in the OT method. HM and HACHTEL methods are a reasonable compromise between the numerical stability and computation efficiency, and both methods have an acceptable implementation complexity, though implementation of the HACHTEL method is a bit more complicated. Some simulation results have shown that the HACHTEL method is more stable and efficient than HM [Hol88], in particular, for large-scale power systems.

As in load flow computation, the utilization of the fast P-Q decoupling technique can improve the computational efficiency of state estimation to a great extent. However, with the fast P-Q decoupling technique, it is difficult to use the abundant ampere measurements for state estimation, since ampere measurements cannot be easily decoupled into two parts. In most circumstances, ampere measurements are only used as a kind of auxiliary information for the topology analysis of state estimation. The ampere measurements are used in the state estimation of distribution systems, where the fast P-Q decoupled method is not applied [Roy93]. The utilization of ampere measurements can not only improve the accuracy of state estimation but also enlarge the observable area especially when the measurements in the system are insufficient.

In 1978, Irving, Owen, and Sterling introduced the first practical weighted least absolute value (WLAV) approach to power system state estimation [Irv78]. The estimation problem was formulated as a linear programming problem in which the objective was to minimize the sum of the residual module. The linear programming (LP) estimator was shown to possess the exceptional property of outright bad data rejection, apparently eliminating the need for expensive error analysis methods. It was stated that the superior robustness of the linear programming estimator stems from its zero-residual property. The LP estimator satisfies a subset of measurement equations exactly, assigning zero residuals to these measurements, while freely rejecting the remaining measurements. As a consequence the LP estimator is insensitive to bad data (it was later discovered that this is true only in the absence of leveraged bad measurements). It was also shown that the filtering capability of the LP estimator is marginally inferior to that of WLS estimators. This property is due to the same zero-residual property that is responsible for the method's excellent bad data rejection.

A generalized WLAV estimator, functionally equivalent to the LP estimator but more efficient owing to its exploitation of the special structure of the power system measurement equations, was presented in [Kot82]. It was graphically shown that the outstanding robustness of WLAV estimation is attributed to its "interpolative" property, by which the most erroneous measurements are completely disregarded in arriving at an estimate of the system states. Further improvements in the computational efficiency of WLAV estimation were suggested in [Abu91], where the WLAV problem is again implemented as a linear programming problem. The Barrodale and Roberts algorithm is improved upon so that the sparsity and particular structure of the measurement equations is fully utilized. The proposed formulation also reduces the dimension of the basis matrix to be factorized and updated at the start of each Simplex iteration.

To reduce the sizable computational burden of the basic LP estimator, Lo and Mahmoud developed a decoupled LP formulation [LoK86]. Along similar lines to the decoupled load flow, the original problem is decomposed into two smaller problems, one P-delta, the other Q-V, which are solved sequentially to obtain the final solution. Furthermore, the elements of the Jacobian matrix are evaluated only at the beginning of the first iteration and held constant for all subsequent iterations, a strategy which avoids the considerable computation required to update the elements at the start of each iteration.

The Dantzig-Wolfe decomposition algorithm was used to devise a multi-area approach to LP power system state estimation [Kei91]. The global LP problem is decomposed into multiple subproblems, one for each designated network area, while the interactions between different areas are accounted for by the master problem. Each subproblem is solved independently before submitting its results to the master problem. The solution of the master problem, in turn, dictates modifications for the subproblem objective functions, and so forth. Parallel processing is proposed as a practical means by which all subproblems may be executed concurrently, so that the execution time of the method is minimized.

Abbasy and Shahidehpour considered the question of optimal measurement sets in mathematical programming state estimation and proposed a method by which the measurements corresponding to redundant constraints, which do not contribute to the resulting state estimate, are identified [Abb88]. The redundant constraints can thus be excluded without negatively affecting the estimation accuracy, leading to an appreciable savings in the computational effort required for the estimation problem. The proposed method can also be used to devise a measurement set which contains a minimal number of measurements and adequate redundancy for estimate accuracy.

7.2.1 Applications of Parallel and Distributed Processing

Since the late 1980s, with the advancement of computer science and network technology, the application of parallel and distributed processing to power systems has become a critical research area. Now, with the emerging ISO/RTOs which may have several interconnected control areas, distributed computing and control has becomes more preferable than completely centralized management. Certainly, the parallel and distributed computation of state estimation is one of the attractive research subjects in electricity restructuring.

The main difficulty in applying parallel and distributed computation to state estimation lies in the decomposition of the system into subareas. Accordingly, parallel and distributed state estimation will be able to converge to the correct solution as its serial counterpart with a reasonable level of data communication among processors. By overlapping boundary buses, [Ebr00] proposed a distributed state estimation method and concluded that this new method is more flexible in system decomposition and has a smaller amount of inter-processor data communications compared with other distributed state estimation methods [Bri82, Cle72, Cut81, Cut83, Kob74, Lin92, Zab80].

Unlike parallel processing where data communications among processors are presumed to be determined and communication delay is almost negligible, distributed computation needs to take the data communication delay into account. This is because in most distributed systems, data communication delays among computers are hard to predict. Based on the analysis of the Jacobian matrix of state estimation, [Wan94,

Wan96] proposed a distributed state estimation method that takes the unpredictable data communication delays among processors into account for state estimation; this method adopts the most flexible system decomposition and may become the most feasible method for practical applications of distributed state estimation.

In most cases parallel processing would definitely expedite the solution process, but distributed processing may not guarantee a faster solution speed. The synchronization penalty could be very high in some distributed systems and under such circumstance the speed of distributed computation could be even slower than that of sequential processing. However, obtaining a fast solution speed is not the only purpose of distributed processing. For a large power system, distributed processing could bring more flexibility and reliability in monitoring and control. Moreover distributed processing could save on large investment in communication networks.

7.3 COMPONENTS OF STATE ESTIMATION

Although state estimation resembles a generalized load flow process, the implementation of state estimation is much more complex than that of the load flow computation. Before state estimation can be computed, some preliminary functions such as measurement filtering, network topology analysis, and observability analysis have to be performed, and the bad data detection and identification have to be processed to improve further the accuracy of the state estimation.

A state estimation program would generally include the four components shown in Figure 7.2. A bad data filter could be added just behind the SCADA system to reject the bad data that are obviously out of bounds. It is easy to design such a filter based on some common logics and the given limits for relative measurements. The logic used in this filter should be very simple; for instance, the voltage measurement for a 110 kv bus cannot exceed 150 kv, or the power measurements at the two terminals of one line should have different signs. The functions of these four components are described as follows:

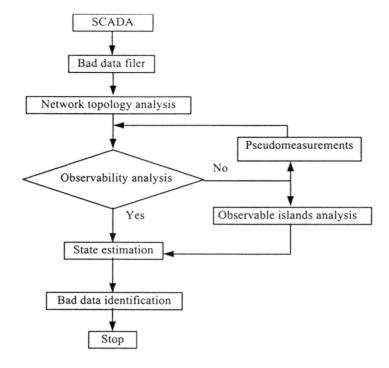

Figure 7.2 Main Components of State Estimation

- **Network topology analysis.** This is used to determine the real-time network structure of the power system based on its acquired digital measurements. The network topology analysis is the basis of the real-time power system analysis and control.

- **Observability analysis.** This is used to determine whether the system is observable based on the usable analog measurements and the results of the network topology analysis. If the power system is not observable, the analysis will further determine where pseudo-measurements will need to be added to make the system observable. It will also determine the portions of the system which constitute observable islands. Only the observable parts of the system can be estimated. There are two kinds of observability analysis methods: the topological method and the numerical method. The topological method uses the information on the physical measurement distribution throughout the system, while the numerical method uses the information provided by the Jacobian matrix of state estimation.

PARALLEL AND DISTRIBUTED STATE ESTIMATION 247

- **State estimation computation.** This is used to estimate the state of the power system according to the usable measurements. The state estimation computation encompasses the bad data detection and identification process shown in Figure 7.3.

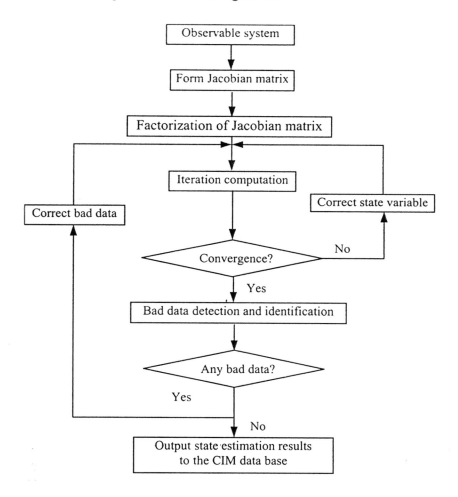

Figure 7.3 Flow Chart of State Estimation

- **Bad data detection and identification.** This is used to detect and identify bad data in the measurement set based on the analysis of measurement residuals. If any bad data are detected in the state estimation process, the algorithm proceeds to identify the bad data. The bad data can contaminate its adjacent measurements severely and thus cause more errors in state estimation. Hence bad data need to be

7.4 MATHEMATICAL MODEL FOR STATE ESTIMATION

The NE method is the most fundamental state estimation method which is used to apply the principles of power system state estimation. The state estimation based on the NE method is widely implemented and used in many commercial EMS packages. In this chapter we use the NE state estimation method to illustrate how the distributed state estimation is designed and implemented; the same distributed processing technique can be applied to other state estimation methods.

Suppose the measurement equation is given by

$$z = h(x) + v \qquad (7.1)$$

where

z = measurement vector (m dimension)
x = state variable vector (2n - 2 dimension)
v = measurement error vector (m dimension) whose elements are random variables with Guassian distribution
h = a relation function of the measurement z and the state variable x.

The weighted least square state estimation would minimize the following objective function:

$$J(x) = [z - h(x)]^T W [z - h(x)] \qquad (7.2)$$

where

W = weight matrix ($m \times m$)

$W = diag[1/\sigma_1^2, \ldots, 1/\sigma_i^2, \ldots, 1/\sigma_m^2]$, and σ_i^2 is the covariance of the ith measurement error.

The measurement residual vector is defined as

$$\Delta z(x) = z - h(x) \qquad (7.3)$$

PARALLEL AND DISTRIBUTED STATE ESTIMATION

Then, the least square state estimation computes the system state variables by iterating the following normal equation:

$$G(x)\Delta x = H^T(x)W\Delta z(x) \qquad (7.4)$$

where

$G(x)$ = gain matrix

$H = \partial h(x)/\partial x$ is the Jacobian matrix of $h(x)$

7.5 PARALLEL STATE ESTIMATION

7.5.1 Data Partition

Say that a parallel machine has N processors for parallel state estimation computation. The first step to implement parallel state estimation computation is to divide the measurement set into N smaller sets so that each processor can compute one subset in parallel. The second step is to minimize the communication overhead among these N parallel processors.

Accordingly the system is divided into N subareas in which each processor is responsible for the state estimation computation of a certain subarea. To implement this step, the measurement data set acquired through SCADA is correspondingly divided into N sets. The mapping of the SCADA measurement data set into subareas is illustrated in Figure 7.4.

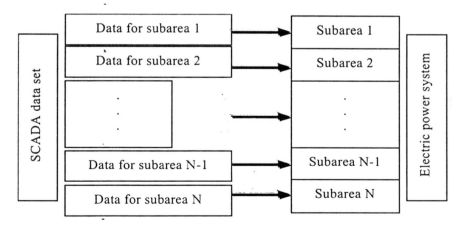

Figure 7.4 System Data Partition

The parallel state estimation computation will be performed based on the partitioning of the measurement set. The proper partitioning of the measurement set and the power system facilitates the parallel computation and expedites the entire solution process by reducing the data communication among processors.

7.5.1.1 Parallel Topology Analysis. The algorithm chooses an initial search point in each subarea, so that the N processors can perform a depth-first search in parallel. Accordingly each processor will form a sub-tree of the system topology based on its own subset of data. The results of the local topology analysis will be merged to form the network topology of the entire system by using the boundary status information.

7.5.1.2 Parallel Observability Analysis. Each processor performs the observability analysis for its own subarea. Either the topology analysis method or the numerical analysis method can be employed for this purpose. The observability analysis for the entire system depends on the results of the observability analyses of subareas and the measurements on the boundary lines and buses. All boundary nodes of the system should be observable using boundary measurements, which include measurements on boundary buses and tie lines.

7.5.1.3 Parallel State Estimation Computation. Each parallel processor calculates the Jacobian matrix for its own subarea. The calculation of Jacobian matrix elements is the same as that for the serial state estimation. Suppose that the ith subarea has n_i measurements. Figure 7.5 shows the Jacobian matrix of the ith subarea that is quite similar to the subarea Jacobian matrix in the distributed load flow computation, $H_i = \partial h(x)/\partial x_i$.

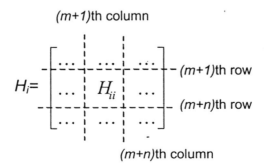

Figure 7.5 Elements of H_i

PARALLEL AND DISTRIBUTED STATE ESTIMATION

Correspondingly, the gain matrix of the ith subarea is calculated as

$$G_i(x) = H_i^T(x)W_i H(x) \qquad i = 1, \cdots, N \qquad (7.5)$$

Then, each processor will iterate the following equation

$$G_i(x)\Delta x_i = H_i^T(x)\Delta z_i(x) \qquad i = 1, \cdots, N \qquad (7.6)$$

where x in (7.5) and (7.6) is the state variable vector for the entire system. However, for the state estimation computation of a specific subarea, only its internal state variable x_i and its subarea boundary state variable, $x_{i,b}$ will be needed. Hence (7.5) and (7.6) are rewritten as

$$G_i(x_i, x_{i,b}) = H_i^T(x_i, x_{i,b})W_i H(x_i, x_{i,b}) \qquad i = 1, \cdots, N \qquad (7.7)$$

Then each processor will iterate the following equation:

$$G_i(x_i, x_{i,b})\Delta x_i = H_i^T(x_i, x_{i,b})\Delta z_i(x_i, x_{i,b}) \qquad i = 1, \cdots, N \qquad (7.8)$$

At every iteration in the parallel computation, each processor has to wait for the boundary information $x_{i,b}$ to be transmitted by the processors that are responsible for neighboring areas. If the computation task is not well balanced, a significant amount of time will be spent on waiting for data communication, which may result in a high synchronization penalty.

7.5.1.4 Parallel Bad Data Detection and Identification.
As we saw earlier, bad data detection and identification is an indispensable module of state estimation because there are measurement errors that can cause incorrect state estimation solutions.

The bad data detection module checks if there are any bad data in the measurement set used for state estimation. This is a difficult process, and so far most detection methods have been based on traditional statistical theories. The most simple and effective approach is the hypothetical testing method, which checks the value of the state estimation objective function $J(x)$. If the value of $J(x)$ exceeds a certain limit, then theoretically there are some bad data in the measurement set, and further identification must proceed. Otherwise, it is supposed that there are no bad data in the measurement set and the state estimation results are credible.

Since the measurement residual is a vector, an efficient way to the residual calculation is for each subarea processor to compute its own sub-vector of the residual. To see this, let us divide the power system into N subareas. Then, the objective function of state estimation can be calculated in parallel as follows:

$$J_i(x) = [z_i - h_i(x)]^T W_i [z_i - h_i(x)], \quad i = 1, \cdots, N \tag{7.9}$$

$$J(x) = \sum_{i=1}^{N} J_i(x) \tag{7.10}$$

Say that the processor k acts as the coordinator for parallel computation, meaning that it collects and summarizes all subarea residuals and determines whether there are any bad data in the measurement set. Bad data identification will find the suspected bad data after the bad data detection shows that some bad data exist in the measurement set.

There are several other methods for bad data identification. In particular, the normalized residual method has received wide application. The principle of the normalized residual method is rather simple. After the computation of the residual vector for all measurements that are utilized in state estimation, the measurement which has the largest absolute residual is taken as the prime suspect for bad data. The normalized residual can be computed in parallel as follows:

$$\Delta z_i = z_i - h_i(x), \; i = 1, \cdots, N \tag{7.11}$$

The processor k, which is the coordinator for parallel computation, will collect $\Delta z_i, i = 1, \cdots, N$, to form the system residual

$$\Delta z = \begin{bmatrix} \Delta z_1 \\ \vdots \\ \Delta z_N \end{bmatrix} \tag{7.12}$$

Δz can be further normalized as

$$\Delta z = \begin{bmatrix} \Delta z_1 / d \\ \vdots \\ \Delta z_N / d \end{bmatrix} \quad (7.13)$$

where $d = \sqrt{(\Delta z_1)^2 + \cdots + (\Delta z_N)^2}$.

The bad data can then be identified one by one according to this normalized residual. To expedite the bad data identification process, at one time b_n measurements, which have the b_n largest absolute residuals, are taken as the primary bad data suspects.

This simple method can be used for theoretical analyses and engineering applications, although it has difficulties in processing correlated bad data. The identified bad data must be purged out or be corrected in time according to certain criteria for the next round of iterations in order to prevent the bad data from contaminating credible data and influencing state estimation results.

7.6 DISTRIBUTED STATE ESTIMATION

Unlike its parallel state estimation where the power system is virtually divided into N subareas, the distributed state estimation will actually divide the power system into several subareas. In other words, distributed state estimation is based on the existing distributed system, which is composed of ISO/RTO computers and satellite control centers.

The control centers signify the physical partitioning of a power system into control area/subareas. In this circumstance distributed computation would specifically be performed with a fixed number of computers. This is unlike the parallel computation in a parallel machine where the number of processors to be employed is conveniently adjusted to meet parallel algorithm requirements.

Each subarea in Figure 7.6 has its own SCADA system, which collects real-time measurement data of its own subarea to be processed by its subarea control center (SACC). If SCC is an RTO, SACCs in Figure 7.6 can be ISOs under this RTO. However, if SCC is an ISO, SACCs can be satellite control centers of this ISO. In such a distributed system, since the system partitioning is already set, each SACC will only perform distributed state estimation for its own subarea.

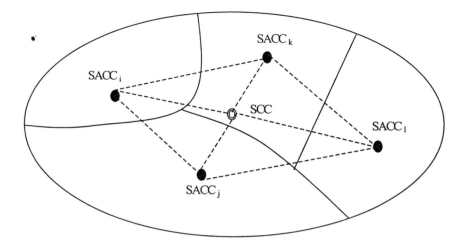

SCC: system control center
SACC: subarea control center
---: telecommunication links between control centers

Figure 7.6 Power System Partition for Distributed State Estimation

As in the parallel state estimation, distributed state estimation requires a coordinator in the distributed computation process that acts as the information collector. There are two possible ways to realize this requirement. One way is to let the SCC be the coordinator and not participate in the distributed computation. Such an SCC may exist in electric power systems, where the main function of SCC is to coordinate the behavior of its SACCs. The other way is that the SCC participates in distributed computation, which theoretically allows anyone of computers to be the coordinator. In any case, since there is actually not much coordination in the distributed state estimation, there is not much need to designate a particular computer to be the coordinator. However, since SACCs have a communication link to SCC, it is better to choose an SCC as the coordinator which is rather more concerned with computation than SACCs.

7.6.1 Distributed Topology Analysis

Since the data partitioning methodology for the distributed state estimation is exactly the same as that of the parallel state estimation, the topology analysis for the distributed state estimation is the same as that for the parallel state estimation.

PARALLEL AND DISTRIBUTED STATE ESTIMATION

Each SACC will first analyze the topology of its own subarea, and then the coordinator determines the topology for the entire system based on the status of tie line measurements. In some sense, the topology of the entire system may not be so necessary for a totally distributed computing system since each subarea is concerned with its local system. However, from the viewpoint of the ISO/RTO, the topological information for the entire system is most important for the real-time monitoring and control of the system. Figure 7.7 depicts the procedure for the topological analysis of two adjacent subareas.

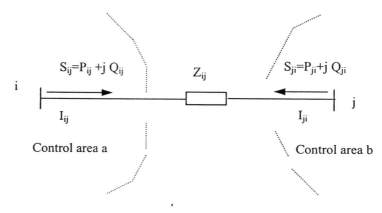

Figure 7.7 Boundary Topology Analyses

The operating status of a tie line is determined with the following criteria as described in Table 7.1. This determination procedure can be further enhanced by making use of ampere measurements when the status of switches at the two ends of a tie line contradicts each other. The same procedure is also used to make a crosscheck for the uncertain status of certain transmission lines. The ampere measurements used here may be replaced with active or reactive power measurements.

Table 7.1 Determination of Tie Line Operation Status

Measurements	Status of T_{ij}
$S_{ij}=0$, $S_{ij}=0$	OPEN
$S_{ij}=NZ$, $S_{ij}=NZ$	CLOSE
$S_{ij}=0$, $S_{ij}=1$, $I_{ij}=NZ$, $I_{ji}=NZ$	CLOSE
$S_{ij}=1$, $S_{ij}=0$, $I_{ij}=NZ$, $I_{ji}=NZ$	CLOSE

Note: NZ stands for a non-zero value; an "OPEN" status means the tie line is out of service, while "CLOSE" means in service.

The SCADA at each SACC will scan its own subarea in a preset cyclic period, and the acquired status and analog data will be collected and stored in the real-time EMS database of SACC.

7.6.2 Distributed Observability Analysis

After each SACC has finished the topological analysis of its own subarea, it will proceed to analyze the observability of its subarea. There are two methods that are used for the observability analysis. One is the numerical analysis method, and the other is the topological analysis method. Either method possesses advantages and disadvantages. The numerical analysis method determines the system observability by investigating whether there are any zero elements in the diagonal positions when the information matrix of state estimation is factorized. This method is widely used in the centralized state estimation computation. In a distributed environment, it is more convenient to determine the observability using the topological analysis method, which investigates whether the valid measurements form a minimum spanning tree.

The precondition for the distributed state estimation is that every subarea should be observable; specifically, the system will become unobservable if any subarea is unobservable. We should note that the reverse assertion does not always hold, meaning that a power system may not be observable if all subareas are observable. The reason for this phenomenon is that observable subareas are to be interconnected by observable boundary measurement deployments in order to make the entire system observable. If some boundary lines or buses are unobservable, the entire system can be separated into several isolated observable islands. In some special cases, even if the system is observable for a serial state estimation, the entire system may be unobservable for the distributed state estimation.

Suppose that the subarea e in Figure 7.8 has a single tie line with its neighboring subareas and is connected to subarea m. If subareas e and m are observable, the entire system will be observable for the serial state estimation based on the measurement deployment shown in the figure. However, because there is no boundary measurement at bus j of subarea e, there is no $x_{e,b}$ in (7.8). In this case subarea e cannot obtain the relevant boundary state variable information from its neighboring subarea m, and the distributed state estimation for the entire system will become unobservable, except that bus j is chosen as the reference bus for the entire system.

PARALLEL AND DISTRIBUTED STATE ESTIMATION

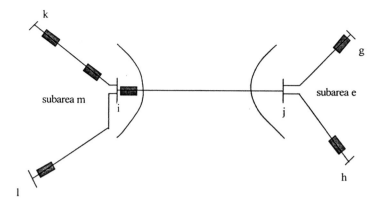

▨ : Effective active and reactive power measurements

Figure 7.8 Unobservable Measurement Dispositions

The example in Figure 7.8 shows that distributed state estimation poses more strict requirements for measurement dispositions at boundary buses and tie lines. The main reason for this is that there is only one reference bus in the entire system for distributed state estimation, and every subarea should refer to this reference bus.

We are not concerned with this issue in the conventional serial state estimation because the observability analysis for the serial state estimation assumes implicitly that all buses refer to the same reference bus. However, for the parallel and distributed state estimation, we need to make sure that every subarea can obtain the reference information, either directly or indirectly. This is a critical condition for the observability of parallel and distributed state estimation.

Theorem 7.1. Every subarea of a system is individually observable. If there are at least $(N - 1)$ tie lines with valid measurements at both ends that interconnect the N subareas. The entire system will then be observable for the distributed state estimation.

Proof. Measurements in each subarea form a minimum spanning tree because subareas are observable. Consider the N subareas as N nodes with $(N - 1)$ lines which form a minimum spanning tree of the N nodes. When at least $(N - 1)$ tie lines interconnect the subareas, there will be a minimum spanning tree for the entire system. Since each of the $(N - 1)$ tie lines has valid measurements at both ends, each subarea can receive the information from at least one of its neighboring subareas and finally get the reference information from the subarea where the reference bus resides. Therefore, the entire system becomes observable for the parallel and distributed state estimation.

Theorem 7.1 is conservative as it offers the sufficient, but not the necessary, condition for calculating the observability of the parallel and distributed state estimation. This issue is illustrated in Figure 7.9. Suppose that the system reference bus is in subarea a and exposed measurements are all valid. Then, the entire system is observable for the parallel and distributed state estimation even if only one end of each of the three tie lines has valid measurements. This is because subareas b, c, and d can either directly or indirectly receive the reference information through the exchange of boundary variables.

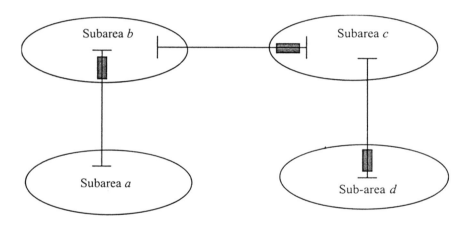

Figure 7.9 Observable System with Fewer Boundary Measurements

The following theorem provides the sufficient and necessary condition for the observability of parallel and distributed state estimation.

Theorem 7.2. Every subarea is observable if all subareas can either directly or indirectly receive the reference information through the exchange of boundary variable information. The entire system is then observable for the parallel and distributed state estimation.

7.6.3 Computation of Distributed State Estimation

7.6.3.1 Component Solution Method for State Estimation. Similar to the distributed load flow computation, distributed state estimation is also based on the component solution method. Each SACC will collect measurements separately and compute the state estimation for the subarea that is under its jurisdiction. The difficulty with employing the component solution method lies in finding an effective way to partition the system's Jacobian matrix H into several H_i for each subarea. Here we adopt a procedure similar to that for the distributed load flow, which can be described as follows: First, as we calculate the elements of the subarea Jacobian matrix H_i all of the state variables of other subareas, including all the state variables of external boundary buses of the ith subarea, are taken as constants, so that all of the elements of H_i corresponding these state variables are equal to zero. Then, we use the real and reactive power measurements of the ith subarea including the measurements on the tie lines of the ith subarea to form the local Jacobian matrix H_i.

Once the data communication delays between SACCs are taken into account, the component solution of the distributed state estimation is described by the following iterative equation:

$$G_i(x^i(t))\Delta x_i = H_i^T(x^i(t))\Delta z_i(x^i(t)) \qquad i = 1,\cdots,N \qquad (7.14)$$

where $x^i(t) = (x_1(\tau^i{}_1(t)), \ldots, x_N(\tau^i_N(t)))$. Compared with the component solution of parallel state estimation (7.6), we find that the only difference between the parallel state estimation and the distributed state estimation lies in the latter taking into account the unpredictable time delays of communications among SACCs.

7.6.3.2 Synchronous Distributed State Estimation. If for any i and j, there exists $\tau^i{}_j(t) = t$, then (7.14) represents the synchronous distributed state estimation. In the synchronous distributed state estimation computation, computers will be synchronized to the same proceeding step. Therefore, faster computing machines will have to wait for slower machines. In this solution process only the state variables of boundary buses are to be exchanged. The computer at each SACC will proceed to its next time of iteration immediately after it has received the boundary variable information from its neighboring SACCs.

7.6.3.3 Asynchronous Distributed State Estimation. When the data communication delays $\tau^i{}_j(t)$ among SACCs meet the partially asynchronous assumption, (7.14) is referred to as the partially asynchronous distributed state estimation. Similarly, when the data communication delays $\tau^i{}_j(t)$ meet the totally asynchronous assumption, (7.14) is referred to as the totally asynchronous distributed state estimation.

7.7 DISTRIBUTED BAD DATA DETECTION AND IDENTIFICATION

Bad data detection and identification is processed after the distributed state estimation computation is completed. The procedure is similar to that of parallel state estimation; the only difference here is that the residual subvectors are sent to a designated computer rather than to a processor. The computer summarizes these subvectors and judges whether there are any bad data in the measurement set used for the distributed state estimation. Hence, for the bad data detection and identification of distributed stated estimation, we refer to the corresponding procedures for the parallel state estimation discussed in Section 7.4.

7.8 CONVERGENCE ANALYSIS OF PARALLEL AND DISTRIBUTED STATE ESTIMATION

In a broad sense, state estimation is a special kind of load flow. Hence, based on the convergence analysis of the distributed load flow given in Chapter 6, we assume that there is an iteration mapping function $\varphi: R^n \to R^n$, where $\varphi = (\varphi_1, \cdots, \varphi_N)$, and n is the dimension of the iteration mapping function φ. Also, $\varphi_i : R^n \to R^{n_i}$, $(i = 1, \cdots, N)$ is the component iteration mapping function of distributed state estimation, and n_i is the iteration mapping function φ_i. Say that the distributed state estimation is formulated as

$$\varphi(x) = x + H^T(x)W(z - h(x)) \tag{7.15}$$

According to (7.14) we further formulate distributed state estimation as

$$\varphi_i(x) = x_i + G_i^{-1}(x)H_i^T(x)W_i\Delta z_i(x) \quad i = 1, \cdots, N \tag{7.16}$$

In the following, by using a procedure similar to the distributed load flow, we analyze the convergence of distributed state estimation. First, we denote

$$g(x) = H^T(x)W(z - h(x)) \tag{7.17}$$

So the weighted least square state estimation has the same form as that of load flow. According to the convergence condition of the distributed load flow, the partially asynchronous distributed state estimation will converge if φ is a nonexpansive mapping function. The totally asynchronous distributed state estimation will converge if the mapping function is a block maximum normal contraction mapping. Hence we have the following convergence theorem for distributed state estimation.

Theorem 7.3. Convergence of Totally Asynchronous Distributed State Estimation If the distributed state estimation iteration mapping function φ is a block maximum norm contraction mapping, the totally asynchronous distributed state estimation will converge to some fixed point of φ.

Since it is difficult to guarantee that the iteration mapping function φ is a block maximum normal contraction mapping, we offer the following theorem.

Theorem 7.4. If the component iteration mapping function φ_i is a block maximum normal contraction mapping with regard to the local area state variable x_i $(i = 1, \cdots, N)$, the totally asynchronous distributed state estimation converges to a fixed point of φ.

The convergence theorem for the partially asynchronous distributes state estimation is given as follows.

Theorem 7.5. (Convergence of the partially asynchronous distributed state estimation) The state estimation iteration mapping function $\varphi : X \to R^n$ is non-expansive if the partially asynchronous assumption holds. Then the sequence $\{x(t)\}$ generated by a partially asynchronous state estimation converges to a fixed point $x^* \in X^*$.

Now let us change (7.15) into the form

$$\varphi_i(x) = x_i - G_i^{-1}(x)H_i^T(x)W_i(h_i(x) - z_i) \tag{7.18}$$

In the following we analyze the convergence condition of (7.16). If we denote $s(x) = G_i^{-1}(x)H_i^T(x)W_i(h_i(x) - z_i)$, as we ignore the second-order derivatives of the $h(x)$ and $H(x)$, we obtain

$$\partial s(x)/\partial x_i = G_i^{-1}(x)H_i^T(x)W_i H_i(x) = E \qquad (7.19)$$

Hence, if we let $D = \partial s(x)/\partial x_i$, then D is diagonal dominant. Suppose that the maximum value of the diagonal elements of D is k; then, according to the Proposition 5.6 in Chapter 5, we arrive at the following conclusion:

Proposition 7.6. If D is diagonal dominant and γ is within the range of (0, 1/k), the totally asynchronous state estimation will converge to a fixed point of $\varphi(x) = 0$.

It is easier to analyze $D = \partial s(x)/\partial x_i$ for the case of P-Q fast decoupled state estimation, where matrices G and H are supposed to be constant. For a more general case, we set $k = 1 + \xi$, where ξ is a small positive factor; then, for a γ in (0, 1/(1 + ξ)), the totally asynchronous fast P-Q decoupled state estimation will converge. The value of ξ will depend on system characteristics.

7.9 CASE STUDIES

7.9.1 Test System

Let us consider the IEEE 118-bus system with the same system partitioning as in Appendix A1. The corresponding distributed system composed of three computers of the SACCs is depicted in Figure 7.10. Since there is no direct physical link between subareas 1 and 2, no information will be exchanged between EMS1 and EMS2 for distributed state estimation. However, there is a possible communication link between EMS1 and EMS2 for certain operational purposes. The system partitioning results are given in Tables 5.1 through 5.4. The measurement sets of the subareas are listed in the Tables B.1 to B.4 in Appendix B.

PARALLEL AND DISTRIBUTED STATE ESTIMATION

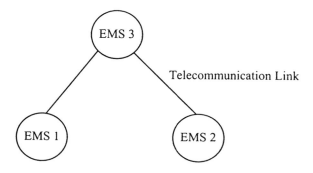

Figure 7.10 Distributed System for State Estimation

7.9.2 Synchronous Distributed State Estimation

The iteration number of the synchronous distributed state estimation versus γ is shown in Figure 7.11. Here the parallel state estimation should have the same kind of convergence curve, though it may have a faster solution speed than that of the distributed state estimation.

7.9.3 Partially Asynchronous Distributed State Estimation

The iteration number of the partially asynchronous distributed state estimation versus the asynchronous measure M is illustrated in Figure 7.12, where we assume that $\gamma = 0.1$.

7.9.4 Totally Synchronous Distributed State Estimation

As for the distributed load flow computation, the totally asynchronous distributed state estimation is realized by using MPI_Isend and MPI_Irecv; however, since the data communication process is completely random, it is difficult for the proposed distributed state estimation to converge. Theoretically, the totally asynchronous distributed state estimation should converge to the solution when γ is small enough.

264   CHAPTER 7

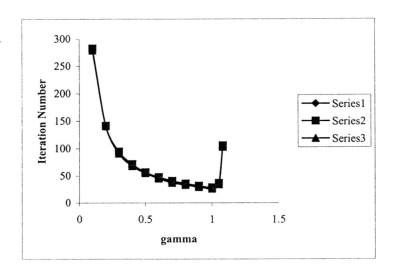

Figure 7.11 Parallel and Synchronous Distributed State Estimation

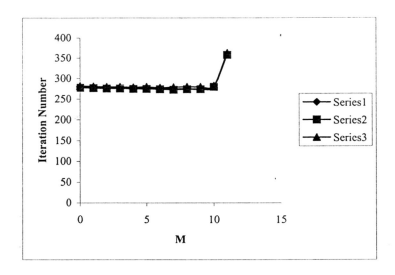

Figure 7.12 Partially Synchronous Distributed State Estimation

Note: series1, 2 and 3 represent computers 1, 2, and 3 respectively.

Chapter 8

Distributed Power System Security Analysis

8.1 INTRODUCTION

Since cascading outages can incur significant financial losses to electric utility companies, an overriding concern in power system operation is the security of the system in contingencies. Security monitoring and control has been the most important task of the system operator in both regulated and restructured power systems. The objective in security analysis is to take preventive or corrective measures if an insecure state of the system is imminent. The results of the security analysis will forewarn the system operator and give it sufficient time to take corresponding remedial measures.

Power system security analysis requires the computation of complex voltages and transmission flows in power systems. Given the different loads and generations at each bus, this calls for simultaneous solutions of a large, sparse and symmetrical system of equations whose topology may change due to changes in the network status [Alv98, Bal92, DyL97, Luk00, Qui02, Sch99, Sid00].

The solution process for large-scale power systems is usually complicated by the consideration of transformer tap-changing and generator voltage regulations. In this regard it is necessary to seek more efficient methods for the power system security analysis. Over the last few decades, various methods based on heuristics, linear approximations, pattern recognition, and artificial neural networks have been applied to the power system security analysis. Even though some methods such as computational intelligence have been utilized to speed up the entire security analysis process, the security analysis has been recognized as a time-consuming and computationally intensive process, especially for large-scale power systems with a large number of possible contingencies.

In general, power system security analysis falls into the two categories: static security analysis and dynamic security analysis, which are described below. Static security (which is also referred to steady state security) is the ability of a power system to reach a new steady state without violating constraints after certain disturbances. The static security analysis is composed of: contingency selection and contingency evaluation. The violations of thermal limits of transmission lines and bus voltage limits are the main concerns for the static security analysis. Static security analysis calculates certain types of performance indexes that can reflect the distance from the current operating state to a potential insecure state. Dynamic security is the ability of a power system to operate consistently within the limits imposed by the system stability phenomena. Generation, demand, and interchanges in a power system can be constrained by its security. [Fer01, Kum97, Kum98].

8.1.1 Procedures for Power System Security Analysis

Conventionally the task of power system security analysis encompasses the following three parts:

- **System monitoring.** Power system monitoring is realized exclusively through the SCADA system installed at control centers. SCADA collects real-time data from RTUs (remote terminal units) that are installed in substations and power plants and distributed throughout the power system. Usually SCADA scans RTUs at a frequency of two to five seconds. The data acquired typically include watts, vars, volts, amps, KWhr, frequency, circuit breaker status, and tap changing transformer settings. These data are transmitted to the system control center and stored in the SCADA/EMS real-time database. The system operator then monitors and controls the power system in real time with the help of a state estimator that can detect and identify the bad data among "raw" measurements and provide more accurate data for EMS applications.

DISTRIBUTED SECURITY ANALYSIS

- **Contingency analysis.** As mentioned earlier, power system security analysis involves the two functions of contingency selection and contingency evaluation. Contingency selection is the first step in both static and dynamic security analyses. There are a large number of components that can cause contingencies. Because the potential contingencies in a large-scale power system can occur at any time, the available window of time for system operators to figure out trouble spots and make up appropriate corrective measures is quite limited. Hence, contingency selection is used for detecting the most possible single and multiple contingencies. The processing speed becomes a main concern in the contingency selection and evaluation when very complex system dynamics are taken into account. If the contingency selection is too conservative, the time needed for contingency evaluation will be long; if the contingency selection is not aggressive enough, it risks overlooking certain vital contingencies that can cause a catastrophe to the power system. Therefore, the dominant analytical approaches to contingency analysis ought to detect potential contingencies with high efficiency and reliability.

- **Corrective action analysis.** Once the potential contingency set is selected, the system operator will embark on determining the preventive or corrective control measures that could partially alleviate or totally eliminate certain contingencies. The corrective actions could range from adjusting control transformers, adjusting the network configuration, and modifying the economic schedule of units to a pre-calculated set of load-shedding alternatives. It is customary for a system operator to obtain decision support by running the EMS static security analysis. However, it is usually more difficult to achieve credible decision support from EMS for the dynamic security analysis of a power system.

8.1.2 Distributed Power System Security

The distributed power system security analysis is the process of evaluating the vulnerability of a restructured power system to any credible set of contingent events in the ISO/RTO territories. Regarding the online security analysis, the following two approaches could be utilized by an ISO/RTO:

- The ISO performs the security analysis for the entire system.

- Each satellite control center of the ISO performs distributed security analysis for its own subarea by making use of the online external equivalents of its neighboring control areas, as shown in Figure 8.1.

Likewise the RTO may use the following two approaches to the security analysis for the system under its jurisdiction:

- The RTO performs the security analysis for the entire system.
- Each subordinate ISO of the RTO performs the distributed security analysis for its own subarea by making use of the online external equivalents of its neighboring control areas, as shown in Figure 8.2.

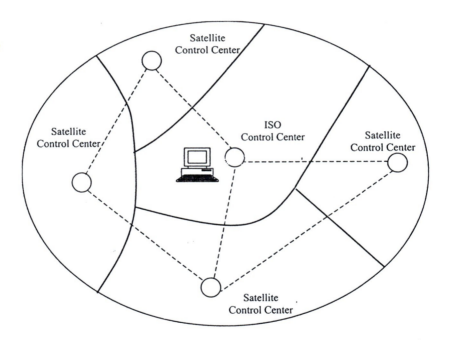

Figure 8.1 Distributed Computing Network of an ISO

According to FERC Orders 888 and 2000, when setting up a control area for an ISO or RTO, boundaries should be set where at least in the normal operation state there is only a weak energy exchange between neighboring control areas. However, this requirement does not mean that tie lines between control areas are to be neglected in the system security analysis. Moreover, in most cases it may be difficult to meet this requirement. According to the FERC Orders, neighboring

DISTRIBUTED SECURITY ANALYSIS

ISOs/RTOs should provide mutual support in cases of emergency, which requires an ISO/RTO to take into consideration the response of its neighboring control areas for security analysis.

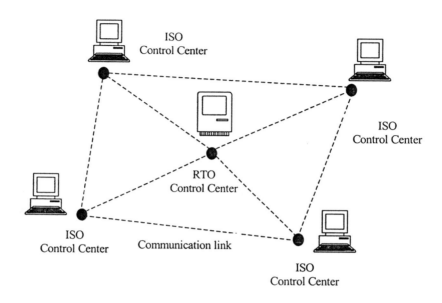

Figure 8.2 Distributed System of RTO

8.1.3 Role of the External Equivalent

The external equivalent is the basis for the online security analysis of a power system, and its computational accuracy can largely influence the security analysis. However, an external equivalent can be compounded at different load levels, network topologies, and operational strategies of the representative system. The traditional Ward equivalent method chooses a base case of the external system. Hence, the equivalent result is naturally related to this pre-selected base case of the power system. However, any change in the operation state of the representative system will inevitably incur computation errors to the security analysis of the internal system, since it does not affect the external equivalent. This is a major shortcoming of traditional external equivalent methods.

The main objective of this chapter is to build a more efficient external equivalent mechanism by making a full use of the existing hierarchical and distributed computing and control systems. The external equivalent plays a very important role in the power system security analysis [Cle99, Mon84]. The problem with applying the external equivalent in previous studies was that the external system was often unobservable to the internal system operator, or the details of the external system were bypassed to obtain a fast solution. The accuracy of the security analysis of an internal system depends on the information obtained from the external system. With the help of the advanced computer and telecommunication techniques, the power system operator can view the external system more closely. In this new situation all control areas of ISOs within one RTO become observable to one another.

The ISO/RTO would be interested in evaluating whether the loss of a certain equipment would cause violations of line thermal limits or bus voltage limits anywhere in its power system. Among the topics that we would like to discuss in this chapter are other approaches for security analysis and control that can be developed based on the hierarchical and distributed ISO/RTO structure. With an efficient distributed computing and control system, an ISO/RTO could monitor and control a power system in a distributed manner. In this chapter we discuss distributed approaches to security analysis based on a distributed system as shown in Figures 8.1 and 8.2. We will propose new schemes for an ISO/RTO to perform distributed security analysis and control. First we explore the approaches for online parallel and distributed external equivalent, which can provide a basis for parallel and distributed security analysis; then we discuss the methods for both the steady and dynamic state security analysis. In our discussion of these schemes, we include possible distributed restorative control approaches for certain contingencies.

8.2 EXTERNAL EQUIVALENCE FOR STATIC SECURITY

In the modeling of the power system external equivalent, a power system is partitioned into three sub-systems: the internal system I, the boundary system B, and the external system E, as shown in Figure 8.3. The internal system is the part that the system operator is interested in; the external system is the part that is usually out of the system operator's control of the designated internal system. When the internal system is an ISO's control area, external systems would be the neighboring ISOs' control areas. The conventional Ward equivalent method eliminates the external system buses from the system equation, and this requires the external state to be known in advance. To utilize

DISTRIBUTED SECURITY ANALYSIS

the Ward method, a base case for the external system needs to be assumed.

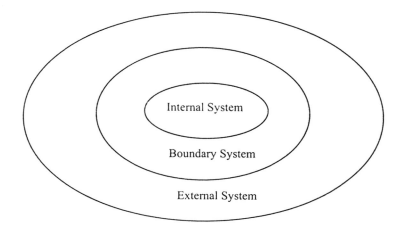

Figure 8.3 System Partition for External Equivalent

The bus voltage matrix of a power system is written as follows

$$[Y][V]=[I] \qquad (8.1)$$

where **Y** is the admittance matrix, **V** is the bus voltage vector, and **I** is the bus injection current vector. If the system is partitioned as in Figure 8.3, (8.1) is rewritten as

$$\begin{bmatrix} Y_{II} & Y_{IB} & 0 \\ Y_{BI} & Y_{BB} & Y_{BE} \\ 0 & Y_{EB} & Y_{EE} \end{bmatrix} \cdot \begin{bmatrix} V_I \\ V_B \\ V_E \end{bmatrix} = \begin{bmatrix} I_I \\ I_B \\ I_E \end{bmatrix} \qquad (8.2)$$

where subscripts *I*, *B*, and *E* represent the internal system, boundary system, and external system, respectively. By deleting the external bus voltage vector V_E in (8.2), we obtain the following bus voltage equation for the equivalent network:

$$\begin{bmatrix} Y_{II} & Y_{IB} \\ Y_{BI} & Y_{BB} - Y_{BE} Y_{EE}^{-1} Y_{EB} \end{bmatrix} \cdot \begin{bmatrix} V_I \\ V_B \end{bmatrix} = \begin{bmatrix} I_I \\ I_B - Y_{BE} Y_{EE}^{-1} I_E \end{bmatrix} \qquad (8.3)$$

It is seen in (8.3) that the Ward external equivalent would only change the topologies of the boundary system and the injection currents of boundary buses. This is because the Ward external equivalent is related to the bus injection currents I_E of the external system. When the state of the external system changes a lot, the external equivalent will reveal to a significant error in the computation.

To get a more accurate external equivalent, new algorithms use a combination of extended Ward and Ward injection techniques to produce an equivalent that retains the effect of external lines and transformer impedances. These algorithms utilize shunt MVA and impedances at the boundary of the retained network. Some selected load and generation buses can be retained as well, thus giving the system operator the freedom to model the effects of nonconforming loads in the external system more accurately for the static security assessment.

8.3 PARALLEL AND DISTRIBUTED EXTERNAL EQUIVALENT SYSTEM

8.3.1 Parallel External Equivalent

We learn from (8.3) that the main computation of external equivalent lies in the multiplication of the boundary nodal admittance matrix and the admittance matrix of the external system. When the boundary conditions are given and the state of the external system is determined, a way to improve the computational efficiency of the online external equivalent is to use parallel computation for this multiplication.

We avoid the computation of the inverse of the external admittance matrix by forming the impedance matrix of the external system directly, because the calculation of the inverse of the external admittance matrix is time consuming. The impedance matrix of the external system will be stored in the control center computer at the internal system and will be modified in time, using the compensation method, as the topology of the external system changes. This parallelization will improve the computational efficiency of the external equivalent.

The conventional formulation of external equivalent is described by (8.3). However, the control area of an ISO/RTO (i.e., the

DISTRIBUTED SECURITY ANALYSIS

internal system depicted in Figure 8.4) has usually more than one neighboring control area (i.e., external system).

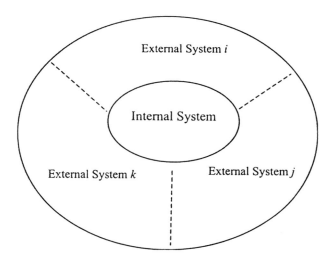

Figure 8.4 Power System with Multiple External Systems

For simplicity, and without losing of generality, we let the internal system have two external systems denoted by E_1, B_1 and E_2, B_2, respectively. According to the second row of (8.3), we have

$$\mathbf{Y}_{BI} \mathbf{V}_I + \left\{ \mathbf{Y}_{BB} - \mathbf{Y}_{BE} \mathbf{Y}_{EE}^{-1} \mathbf{Y}_{EB} \right\} \mathbf{V}_B = \mathbf{I}_B - \mathbf{Y}_{BE} \mathbf{Y}_{EE}^{-1} \mathbf{I}_E \quad (8.4)$$

which can be further expanded as

$$\begin{bmatrix} \mathbf{Y}_{B_1 I} \\ \mathbf{Y}_{B_2 I} \end{bmatrix} \cdot [V_I] + \left\{ \begin{bmatrix} \mathbf{Y}_{B_1 B_1} & 0 \\ 0 & \mathbf{Y}_{B_2 B_2} \end{bmatrix} - \begin{bmatrix} \mathbf{Y}_{B_1 E_1} & 0 \\ 0 & \mathbf{Y}_{B_2 E_2} \end{bmatrix} \cdot \right.$$
$$\left. \begin{bmatrix} \mathbf{Y}_{E_1 B_1} & 0 \\ 0 & \mathbf{Y}_{E_2 B_2} \end{bmatrix} \right\} \cdot \begin{bmatrix} \mathbf{V}_{B_1} \\ \mathbf{V}_{B_2} \end{bmatrix} = \begin{bmatrix} \mathbf{I}_{B_1} \\ \mathbf{I}_{B_2} \end{bmatrix} -$$
$$\begin{bmatrix} \mathbf{Y}_{B_1 E_1} & 0 \\ 0 & \mathbf{Y}_{B_2 E_2} \end{bmatrix} \begin{bmatrix} \mathbf{Y}_{E_1 E_1} & \mathbf{Y}_{E_1 E_2} \\ \mathbf{Y}_{E_2 E_1} & \mathbf{Y}_{E_2 E_2} \end{bmatrix}^{-1} \begin{bmatrix} \mathbf{I}_{E_1} \\ \mathbf{I}_{E_2} \end{bmatrix} \quad (8.5)$$

Let us assume that there are no tie lines between these two external systems, which means that $Y_{E_1E_2} = Y_{E_2E_1}^T = 0$. We use the following parallel computation for the equivalents of two external systems:

$$\left[Y_{B_KB_K} - Y_{B_KE_K}Y_{E_KB_K}^{-1}Y_{E_KB_K}\right]\cdot\left[V_{B_K}\right] + \left[Y_{B_KI}\right]\left[V_I\right] =$$

$$\left[I_{B_K} - Y_{B_KE_K}Y_{E_KE_K}^{-1}I_{E_K}\right], \quad k = 1,2\ldots \tag{8.6}$$

Used in this way, the two external equivalents are computed in parallel for the same internal system.

Now let us extend the simple example to a general case. Suppose that the internal system has N external systems that are not interconnected. Hence, for any two of the N external systems, say k and m systems, we have $Y_{E_kE_m} = 0$. Suppose that the bus sets for external and boundary systems are $E_1, B_1, E_2, B_2, \cdots, E_N, B_N$ respectively, and $\bigcup_{k=1}^{N} B_K = B$, and $\bigcup_{k=1}^{N} E_K = E$. Then the ISO of the internal system computes its external network equivalents in parallel according to the following formulations:

$$Y_i = Y_{B_iB_i} - Y_{B_iE_i}Y_{E_iE_i}^{-1}Y_{E_iB_i}, \quad i = 1, \cdots, N \tag{8.7}$$

$$I_i = I_{B_i} - Y_{B_iE_i}Y_{E_iE_i}^{-1}I_{E_i}, \quad i = 1, \cdots, N \tag{8.8}$$

where Y_i and I_i are the correction item of the admittance matrix $Y_{B_iB_i}$ and the correction item of the injection current vector of the corresponding boundary system, respectively.

The computation of (8.7) and (8.8) decomposes the admittance matrix of all external systems into N small matrices, which are merely composed of the nodes of one external system. From the viewpoint of matrix computation, the parallelization of external equivalent can greatly improve the computation efficiency.

When the ISO of the internal system is to compute the external equivalent, its neighboring ISOs will transmit the relevant information to the ISO of the internal system as required by the ISO of the internal

DISTRIBUTED SECURITY ANALYSIS 275

system. To expedite the computation, the ISO of the internal system will compute its external equivalents through one of the following two ways:

- **Employing a parallel machine.** Each processor calculates Y_i and I_i of one external system according to (8.7) and (8.8). Here no data communication is necessary among processors because there is no interconnection with the external systems.

- **Employing system area network (SAN).** The ISO can simply realize this parallel computation on a loosely coupled SAN because there is no data communication in the parallel computation; each computer of SAN calculates Y_i and I_i of one external system according to (8.7) and (8.8).

The additional advantage of the parallel computation of external equivalent is that the ISO of the internal system can make certain modifications on the external systems' data as needed. As for the static external equivalent computation, the external system's base case can be pre-selected according to the internal system's study requirements. Although there is no rigid time requirement for an offline external equivalent computation, (8.7) and (8.8) also provide a means of the parallel offline external equivalent computation.

8.3.2 Online Distributed External Equivalent

Unlike the parallel external equivalent computation in which the external system's data will have to be transmitted to the designated internal system for centralized processing, the distributed external equivalent will process the external data in its original location. Instead of transmitting the external equivalent's data to the ISO of the internal system, the ISOs of external systems will retain the relevant data at their own control centers and participate in the distributed computing of the designated internal system's equivalent.

When requested by the ISO of the internal system, the ISOs of external systems will compute their own Y_i and I_i according to (8.7) and (8.8) and transmit their computation results to the internal system ISO. The ISO of the internal system will inform the ISOs of external systems of its requirements for the external equivalent so that the ISOs of external systems can make the necessary modifications before they start the distributed computation of the external equivalent. Though certain information needs to be exchanged between ISOs in both parallel and distributed external equivalent computations, the amount of data exchanged in distributed computation is much less than that in parallel computation.

8.3.3 External Equivalent of Interconnected External Systems

In many cases, there are tie lines between two adjacent control areas. However, it may be difficult to satisfy $Y_{E_iE_j} = 0$ as we assumed for the parallel and distributed external equivalent computation discussed in the last section. Two approaches can be utilized to deal with this case. One is to simply model tie flows as constant loads or injections at the corresponding terminal buses, and assume $Y_{E_iE_j} = 0$ by neglecting the tie flows. This can be done with the aid of online state estimation. The second approach is to merge the external systems into a larger system in which the ISO collects all the necessary data from the external areas and then compute the external equivalents on behalf of all external areas. Comparatively, the first approach is simpler and easier to implement in a distributed computing environment. However, in utilizing the equivalent power loads and injections, we assume no tie lines in the external system. We have to make sure that the assumption will not have a significant bearing on the security analysis of the internal system.

The distributed external equivalent computation can be performed by the ISO components of an RTO, or by the satellite control centers of an ISO. Both parallel and distributed external equivalent computations use real-time data; hence, they can overcome the shortcoming of relying on pre-selected base cases in the traditional external equivalent method. This attribute is important for the online security analysis as it guarantees that the analysis of the internal system will be least affected by changes in the external systems. On the other hand, since the proposed parallel and distributed external equivalent method does not rely on a single base case, the parallel and distributed external equivalent computation described above could use the distributed state estimation results at individual control centers to avoid a myriad of data communication between control centers.

8.4 EXTENSION OF CONTINGENCY AREA

A contingency area is defined as the endangered part of an internal system due to a contingency. There is one contingency area in the system, when there is only one contingency in the internal system. The size of a contingency area depends on the severity of the contingency which could be smaller than the internal system. There could be multiple contingency areas within the internal system when multiple outages occur simultaneously in the internal system. The system operator might prefer to use the contingency area rather than the entire internal system for analyzing the impact of a contingency.

The extension of a contingency area is important in distributed contingency analysis [Alo99]. For instance, when a contingency occurs near the boundary of a control area, the area control center needs to know the impact of the contingency on the lines and buses of its neighboring areas. The information provided by distributed external equivalents does not suffice for this purpose, as the area control center requires more information for the contingency analysis. Usually it is not practical for an area control center to get much detailed information about its neighboring systems. Hence, it is crucial to determine the type of information that will be needed for the contingency analysis of the internal area.

Many techniques were proposed for the contingency analysis, and were used in defining the most endangered system components in a contingency. To reduce the computational time for the contingency analysis in large-scale power systems, several techniques resort to dc power flow for analyzing the impact of a contingency. The ac-based techniques, however, involve fast-decoupled power flow analyses. In addition, bounding methods were used for contingency analyses to identify the most endangered branches. The bounding methods were inspired by the fact that an outage has a limited geographical effect on its surrounding. Accordingly, the efficient bounding methods proposed additional efficiency by solving a small subarea instead of the entire area, in which branch flow limits were checked for violations. The "incremental angle" criterion was introduced in efficient bounding methods to detect branch flow violations by establishing an upper bound on changes in the angular spread based on the solution of a subarea. In these methods, certain bounding criteria were used to extend the interest subarea of a contingency, starting form a subset of the area around the contingency location.

The demand on speed and accuracy of computation increases as additional lines and generators are added, the system under study gets larger, and multiple outages are possible. For these reasons, a straightforward algorithm is necessary to simulate single line outages (SLOs) as well as multiple-line outages (MLOs). This can be achieved by dealing with MLOs as successive SLOs. We will perform the inverse of a sparse Jacobian matrix under SLO by way of the Woodbury formula [Cam95]. We will also make use of this technique to formulate a solution for MLOs. The Woodbury algorithm will help us find the exact distribution factors and enlarge the initial interest subarea of the contingency in the correct direction. The algorithm is used to define the exact locations of most endangered branches in the system. The total distribution factor is a new contingency term that is used for specifying the direction of the exact expansion of the interest subarea for the most endangered branches. Once starting from an initial interest subarea around an outage, the exact direction for enlarging the interest subarea will be based on total distribution factors of buses inside the interest subarea.

A new technique for the derivation of exact distribution factors will be discussed in this section. Accordingly we will be able to calculate calculate the boundaries of the subarea with violated limits more accurately, and this will also save on the computation time.

8.4.1 Distribution Factors

Suppose that the base case dc power flow equation is given by the following sparse linear equation

$$\mathbf{B}\,\delta_B = \mathbf{P} \tag{8.9}$$

The equation representing the system after the occurrence of outage(s) is given by

$$\mathbf{B}_m \delta = \mathbf{P}_m \tag{8.10}$$

After adding $-\mathbf{B}_m \delta_B$ to both sides of (8.10) we get

$$\mathbf{B}_m \delta - \mathbf{B}_m \delta_B = \mathbf{P}_m - \mathbf{B}_m \delta_B \tag{8.11}$$

which can be rewritten as

DISTRIBUTED SECURITY ANALYSIS

$$\mathbf{B}_m \Delta\delta = \mathbf{L} \tag{8.12}$$

where $\mathbf{L}$ is the incremental injection vector of the multiple-line outage.

For a single-line outage between buses r and s, the correction to the base case dc power flow Jacobian matrix is given by

$$\mathbf{B}_m = \mathbf{B} + \mathbf{U}_{rs}\mathbf{R}_{rs}^T \tag{8.13}$$

Vectors $\mathbf{U}_{rs}$ and $\mathbf{R}_{rs}$ have nonzero elements only in the r and s positions, and zeros elsewhere, where $\mathbf{R}_{rs} = (1/x_{rs})\,\mathbf{U}_{rs}$. For a general case with any number of line outages, (8.13) can be rewritten as

$$\mathbf{B}_m = \mathbf{B} + \mathbf{U}\mathbf{R}^T \tag{8.14}$$

and (8.12) becomes

$$(\mathbf{B} + \mathbf{U}\mathbf{R}^T)\,\Delta\delta = \mathbf{L} \tag{8.15}$$

Using the Woodbury formula, we express the inverse of the modified Jacobian matrix $\mathbf{B}_m$ in (8.14) as

$$\mathbf{B}_m^{-1} = [\mathbf{B} + \mathbf{U}\mathbf{R}^T]^{-1} \tag{8.16}$$

$$\mathbf{B}_m^{-1} = \mathbf{B}^{-1} - [\mathbf{B}^{-1}\mathbf{U}(\mathbf{I} + \mathbf{R}^T\mathbf{B}\mathbf{U})^{-1}\mathbf{R}^T\mathbf{B}^{-1}] \tag{8.17}$$

Using this convention, we write the following equations:

$$\Delta\delta = \mathbf{B}_m^{-1}\,\mathbf{L} \tag{8.18}$$

$$\Delta P_{ij} = (\Delta\delta_i - \Delta\delta_j)/x_{ij} \tag{8.19}$$

$$\Delta P_{ij} = \mathbf{R}_{ij}^T\,\Delta\delta \tag{8.20}$$

The distribution factors are given by:

$$\rho_{ij} = \mathbf{R}_{ij}^T\,\Delta\delta / P_L \tag{8.21}$$

$$\rho_{ij} = R_{ij}^T B_m^{-1} L / P_L \tag{8.22}$$

where P_L is the pre-contingency power flow on the outaged line.

For multiple-line outages, UR^T in (8.14) is reformulated as follows:

$$U R^T = [U_1 \quad U_2 \quad ... \quad U_q][R_1^T \quad R_2^T \quad ... \quad R_q^T]^T \tag{8.23}$$

where q is the number of line outages. For each $i \in 1,2,..q$, U_i is a vector with dimension n (the number of buses), which possess two elements 1 and -1 and the rest of its elements are zero. We recall that $R_i = (1/x_{rs}) U_i$, with r and s representing buses where the line outage has occurred. For the case of q multiple-line outages, vector L is given as

$$L = \sum_{i=1}^{q} L_i \tag{8.24}$$

In this case the system equations are solved by any of the following two methods:

- By finding $U = [U_1 \quad U_2 \quad ... \quad U_q], R = [R_1 \quad R_2 \quad ... \quad R_q]$, and then solving (8.17) and (8.18).
- By successive solutions for a single line outage, matrices U and R become vectors and the identity matrix I in (8.17) will be 1. Then, (8.17) will be written as:

$$B_m^{-1} = B^{-1} - [B^{-1} U (1 + R^T B U)^{-1} R^T B^{-1}] \tag{8.25}$$

We define $B_1, B_2, ..., B_q$ for line outages 1,2...q, given by:

$$B_1 = B + U_1 R_1^T, \quad B_2 = B + U_2 R_2^T, \quad B_3 = B + U_3 R_3^T \tag{8.26}$$

for outage q, we have

DISTRIBUTED SECURITY ANALYSIS

$$B_q = B_{q-1} + U_q R_q^T \qquad (8.27)$$

$$B_q^{-1} = [B_{q-1} + U_q R_q^T]^{-1} =$$

$$B_{(q-1)}^{-1} - [B_{(q-1)}^{-1} U (1 + R^T B_{(q-1)} U)^{-1} R^T B_{(q-1)}^{-1}] \qquad (8.28)$$

We find B_1^{-1} using the inverse in (8.25) and then use this inverse to find B_2^{-1} and the next inverse, and so on, until we reach $B_{(q-1)}^{-1}$ which is used to find B_q^{-1}.

To make computations of the second methodology more straightforward and computationally efficient, we present the equations in a more systematic form as stated below. Suppose that the inverse of the base case Jacobian matrix B^{-1} was calculated initially and given by the following matrix:

$$B^{-1} = Z = \begin{bmatrix} z_{11} & z_{12} & z_{13} & \cdots & z_{1n} \\ z_{12} & z_{22} & z_{23} & \cdots & z_{2n} \\ z_{13} & z_{23} & z_{33} & \cdots & z_{3n} \\ \cdot & \cdot & \cdot & & \cdot \\ \cdot & \cdot & \cdot & & \cdot \\ z_{1n} & z_{2n} & z_{3n} & \cdots & z_{nn} \end{bmatrix} \qquad (8.29)$$

For a single-line outage between buses (i, j), we have

$$B_1^{-1} = (B + U_1 R_1^T)^{-1} \qquad (8.30)$$

The left side of the last equation can be calculated as follows:

- For an outage between buses i and j define the modification factor Δ_{ij} as

$$\Delta_{ij} = (x_{ij} + z_{ii} + z_{jj} - 2z_{ij}) \qquad (8.31)$$

Then, each kl^{th} entry of the matrix B_1^{-1} is given by:

$$(\mathbf{B}_1^{-1})_{kl} = z_{kl} - \frac{1}{\Delta_{ij}}(-z_{ik} + z_{jk})(-z_{il} + z_{jl}) \quad (8.32)$$

- For a double-line outage (i-j,u-v), any kl^{th} element of $\mathbf{B}_2^{-1}$ is given by

$$(\mathbf{B}_2^{-1})_{kl} = (\mathbf{B}_1^{-1})_{kl} - \frac{1}{\Delta_{uv}}\{-(\mathbf{B}_1^{-1})_{uk} + (\mathbf{B}_1^{-1})_{vk}\} \quad (8.33)$$
$$\{-(\mathbf{B}_1^{-1})_{ul} + (\mathbf{B}_1^{-1})_{vl}\}$$

and for the qth line outage between buses (a, b)

$$(\mathbf{B}_q^{-1})_{kl} = (\mathbf{B}_{q-1}^{-1})_{kl} - \frac{1}{\Delta_{ab}}\{-(\mathbf{B}_{q-1}^{-1})_{ak} + (\mathbf{B}_{q-1}^{-1})_{bk}\} \quad (8.34)$$
$$\{-(\mathbf{B}_{q-1}^{-1})_{al} + (\mathbf{B}_{q-1}^{-1})_{bl}\}$$

Superposition is applied here to find the equivalent incremental injection vector $\mathbf{L}_{eq}$, which is equivalent to q line outages if we use the successive solution methodology:

$$\mathbf{L} = \mathbf{L}_{eq} = \mathbf{L}_1 + \mathbf{L}_2 + \ldots + \mathbf{L}_q \quad (8.35)$$

Once we determine the inverse of the Jacobian matrix and changes in the bus angle vector of (8.12), the change in the power flow between buses h and k is given by

$$\Delta P_{hk} = \frac{1}{x_{hk}} \mathbf{U}_{hk}^T \Delta\delta = \mathbf{R}^T \Delta\delta \quad (8.36)$$

where $\mathbf{U}_{hk}$ has only two nonzero elements 1 and -1 in positions h and k, respectively. For a line outage between buses (i, j), the distribution factor of the line between buses (r,s) is

$$\rho_{ij,rs} = (-z_{ij}/z_{rs})(-z_{ir} + z_{jr} + z_{is} - z_{js})/\Delta_{ij} \quad (8.37)$$

For multiple-line outages the distribution factor for q simultaneous outages is given by the last equation in which buses (i, j) correspond to the last outage, and Δ_{ij} is calculated for the last outage considering that modifications are done for the rest of outages as shown

DISTRIBUTED SECURITY ANALYSIS

before. By using successive solutions of system equations, we only need a single set of distribution factors for multiple contingencies.

8.4.2 Extension of Subarea

At this stage we utilize the concept of the bus total distribution factor (TDF) (ρ_i). For any bus i, TDF (ρ_i) is the sum of all distribution factors of lines connected to that bus. Assume for buses i, j in Figure 8.5 that

$$\rho_i = \sum_{x \in R_i} \rho_{i-x} \tag{8.38}$$

$$\rho_j = \sum_{x \in R_j} \rho_{j-x} \tag{8.39}$$

where R_i is the set of buses connected to i and R_j is the set of buses connected to j.

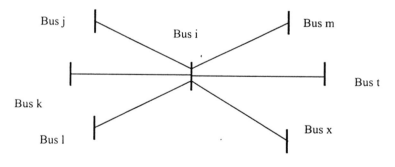

Figure 8.5 Illustration of TDF Concept

    TDF will be used to define the exact direction for extending the initial interest subarea as it represents the most endangered branches due to an outage. In other words, boundaries of the interest area should enclose the most endangered branches. By applying this concept, we introduce the following algorithm for expanding the initial interest subarea for a line outage located between buses i and j:

Step 1. Solve for $\Delta\delta$ in (8.18) using the inverse of the modified Jacobian matrix given by (8.25).

Step 2. Initialize the TDF vector of order n, which has 1 in position i, -1 in position j, and zero elsewhere.

Step 3. Define all buses that are directly connected to buses i and j. This can be done by checking the non-zero reactance of different lines of the original system. The initial subarea of interest (subarea 0) includes these buses along with outaged line buses.

Step 4. For each bus x specified in Step 3, find the distribution factor of the line connecting buses x and i, and update the TDF of x using the updating formula $\rho_x = \rho_x + \rho_{x\text{-}i}$. Then update the TDF of bus i using the updating formula $\rho_i = \rho_i + \rho_{i\text{-}x}$ and apply the same procedure to j and the nodes connected to bus j.

Step 5. Define the bus that has the absolute maximum TDF. Let this bus be called g.

Step 6. Check if the absolute value of the maximum of TDF is less than a specified value ε (note that $\text{TDF} \times P_L$ is the power flow which can be compared with ε). If yes go to the last step (Step 9).

Step 7. Specify branches that are connected directly to the node g from the previous step. Include in the new expansion all nodes that are connected to the ends of branches that stem out of node g.

Step 8. Set $i = g$ and go to step 4.

Step 9. End of the algorithm.

The most endangered area is specified by Step 7. This algorithm is shown by the flowchart in Figure 8.6. To illustrate the concept of TDF and initial contingency area that will be expanded, we use a portion of a network around the outage, as shown in Figure 8.7.

Initially we define the TDF vector ρ to be a zero vector of order n. There is a line outage between buses (i, j), while buses e, f and w are directly connected to i, and buses r and s are directly connected to j. Then from (8.20) we have

$$\Delta P_{ie} = R_{ie}^T \Delta \delta$$
$$\Delta P_{iw} = R_{iw}^T \Delta \delta$$
$$\Delta P_{if} = R_{if}^T \Delta \delta$$
$$\Delta P_{jr} = R_{jr}^T \Delta \delta$$
$$\Delta P_{js} = R_{js}^T \Delta \delta$$
$$\rho_i = \rho_{i\text{-}e} + \rho_{i\text{-}w} + \rho_{i\text{-}f} \tag{8.40}$$

DISTRIBUTED SECURITY ANALYSIS 285

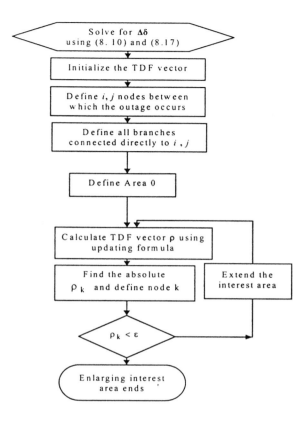

Figure 8.6 Flowchart for Contingency Area Extension

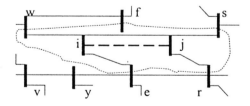

Figure 8.7 Portion of the Network around the Outage

From the last equation of (8.40), we can write ρ_i as

$$\rho_i = (R_{ie}^T + R_{iw}^T + R_{if}^T) \Delta\delta / P_L \tag{8.41}$$

Also, for the nodes connected to i, we can write TDFs as

$$\rho_e = \rho_{e-i} = -\rho_{i-e}$$
$$\rho_w = \rho_{w-i} = -\rho_{i-w}$$
$$\rho_f = \rho_{f-i} = -\rho_{i-f}$$

For node j, we write

$$\rho_j = (R_{jr}^T + R_{js}^T + R_{jf}^T) \Delta\delta / P_L \tag{8.42}$$
$$\rho_r = \rho_{r-j} = -\rho_{j-r}$$
$$\rho_s = \rho_{s-j} = -\rho_{j-s}$$
$$\rho_f = \rho_f + \rho_{f-j} = -\rho_{f-i} - \rho_{j-f}$$

If we define ρ as

$$\rho = \mathbf{R}_{tot}^T \Delta\delta / P_L \tag{8.43}$$

Equations (8.40)-(8.42) result in $\mathbf{R}_{tot}$ in (8.43). The sum of vectors between brackets in (8.41) gives a vector of dimension n, which has the following entries: sum of $(1/x_{ie} + 1/x_{iw} + 1/x_{if})$ in position i, $-1/x_{ie}$ in position e, $-1/x_{iw}$ in position w and $-1/x_{if}$ in position f. In each position k, where $k \in$ nodes connected to the outaged line, we insert the negative of $1/x_{ik}$, and in position i we insert the positive of summation of all other entries. Also, the same idea is applied to (8.42). It is easily seen that $\rho_i = 1$ which comes from the fact that the sum of the power entering a certain node is equal to the sum of the power leaving it, as node i is at one of the terminals of the outaged line, and for the other terminal of the outaged line, $\rho_j = -1$.

8.4.3 Case Study

We consider the IEEE 30-bus system shown in Figure 8.8 as a test system. We study a line outage between buses 2 and 4, as shown by the dashed line in the figure. Suppose that Area 0, in Figure 8.9, is our initial contingency area for this outage.

DISTRIBUTED SECURITY ANALYSIS

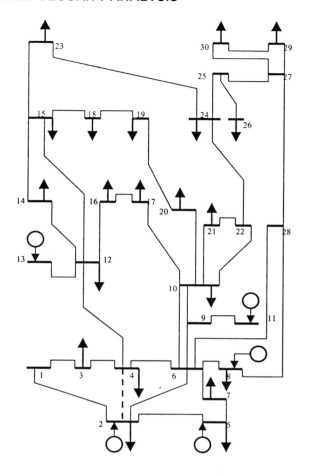

Figure 8.8 IEEE 30-Bus System

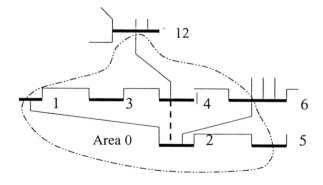

Figure 8.9 Initial Interested Contingency Area around Outaged Line 2-4

For Area 0, the initial total distribution factors are given in Table 8.1. The distribution factors for the initial contingency area are shown in Table 8.2, and the TDF values corresponding to buses in this contingency area are shown in Table 8.3, in which the values are calculated using the following relations:

$$\rho_k = \rho_k + \sum_{k \in R_{ij}} \rho_{k-t} \qquad (8.44)$$

or

$$\rho_k = \rho_k + \sum_{k \in R_{ij}} -(\rho_{t-k}) \qquad (8.45)$$

where R_{ij} is set of buses connected to either i or j, and k and t are any buses in R_{ij} in the area.

Table 8.1 Initial Total Distribution Factors of Area 0

ρ_1	ρ_2	ρ_3	ρ_4	ρ_5	ρ_6	ρ_{12}
0.0	+1.0	0.0	-1.0	0.0	0.0	0.0

Table 8.2 Distribution Factors of Lines in the Initial Interest Subarea

ρ_{2-1}	ρ_{2-5}	ρ_{2-6}	ρ_{3-1}	ρ_{4-3}	ρ_{4-6}	ρ_{4-12}
-0.3673	-0.1942	-0.4385	0.3673	0.3673	0.5911	0.0416

Table 8.3 Total Distribution Factors of Area 0

ρ_1	ρ_2	ρ_3	ρ_4	ρ_5	ρ_6	ρ_{12}
0.0	0.0	0.0	0.0	0.1942	-0.1526	-0.0416

According to Table 8.3, for buses 5, 6, and 12 at the borderline of the interest subarea, $\rho_5 + \rho_6 + \rho_{12} = 0.0$ or $\rho_5 = -(\rho_6 + \rho_{12})$, which means the net power that enters the interest subarea is equal to the net power that exits the area. The important note here is that the net power the exits the interest subarea mainly comes from buses with maximum TDFs. Another note is that ρ_1, ρ_2, ρ_3, and ρ_4 are all zeros because none of these buses are connected to any other buses outside the interest

DISTRIBUTED SECURITY ANALYSIS

subarea, which means that the sum of flows entering any node is equal to the sum of exiting flows.

The idea is as follows: buses at the borderline of the interest subarea are connected to buses that are located inside and outside the interest subarea. The sum of the power entering any of these buses should be equal to sum of power exiting that bus. The entering and exiting power can be calculated by distribution factors as stated above. The bus that has the maximum TDF, gives us an indication on how much of the outaged line power will flow outside the interest subarea through the lines connected to that bus. For our case, $\rho_5 P_L$ is the change in power flow from bus 5 to the interest subarea. This flow should be compensated with the change in entering power to bus 5 from branches outside the interest subarea, as bus 5 has the maximum TDF. This means that we consider the bus as the one that can be checked for the maximum power flowing outside the interest subarea.

The change in power flow in any branch outside the interest subarea can not be greater than $\rho_k P_L$, where ρ_k is the TDF of bus k with the maximum TDF. So we consider the maximum TDF inside the interest subarea as a bounding criterion for line violations. When $\rho_k P_L$ is less than a predefined limit (ε), we stop the process of extending the interest subarea. Accordingly, we choose ε to be the minimum loadability margin on the list of lines. In each enlargement of the interest subarea, we check whether $\rho_k P_L$ is less than ε. If yes, we have specified the most endangered area around the outage. Note that as the interest subarea enlarges, the maximum TDF gets smaller

Now we return to our example. In Table 8.3 the maximum TDF is $\rho_5 = 0.1942$, which means that bus 5 will be the origin of the next extension of the interest subarea. Figure 8.10 illustrates the next enlargement of the area. Distribution factors of new lines that appeared in Area 1 are shown in Table 8.4.

We calculate TDFs of Area 1 as follows with results shown in Table 8.5.

$$\overline{\rho}_5 = \rho_5 + \rho_{5-7}$$
$$\overline{\rho}_6 = \rho_6 + \rho_{6-7}$$
$$\overline{\rho}_7 = \rho_7 + \rho_{7-5} + \rho_{7-6}$$

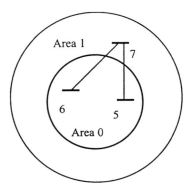

Figure 8.10 First Enlargement of the Interest subarea

Table 8.4 New Distribution Factors in Area 1

ρ_{5-7}	ρ_{6-7}	ρ_{1-2}
-0.1942	0.1942	-0.0416

Table 8.5 Total Distribution Factors of Area 1

ρ_1	ρ_2	ρ_3	ρ_4	ρ_5	ρ_6	ρ_7	ρ_{12}
0.0	0.0	0.0	0.0	0.0	0.0416	0.0	-0.0416

At this point, for a base case flow of 0.2974 p.u. ($P_{base}^{2-4} = 0.2974$) on line 2-4, we note that $0.0416 \times P_{base}^{2-4} = 0.012$ which is too small to affect any line flows. So, we stop enlarging the interest subarea. In case that additional extension is needed, we choose either bus 6 or bus 12 (because they have the same TDFs) for the next extension. Let us choose bus 6 as the point for extending the next subarea. In Figure 8.8, bus 6 is connected to external buses 8, 9, and 28; thus, the newly extended contingency subarea 2 is as shown in Figure 8.11.

TDFs of these newly represented buses are calculated according to the following formulations with the results shown in Table 8.6:

$$\rho_6 = \rho_6 + \rho_{6-8} + \rho_{6-9} + \rho_{6-10} + \rho_{6-28}$$
$$\rho_8 = \rho_8 + \rho_{8-6}$$

DISTRIBUTED SECURITY ANALYSIS

$p_9 = p_9 + p_{9-6}$
$p_{10} = p_{10} + p_{10-6}$
$p_{28} = p_{28} + p_{28-6}$

The same process for the extension of the contingency subarea is applied to bus 12. However, according to Table 8.6, the maximum DTF decreases fast as the interest subarea is expanded, which means that a preset limit (ε) for $\rho_k P_L$ will be quickly satisfied and the extension of the initial interested subarea determined quickly. After specifying the most endangered subarea, we apply the contingency analysis and check for each line violation inside that subarea. This is done by comparing the loadability margin of each branch with line flow changes based on distribution factors.

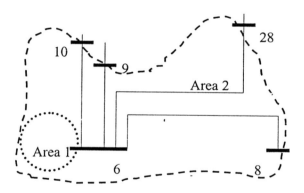

Figure 8.11 Second Enlargement of the Interest Subarea

Table 8.6 Total Distribution Factors of Area 2

Bus	ρ	Bus	ρ
1	0.0000	7	0.0000
2	0.0000	8	0.0017
3	0.0000	9	0.0219
4	0.0000	10	0.0122
5	0.0000	12 -	-0.0416
6	0.0000	28	0.0058

8.4.4 Contingency Ranking Using TDF

The performance index (PI) is used widely for the purpose of contingency ranking. PI is a scalar function of network variables, such as voltage magnitude and real and reactive power flows. It differentiates between critical and non-critical outages, and predicts the relative severity of critical outages. It is perceived that the few contingencies at the top of the contingency list have the most effect on the system security, and so should be analyzed in more detail. In the following analysis, we use the real power performance index given by (8.46):

$$PI_{MW} = \sum_{\ell=1}^{NL} \frac{w_\ell}{2n} \left(\frac{P_\ell}{P_\ell^{lim}}\right)^{2n} \qquad (8.46)$$

The rankings of the first five single line outages using PI are shown in Table 8.7. This table includes only the worst five outages that are ranked at the top of the contingency list according to their PI values. We assumed in the PI equation that the importance factor is 1.0 if a line is overloaded and 0.0 otherwise; n is equal to 1.

We start with the outage that is at the top of the contingency ranking list (i.e., the line connecting buses 2 and 5) and has the largest PI value in Table 8.7. TDFs are given in Table 8.8 and the initial interest subarea is shown in Figure 8.12. Now we extend this initial area (i.e., subarea 0) by starting from bus 1, which has the largest TDF in this subarea. Accordingly this is the subarea 1 as shown in Figure 8.13. We continue the procedure until no further expansion is required. Then we repeat the same procedure for other outages on the list.

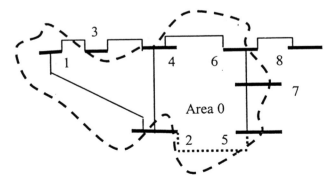

Figure 8.12 Initial Interest Subarea of the Outaged Line 2-5

DISTRIBUTED SECURITY ANALYSIS

Table 8.7 Ranking of the Worst Five Single Line Outages

Bus outaged Line From	Bus outaged Line To	Number of Violated Lines	PI	Rank
2	5	8	23.609	1
9	10	14	7.5308	2
4	12	14	6.8611	3
6	7	2	5.6370	4
28	27	11	5.1086	5

Table 8.8 TDFs of Subarea 0 for the Outage of Line 2-5 with $P_{base}^{2-5} = 0.4347$

Bus i	ρ_i	$\rho_i \times P_{base}^{2-5}$
1	0.211642	-0.0152145
2	0.000000	0.0000000
4	-0.176740	-0.0768280
5	0.000000	0.0000000
6	-0.035008	0.0920007
7	0.000000	0.0000000

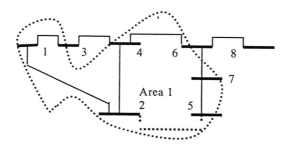

Figure 8.13 First Extension of the Interest Subarea of the Outage of Line 2-5

8.5 DISTRIBUTED CONTINGENCY SELECTION

A static security analysis is an important function of EMS that is performed periodically at the ISO/RTO control center. Static security is the capability of a power system to reach a steady state operating point upon the occurrence of disturbances, and without violating any system constraints including limits on bus voltages and thermal bounds of transmission lines. By assessing the security of a power system, the ISO can enact appropriate corrective measures to reduce the vulnerability of power systems.

For a large-scale power system, security analysis is time-consuming and computationally intensive. This is because a large number of potential contingencies have to be assessed at each pass. In this section we discuss the applications of distributed processing techniques in a static security analysis of power systems.

Contingency selection is the first step in the power system static security analysis. One of the tough issues that the ISO is facing, in this regard, is the massive number of potential contingencies and the consecutive assessment of selected contingencies. Contingency selection is a procedure that attempts to discover the most severe emergency cases from among all the possible contingences.

There are two approaches to the contingency selection. One is the performance index (PI) method and the other is the network solution method. In practice, the PI method has received wider applications than the network solution method.

The PI method checks the impact of contingencies by executing a simple index based on the first or the second iteration of the power flow calculation for each case. When the power system is large, the computation of PI may require an excessive amount of time. PI for power flow can be formulated as

$$PI^{\delta} = \sum_{i}^{N_l} \frac{w_{\delta}^{i}}{2} \left(\frac{p_i}{p_i^s} \right)^2 \qquad (8.47)$$

where

p_i^s = Power flow limit of the ith line

p_i = Power flow on the ith line

w_{δ}^{j} = Non-negative weighting factor for the ith line

N_l = Number of lines of the system.

DISTRIBUTED SECURITY ANALYSIS

PI for voltages is calculated as

$$PI^v = \sum_{i}^{N_b} \frac{w_v^i}{2} \left(\frac{|v_i| - |v_i^s|}{\Delta v_i^s} \right)^2 \tag{8.48}$$

where

v_i^s = Voltage limit at bus i
v_i = Voltage at bus i
Δv_i^s = Maximum voltage variation at bus i
w_v^i = Non-negative weighting factor at bus i
N_b = Number of buses of the system.

A large PI value implies that the current system is operating in a relatively tense state (e.g., during peak hours). Once the security analysis results indicate the possibility of network violations, the ISO/RTO will take appropriate measures such as generation dispatch, LTC, or phase shift transformer tap adjustments, and capacitor or line switching, load shedding, and so on, to curtail the violation. Such actions can be taken immediately on the completion of security analysis (i.e., preventive action) to prevent violations as early as possible, or can be carried out as a corrective strategy as the contingencies occur.

Since the ISO/RTO has a distributed system, we explore here the possibility of using the distributed processing for expediting the computation of PIs. In the following, we discuss two different methods for the distributed computation of PI. The first method utilizes the distributed power flow solution. The second method is based on the distributed external equivalent calculation.

8.5.1 Distributed Computation of PI Based on Distributed Load Flow

As we discussed previously, the ISO and its subarea control centers (SACCs) can perform the online distributed load flow computation. Therefore, once the distributed power flow computation is done, each subarea control center provides a security analysis for its own subarea according to (8.47) and (8.48). Then the ISO can compute the performance indexes for the entire system according to

$$IP^\delta = \sum_{k=1}^{p} IP_k^\delta \qquad (8.49)$$

$$IP^v = \sum_{k=1}^{p} IP_k^v \qquad (8.50)$$

where IP_k^δ and IP_k^v are PIs of the *kth* subarea. The above procedure is demonstrated in Figure 8.14. This distributed approach will speed up the PI index computation for the entire system, which can save time in executing the online security.

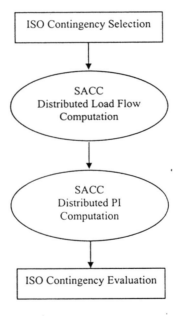

Figure 8.14 Distributed Computation of PI Based on Distributed Load Flow

8.5.2 Distributed Computation of PI Based on Distributed External Equivalent

The ISO/RTO is responsible for the reliable and secure operation of the entire system; as the components of the ISO, SACCs could be required to help accomplish security analysis when necessary. Similarly, as the components of the RTO, ISOs could also be required to help accomplish the system security analysis. In this sense, we assume that

DISTRIBUTED SECURITY ANALYSIS

SACCs of the ISO can provide online external equivalent for other SACCs. Similarly, we can assume that all ISOs under the control of the RTO will provide their neighboring ISOs with online external equivalent for the purpose of security analysis.

Suppose in Figure 8.15 that SACCs are provided with online external equivalent of their neighboring control areas, and SACCs complete the performance index computation of their own control areas separately. Here a, b and c are the online external equivalents of the neighboring control areas of subarea k. Accordingly, Figure 8.16 shows that the power system security analysis can be realized using a distributed processing, which improves the computational efficiency immensely especially for large-scale power systems.

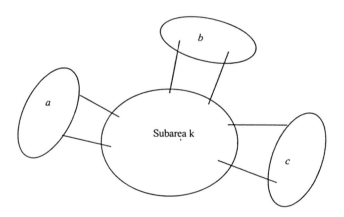

Figure 8.15 Distributed External Equivalent for a Subarea

8.5.3 Comparison of the Two Methods for PI Computation

The efficiency of distributed computation depends heavily on the characteristics of the distributed system, which is composed of computers at the ISO and its SACCs. If the computers have high efficiency, the external equivalent method will be faster because this method requires more computation and less communication. On the other hand, if for some reasons, including the narrow bandwidth of the communication links of the distributed computing system and the overburdened computers at control centers, the distributed power flow computation is not be able to meet the real-time requirements for the online security analysis, then it is better to let SACCs compute their own power flows individually based on the online external equivalents of their neighboring subareas.

The distributed load flow method will be feasible if data communication among computers has a high efficiency. As the computational efficiency of the distributed system is improved more contingencies can be studied in the online security analysis. Both methods discussed above can be applied using online state estimation. The previous methods such as dc power flow, fast decoupled power flow, and the 1P-1Q power flow solution can still be used in the distributed environment of an ISO.

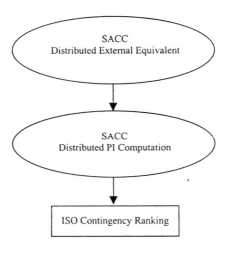

Figure 8.16 Distributed Computation of PI Based on Distributed External Equivalent

8.6 DISTRIBUTED STATIC SECURITY ANALYSIS

Two approaches can be utilized for power system online security analysis based on the distributed computing of an ISO/RTO. The first approach is based on the distributed online load flow computation, and the other is based on the distributed external equivalent computation. These methods are presented next.

8.6.1 Security Analysis Based on Distributed Load Flow

As we mentioned in Chapter 5, the ISO/RTO that can compute the distributed load flow, can use a similar approach for security analysis.

DISTRIBUTED SECURITY ANALYSIS

In this case, each subarea control center will compute the security performance index for its own control area. Then each subarea control center sends its computation results to the ISO/RTO, which can get the security performance index for the entire system by summarizing all the security indexes for individual subareas.

8.6.2 Security Analysis Based on Distributed External Equivalent

Time is the most overriding factor in power system online security analysis. If a severe contingency is recognized in a lengthy time period, there will not be a sufficient time for the ISO/RTO to take effective corrective control measures to prevent the severe consequences. The computation burden can be greatly reduced when the external equivalent is utilized. A security analysis based on the distributed external equivalent can provide a new and effective solution to this problem. Figure 8.17 illustrates this approach with the following modules:

1: SCADA telemetry
2: Observability analysis
3: Distributed state estimation
4: Distributed external equivalent
5: Distributed power flow computation and security analysis

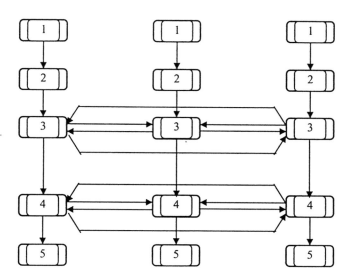

Figure 8.17 Distributed External Equivalent and Security Assessment

An ISO's subarea control center would require its neighboring control areas to provide their online equivalents for distributed security analysis. As Figure 8.18 indicates one of the functions of EMS deployed at the ISO/RTO control center is to calculate the online distributed external equivalent. This proceeds as follows:

- Compute the online external equivalent and transmit the results to neighboring ISOs/RTOs.

- Obtain external equivalents from neighboring ISOs/RTOs and form the admittance matrix of the control subarea for security analysis by taking the external equivalents into consideration.

In this case the distributed computing and control system of an ISO will have a structure illustrated in Figure 8.19.

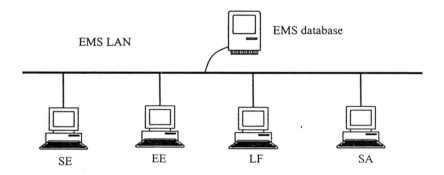

SE: State estimation, EE: External equivalent, LF: Load flow, SA: Security analysis

Figure 8.18   Function Distributed EMS

8.6.3 Enhanced Online Distributed Static Security Analysis and Control

Power system static security analysis is based on contingency selection and evaluation. To reduce the computational burdens so that the ISO has more time for making corrective decisions, the online contingency selection and evaluation should be separated and computed simultaneously with two dedicated computers at the system control center or subarea control center as illustrated in Figure 8.20.

DISTRIBUTED SECURITY ANALYSIS

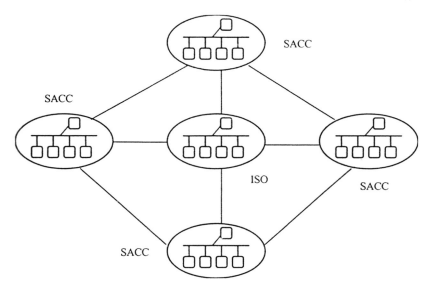

Figure 8.19 Distributed System of the ISO with Distributed EMS

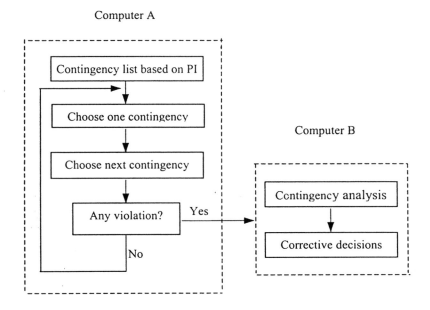

Figure 8.20 Enhanced Online Distributed Security Analysis and Control

In this scheme, computer A is used only to evaluate the contingencies on the list, and computer B is used to make corrective decisions for the contingencies that would cause severe violations. In computer A, the contingency list is created offline (e.g., based on the historical data) and is used by computer A for online evaluation as system operating conditions change. Compared with the traditional scheme that uses only one computer, this distributed scheme can save much time for the ISO to make corrective decisions.

8.7 DISTRIBUTED DYNAMIC SECURITY ANALYSIS

The duration of power system dynamic process could be too short for the ISO to detect a contingency and to make the corresponding remedial measures online. The previous framework for the analysis of power system dynamic security was exclusively based on offline studies in which a group of pre-chosen contingencies in various sequences of events, different initial operating conditions, and different system configurations were analyzed. From these analyses, the ISO can obtain a set of limitations for the secure operation of the power system. These limitations are usually expressed in terms of system operating parameters such as the generation output of certain power plants, and the energy exchange limits of certain tie lines, which can be used as guidance for operating the system under similar conditions.

A power system becomes more vulnerable when it is operating closely to its physical limitations. Hence, the dynamic security analysis becomes a prominent task for those stability-limited power systems. The dynamic security analysis should be based on the real-time operating state of the power system. The online power system dynamic security analysis follows a procedure similar to that of online static security analysis and is comprised of an external network equivalent, contingency selection, contingency evaluation, and corrective control measures. In this section, we explore the implementation of the dynamic security analysis based on the distributed computing and control system of an ISO/RTO.

8.7.1 External Equivalent

Unlike the external equivalent for static security assessment, the external equivalent for dynamic security analysis must include the impact of all generators in the system [Mac85, Mac88]. There have been two approaches to calculating the external equivalent for the system dynamic performance analysis. One is the coherence-based method and the other is modal-based method. In the late 1980s these methods were extended to a more general and systematic approach.

8.7.2 Contingency Selection

At the early stages, contingency selection for dynamic security analysis relied largely on engineering experience and subjective judgment. The most often used screening tool for contingency selection was the direct method for stability analysis, which included a simplified model along with a fast method for evaluating the dynamic behavior of the contingency. With the help of the distributed computing system of an ISO/RTO, the neighboring subareas can provide a detailed online external equivalent; hence, the results of dynamic security analysis can be more accurate. In addition, the distributed processing can improve the computational efficiency of contingency selection and evaluation.

8.7.3 Contingency Evaluation

The evaluation of system dynamics has two objectives. One is to evaluate the system response to some small disturbance such as load fluctuation, and the other is to evaluate the system response to certain disturbance such as short circuit faults. The contingency evaluation methods fall into two categories: one is the numerical simulation method, and the other is the direct method. The numerical simulation method is the traditional approach to system dynamic performance evaluation. For instance, eigenvalue calculation is usually used for small disturbance analysis. The numerical integration method is used for transient stability simulation and long-term dynamic analysis. As these methods are very time-consuming, parallel and distributed processing can improve the computation efficiency. As the direct method determines the system stability without solving the system dynamic equations; hence, parallel and distributed computation can also be applied to improve the solution efficiency.

8.8 DISTRIBUTED COMPUTATION OF SECURITY-CONSTRAINED OPF

As an approach to the real-time security analysis, security-constrained OPF (SCOPF) plays a very important role in power system operation. Traditionally, SCOPF is used in studying outages of transmission lines and generating units. Before distributed computation techniques can be utilized for the real-time power system security analysis and control, the SCOPF computation is performed at the control center by calculating the corrections of the factorization table of the Jacobian matrix of load flow equation, in each pass of the execution of SCOPF. For a large-scale power system, this centralized sequential SCOPF computation is time-consuming because of the large number of transmission lines and generators in the system.

The distributed computing system of the ISO/RTO can provide distributed external network equivalents that obtain fast online SCOPF computation. In practice, the computation of PI is related to the SCOPF computation, since in most cases the PI that is used for the selection of contingency, and so could be further studied based on SCOPF for security analysis.

8.8.1 SCOPF Techniques

As in PI computation there are two approaches to the SCOPF computation based on the distributed system of the ISO/RTO. One is based on the distributed load flow computation. In this method, the effect of line outages on the system can be clearly observed. The computational efficiency can be improved in such a way that associated elements in the factorization table of the Jacobian matrix are the only terms that need to be revised at each stage.

The second approach is based on the distributed external equivalents and can be used for studying less critical transmission lines whose effect of outage is not supposed to be propagated across the entire system but restricted to a limited area. This kind of line SCOPF can be computed by a SACC based on online external equivalents. Each SACC will perform its local computation separately. Because the scale of a subarea is much smaller than the entire system, the correction of the factorization table of the Jacobian matrix for load flow computation goes much easier and faster. Most important, all SACCs can apply the distributed computation to SCOPF, which can greatly improve the computation efficiency. The feasibility of this approach depends on whether the effect of the line outage can be restricted to a limited transmission area. It is perceived that the control area of an ISO is large enough for such an SCOPF computation, and there should be no need to enlarge the distributed load flow computation to cover all ISOs under the RTO for the SCOPF computation.

In SCOPF studies, a generation outage is treated different from a line outage in that the power shortage caused by generation outage must be compensated by all the designated slack buses in the power system according to their distribution factors. This requires that the external response to this generation outage be taken into consideration. When the distributed external equivalent method is used, the following two approaches can be applied:

- The power shortage caused by a designated generation outage is allocated to slack buses according to their power distribution factors in order to keep the entire system in balance.

- The power shortage caused by the designated generation outage is allocated to some of the equivalent generators in the external equivalent. This approach will be more suitable for online application.

In the preceding discussion, we implicitly assumed that the effect of the outage of a line or generator in a subarea on its adjacent subareas can be neglected and hence we could use the external equivalents. If the reactions of some neighboring control areas need to be taken into consideration, then we need to enlarge the realm of the internal system by virtually merging adjacent subareas into one. This way, SACC of the internal subarea will take over the responsibility of SCOPF computation for the virtually merged area.

8.9 SUMMARY

Serial processing of one computer at the system control center does not easily meet the needs of power system real-time monitoring and security analysis. In this chapter we proposed an online distributed external equivalent method, and studied the distributed power system security analysis method based on this equivalency method. The online distributed external equivalent is more accurate than the traditional Ward method, which is based on a pre-chosen base case. The performance index computation can be executed in a distributed manner for a more accurate and faster solution.

Chapter 9

Hierarchical and Distributed Control of Voltage/VAR

9.1 INTRODUCTION

Whether bus voltages are within their operational limits is not only a crucial performance index for power quality but also a major concern for power system security. Unlike the power system frequency, an outstanding feature of power system voltage/VAR optimization and control is its local property. This means that the voltage profile of a local control area in the power system can mainly be regulated through adjustments of reactive power sources in that specific control area. In most cases the impact of remote reactive power sources on the local voltage profile can be neglected in the normal operating conditions. This property of power systems indicates that the voltage/VAR optimization and control of a large-scale system can be better implemented in a distributed manner and based on a hierarchical structure.

The hierarchical and distributed voltage/VAR optimal control is usually based on a system partitioning scheme that considers the minimum power loss as its objective, and constrained by bus voltages that would be within their secure limits. The hierarchy here refers to a certain kind of coordination among upper and lower level control centers for the purpose of voltage/VAR optimization and control.

Recent developments in the restructuring of power systems, and the emerging of distributed and renewable generation, show voltage/VAR optimization and control of a large-scale power system to be an important issue. With so many independent entities appearing in a restructured power system, traditional centralized optimization and control approaches cannot easily meet the various strict requirements for online voltage/VAR optimization and control. The rapid development of computer network technology and the application of high-speed, broadband telecommunication techniques have provided new opportunities for power system voltage/VAR optimization and control. The economical and reliable merits of voltage/VAR optimization and control scheme as proposed in this chapter indicate that distributed and hierarchical generation will be the trend for large-scale power system voltage/VAR optimization and control.

9.2 HIERARCHIES FOR VOLTAGE/VAR CONTROL

Presently, two models are used for the voltage/VAR optimization and control of power systems, which are discussed in the following.

9.2.1 Two-Level Model

The two-level model bypasses secondary control, and the results of global optimization for the whole system take effect directly on primary control [Denz88]. The voltage/VAR control is executed at the control center based on the online OPF as a senior EMS application. The system operator sends out control signals, according to the OPF results, to regulate reactive power resources including generators, tap-changing transformers, FACTS, and so on. This two-level model adapts an open-loop' voltage/VAR control scheme, which can be executed several times a day. Although it can meet the requirements of most electric companies, this two-level model has following disadvantages:

- Performance of OPF may not be reliable as there are many control variables in the computation, and many of them have to be adjusted throughout the power system at the same time. The open-loop control scheme requires an excessive processing time for the global optimization of the entire system. This approach places the burden on the system operator and so requires extensive experience to achieve proper results.

- Because of its complexity, the OPF usually requires long processing time. Even though the closed-loop control can be attained, the control quality is difficult to guarantee.

- Reactive power resources and their control facilities are usually scattered in different parts of the system at different voltage levels, and there are significant differences among their control characteristics.

DISTRIBUTED CONTROL OF VOLTAGE/VAR

These differences lie in continuous and discrete adjustments, limits for maximum adjustments, voltage levels, and so on.

9.2.2 Three-Level Model

The three-level hierarchical voltage/VAR optimization and control scheme [Cors95, Paul87, Thoy86, Ilic95], which is achieved in a centralized processing manner, can be affected in a distributed manner for both vertically integrated and restructured power systems. After partitioning a power system into a number of autonomous control areas, the secondary voltage/VAR optimization and control can be implemented in a distributed manner. However, in a restructured power system, the ISO may have to acquire reactive power from ancillary services providers, and thereby the cost of the voltage/VAR optimization and control becomes a concern. Consideration of cost becomes an additional requirement in voltage/VAR optimization and control strategies.

The three-level model the voltage/VAR optimization and control is consists of the following three stages.

- **Primary control.** Based on the online and on-site sampling signals, automatic control devices take prompt action in responding to instant and large changes at bus voltages. The primary control is designed to be closed-loop and automatic with a quick response, and its control process usually last from several seconds to several minutes. A typical control device for this purpose is the automatic voltage controller (AVC) of a generating unit.

- **Secondary control.** Based on online-metered voltage of the pilot bus, the operator maintains the system voltage profile within a permissible level by readjusting available reactive power resources in the system.

- **Tertiary optimization.** Aimed at satisfying a certain objective, the operator optimizes the system voltage profile and provides reference values for the secondary voltage control.

The three-level voltage/VAR optimization and control scheme is depicted in Figure 9.1.

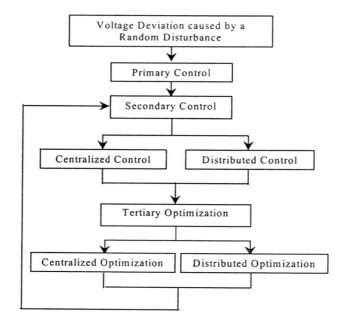

Figure 9.1 Hierarchical and Distributed Voltage/VAR Optimization and Control

There are two approaches for the secondary and tertiary voltage/VAR optimization and control. However, distributed processing, which is based on the system partitioning, is more flexible and reliable. Furthermore, the intrinsic property of voltage/VAR control emphasizes the effectiveness of local reactive power resources because the impact of a reactive power resource is closely related to the electrical distance between reactive power resources and the specified pilot bus. This condition implies that the distributed processing suits the local voltage/VAR optimization and control of power systems.

9.2.3 Hierarchical Control Issues

Hierarchical control has introduced multiple objectives that are attained separately at each control level. Although the three-level control model is relatively complex and difficult for practical implementation, engineering experience tends to adopt the three-level model for control [Glav90, CIGR92]. As local voltage controllers are designed based on local characteristics of the power system, the voltage/VAR control scheme of adjacent control areas is closely tied and the overall control quality depends on the coupling constraints of adjacent control areas.

DISTRIBUTED CONTROL OF VOLTAGE/VAR

The response time of each level in the three-level voltage/VAR control scheme is reasonably set so that voltage violations can be corrected with fast actions provided by the primary and secondary controls. In the decoupling of response time of the three levels, different optimization objectives can be realized, and the control interference among different levels can be avoided. The economy, second to security, can be guaranteed within a relatively longer response by tertiary control. Figure 9.2 depicts the response time of various components in a three-level model.

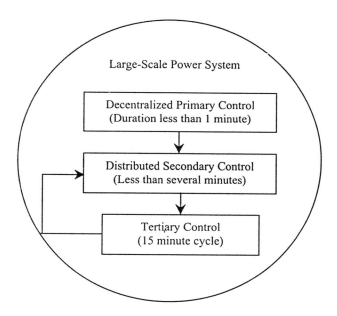

Figure 9.2 Hierarchical and Distributed Voltage/VAR Optimization and Control

When the voltage of a pilot bus in a certain control area deviates from norm because of random disturbances, the primary voltage control will take action. If the voltage of the pilot bus cannot be restored to the preset value, the secondary control will be triggered.

In the secondary voltage control, the system operator first identifies the control area and then makes appropriate reactive power adjustments in that area. In most cases, the voltage of the pilot bus will be restored by redispatching internal reactive power resources. This is because in the majority of contingency cases, it is assumed that each control area has sufficient reactive power resources, so there will not be a need for an external reactive power support. If several pilot buses are deviating from their optimal values, control restorations can be performed independently and based on distributed processing in each control area.

The three-level control scheme utilizes SVC (static voltage controller), which is an area voltage controller [Paul87]. Each SVC is responsible for the coordination of the first-level voltage/VAR control in an independent area. Its control is realized through primary control adjustments to maintain the voltage level of the pilot bus. In this case, the closed-loop control will be realizable with a response time of one to several minutes. The operation experience has shown that the secondary voltage/VAR optimization and control scheme based on SVC can often secure the voltage/VAR operation. Currently, SVC designers focus their analyses on the following issues:

- Methods for selecting the pilot bus
- Methods for partitioning the system
- Control strategies
- Design and implementation of controllers

Generally, tertiary control is implemented through manual operation with a typical response time of about 10 minutes. Closed-loop tertiary voltage control can be used for the coordination of reactive power scheduling and hierarchical voltage control. In such a closed-loop tertiary control model, the optimal voltage of the pilot bus and the optimal reactive power output of generators are first calculated every 30 minutes.

Reactive power scheduling is designed to take the minimum power loss as its objective function, and the optimized bus voltages and reactive power outputs of generators are set for the secondary voltage/VAR control. Alternatively, the tertiary voltage/VAR optimization can be implemented using reactive power optimization within EMS, which takes into account all the prevailing constraints. The cycle for this optimization of voltage/VAR is about 15 minutes, and the optimization results are used for the secondary voltage/VAR control.

9.3 SYSTEM PARTITIONING

The complexity of the voltage/VAR optimization and control process in a large-scale power system can be enhanced by system partitioning. The coupling of adjacent control areas can become very strong, so it is impractical to neglect the interaction of control areas.

DISTRIBUTED CONTROL OF VOLTAGE/VAR

9.3.1 Partitioning Problem Description

A large-scale power system can be partitioned into smaller size control areas for the purpose of voltage/VAR optimization and control, since the reactive power resources possess localized control characteristics. Unlike the partitioning methods for parallel and distributed computation, the system partitioning for voltage/VAR optimization and control stresses the feasibility of the optimization and control process in each area. Hence, each control area will have sufficient reactive power to reduce the interaction between neighboring control areas and affect voltage/VAR control.

In general, the partitioning of a power system for the hierarchical and distributed voltage/VAR optimization and control should comply with the following three requirements:

- The voltage at the pilot bus must reflect the voltage level at the entire control area.

- Each area should have sufficient reactive power resources to maintain the voltage level in its own control area.

- The interaction between neighboring areas should be reduced to a minimum level.

The problem is that the number of areas and the position of the pilot bus in each area are unknown, so it is difficult to partition the system into control areas that can satisfy the three requirements. In some cases it is almost impossible to partition a system without the human operator's involvement. In the following section we propose an algorithm for automatic power system partitioning that combines the experience of the system operator with analytical results.

9.3.2 Electrical Distance

The electrical distance represents a measure of the physical relationship between any two buses in a power system. In the fast P-Q decoupled load flow computation, the voltage solution is calculated based on the following equation:

$$[\Delta \mathbf{Q}] = -[\partial \mathbf{Q}/\partial \mathbf{V}][\Delta \mathbf{V}] \tag{9.1}$$

where $[\partial \mathbf{Q}/\partial \mathbf{V}]$ is the Jacobian matrix of reactive power. By turning all PV buses into P-Q buses (i.e. by including all generators in 9.1), we have

$$[\Delta V] = -[\partial V / \partial Q][\Delta Q] \tag{9.2}$$

where, $[\partial V / \partial Q]$ is the inverse of the Jacobian matrix $[\partial Q / \partial V]$ with similar properties to an impedance matrix.

We assume incremental bus voltage changes are within a small range [Stan91], since the secondary voltage control is relatively slow and bus voltage increments are small, that is,

$$\Delta V_i = \alpha_{ij} \Delta V_j \tag{9.3}$$

where $\alpha_{ij} = [\partial V_i / \partial Q_j]/[\partial V_j / \partial Q_j]$.

Correspondingly, $[\partial V_i / \partial Q_j] = [\partial V_j / \partial Q_i]$ although, in most cases, $[\partial V_i / \partial Q_i] \neq [\partial V_j / \partial Q_j]$. In other words,

$$\alpha_{ij} \neq \alpha_{ji} \tag{9.4}$$

Since the electrical distance between two buses should be unique, we use the following function to map the sensitivities in (9.3) to electrical distances [Lago89]:

$$ed_{i,j} = -\log_{10}(\alpha_{ij} * \alpha_{ji}) \tag{9.5}$$

9.3.3 Algorithm for System Partitioning

Given a parameter r_d for the bus set S of the system, we intend to find out whether the distance between any two buses is within the range of r_d. If it is, we can combine the two buses into one virtual bus, which will take part in the next round of computation of electrical distance. By increasing r_d, all buses will eventually be merged into certain virtual bus sets. One exception is that a bus cannot be merged with any other buses if its electrical distance to all other buses is large. As shown in Figure 9.3, a nuclear power plant that is located at a long distance from the metropolitan area cannot be easily merged with other buses.

DISTRIBUTED CONTROL OF VOLTAGE/VAR

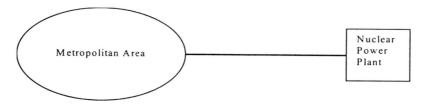

Figure 9.3 A Remote Nuclear Power Plant

The electrical distance between two natural buses is calculated according to (9.5). However, the electrical distance between a natural bus and a virtual bus is calculated according to

$$ed_{i,j} = (ed_{i,j}^{\min} + ed_{i,j}^{\max})/2 \tag{9.6}$$

where $ed_{i,j}^{\min}$ and $ed_{i,j}^{\max}$ are minimum and maximum distances between the single bus i and the virtual bus j. A natural bus will be merged into the virtual bus with which it has the minimum electrical distance. The electrical distance between two virtual buses is calculated according to (9.6), where $ed_{i,j}^{\min}$ and $ed_{i,j}^{\max}$ are the minimum and maximum distances between the two virtual buses. By increasing the distance with the increments of r_d, the number of single buses in S will gradually decrease.

Each virtual bus will form a control area when conditions in Section 9.3.1 are met. We state that the system partitioning can only be based on the electrical distance, as pilot buses are not be chosen before the system is partitioned into areas. To guarantee that there are sufficient reactive power resources in each area, the system should first divide generator buses and then load buses into areas using the method described above. In cases similar to that of Figure 9.3, expert knowledge would be required to put certain special buses in the proper control areas.

9.3.4 Determination of the Number of Control Areas

Theoretically, if we take m (i.e., the number of control areas) to be a function r_d, which is the radius of control area for an n-bus system, the ratio of m to r_d should be inversely proportional the corresponding points on the curve that connects (n, ed_{min}) and (1, ed_{max}), where ed_{min} and ed_{max} are the minimum and the maximum electrical distances of the system. Then, only when r_d becomes sufficiently large can the system be partitioned into feasible control areas. This is because if r_d is relatively small, there will be many control areas in the system, which is not practical. The number of control areas can only be determined after r_d is set. Below, we give a heuristic method to determine r_d.

Let us assume that bus k is the pilot bus of a control area. According to (9.3), α_{kj} is the ratio of voltage changes at buses k and j, and thus represents the ability of the pilot bus k to control the voltage at bus j. If the system operator can approximate α_{kj}, the radius of the control area r_d is determined according to

$$r_d = -2\log_{10}\alpha_{kj} \tag{9.7}$$

Then, using r_d, we can proceed to find the number of control areas in the system.

9.3.5 Choice of Pilot Bus

The voltage/VAR control of an area is enacted by the monitoring and control of the voltage level of the pilot bus in that control area. Hence, the voltage level of the pilot bus must reflect the voltage profile of the entire control area. Furthermore, the voltage of the pilot bus must be controlled properly by regulating reactive power resources near the pilot bus. The usual way to accomplish this is to choose a large generator bus, capacitor bank, or reactor to be the pilot bus of an area. However, the feasibility of the result depends on the location of resources.

Since the electrical distance represents the relations among bus voltage increments including that of the pilot bus, we choose a pilot bus based on the electrical distance. We select a bus k as the pilot bus of an area that satisfies the following criterion:

$$\min_{k} \sum_{i,k \in S_e} ed_{i,k} \tag{9.8}$$

where S_e is the set of buses in the area.

DISTRIBUTED CONTROL OF VOLTAGE/VAR

The reactive power resources available for the voltage control in an area depend on the current operating state that can be represented by the voltage level of the pilot bus. We define the following two indexes to measure the available reactive power resources in a control area:

$$u = \frac{V_p^u - V_p^s}{V_p^s} \times 100\% \tag{9.9}$$

$$l = \frac{V_p^s - V_p^l}{V_p^s} \times 100\% \tag{9.10}$$

where V_p^s represents the voltage level of the pilot bus at the current operating state, V_p^u the voltage of the pilot bus when area generators operate at the feasible upper limits of their reactive power generation, and V_p^l the voltage of the pilot bus when area generators operate at the feasible lower limits of their reactive power generation. The values of u and l reflect the availability of reactive power resources based on the voltage level of the pilot bus. These values should be adjusted within the preset range for every control area

9.3.6 Corrections Based on Expert Knowledge

Due to the complexity of the power system structure, any approach to the automatic system partitioning would have difficulty in satisfying all practical requirements. Hence, it is necessary to modify the system partitioning results based on the system operator's experience on the following issues:

- Determination of the number of control areas
- Determination of indices for reactive power resources
- Evaluation of the feasibility of pilot buses in the system

9.3.7 Algorithm Design

The partitioning algorithm for the hierarchical and distributed control of voltage/VAR in power systems is summarized as follows:

i) Calculate electrical distances between any two buses
ii) Integrate buses based on their electrical distances
iii) Determine the radius of control areas or the number of areas
iv) Choose the pilot bus for each area
v) Check the reactive power generation in each area
vi) Modify the partitioning results based on the operator's experience

The proposed algorithm calculates the number of areas and positions for pilot buses of a specific power system. The problem will become easier if the number of areas or the radius of each area is estimated in advance. In addition, if positions of pilot buses are already determined, we can partition the system more easily into control areas.

9.3.8 Case Studies

To illustrate the procedure of the automatic system partitioning process proposed above, we use a simple test system as shown in Figure 9.4. This system has 10 buses and 17 lines, and the network parameters are listed in Table 9.1. Table 9.2 shows the electrical distances between buses in which the minimum and maximum distances are 0.0504 and 0.4966, respectively. So, when the area radius r_d is less than 0.0504, no bus will be merged and each bus would represent an independent control area.

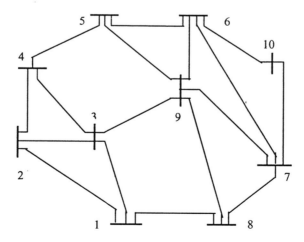

Figure 9.4 10-bus System

DISTRIBUTED CONTROL OF VOLTAGE/VAR

Table 9.1 10-bus System Parameters

i	j	r	x	y_c	K
1	2	0.0119	0.054	0.0071	0
2	3	0.0046	0.0208	0.0027	0
3	4	0.0024	0.0305	0.5810	0
4	5	0.0000	0.0267	0.0000	0.99
1	3	0.0026	0.0322	0.6150	0
1	8	0.0229	0.1888	0.0087	0
3	9	0.0203	0.2682	0.0087	0
5	6	0.0260	0.1196	0.0025	0
6	7	0.0187	0.0616	0.0079	0
7	8	0.0484	0.1600	0.0203	0
8	9	0.0086	0.0340	0.0044	0
6	9	0.0223	0.0731	0.0094	0
7	9	0.0215	0.0707	0.0091	0
5	9	0.0744	0.2444	0.0313	0
2	4	0.0595	0.1950	0.0251	0
6	10	0.0384	0.0760	0.0302	0
7	10	0.0224	0.1500	0.0423	0

Table 9.2 Electrical Distances between Buses

Bus	1	2	3	4	5	6	7	8	9
1	0.0000	0.0814	0.0676	0.1420	0.2111	0.4334	0.4841	0.3405	0.3314
2	0.0814	0.0000	0.0504	0.1196	0.1929	0.4364	0.4966	0.3691	0.3445
3	0.0676	0.0504	0.0000	0.0827	0.1564	0.4021	0.4637	0.3400	0.3111
4	0.1420	0.1196	0.0827	0.0000	0.0835	0.3812	0.4639	0.3619	0.3202
5	0.2111	0.1929	0.1564	0.0835	0.0000	0.3449	0.4486	0.3694	0.3138
6	0.4334	0.4364	0.4021	0.3812	0.3449	0.0000	0.2654	0.3335	0.2435
7	0.4841	0.4966	0.4637	0.4639	0.4486	0.2654	0.0000	0.2890	0.2279
8	0.3405	0.3691	0.3400	0.3619	0.3694	0.3335	0.2890	0.0000	0.1339
9	0.3314	0.3445	0.3111	0.3202	0.3138	0.2435	0.2279	0.1339	0.0000

When r_d is greater than or equal to 0.0504, the system partitioning process is described as follows: When $r_d = 0.0504$, buses 2 and 3 are merged into a virtual bus v_{23} as shown in Figure 9.5. The electrical distances from bus v_{23} to its neighboring buses are

$$ed_{1,v_{23}} = (0.0814+0.0676)/2 = 0.0745$$
$$ed_{4,v_{23}} = (0.0827+0.1196)/2 = 0.10115$$

When $r_d = 0.0835$, buses 4 and 5 are merged into a virtual bus v_{45}, buses 1 and v_{23} are merged into a virtual bus v_{123}, and the system partitioning is as shown in Figure 9.6.

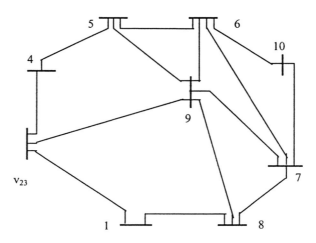

Figure 9.5 System Partitioning Results after Bus 2 and Bus 3 Are Merged

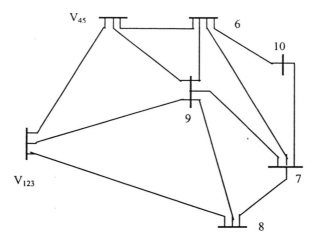

Figure 9.6 System Partitioning Results after Bus 4 and Bus 5 Are Merged

DISTRIBUTED CONTROL OF VOLTAGE/VAR

In Figure 9.6 the electrical distances from buses v_{123}, v_{45} to their neighboring buses are

$$ed_{v_{123},v_{45}} = (0.0827 + 0.2111)/2 = 0.1469$$

$$ed_{8,v_{123}} = (0.3400 + 0.3691)/2 = 0.3546$$

$$ed_{9,v_{123}} = (0.3445 + 0.3111)/2 = 0.3278$$

$$ed_{6,v_{45}} = (0.3812 + 0.3449)/2 = 0.36305$$

$$ed_{9,v_{45}} = (0.3202 + 0.3138)/2 = 0.3170$$

When $r_d = 0.1339$, buses 8 and 9 are merged into a virtual bus v_{89}, the system partitioning result is as shown in Figure 9.7. The electrical distances from buses v_{89} to their neighboring buses are

$$ed_{v_{123},v_{89}} = (0.3111 + 0.3445)/2 = 0.3278$$

$$ed_{v_{45},v_{89}} = (0.3138 + 0.3694)/2 = 0.3416$$

$$ed_{6,v_{89}} = (0.3335 + 0.2435)/2 = 0.2885$$

$$ed_{7,v_{89}} = (0.2890 + 0.2279)/2 = 0.2585$$

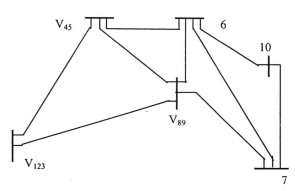

Figure 9.7 System Partitioning Results after Bus 8 and Bus 9 are Merged

When $r_d = 0.1469$, bus v_{123} and bus v_{45} are merged into a new big virtual bus v_{12345}, the system partitioning result is as shown in Figure 9.8. The electrical distances from buses v_{12345} to their neighboring buses are

$$ed_{6,v_{12345}} = (0.3449 + 0.4364)/2 = 0.3907$$
$$ed_{v_{89},v_{12345}} = (0.3111 + 0.3694)/2 = 0.34025$$

When $r_d = 0.2585$, bus 7 and bus v_{89} are merged into a virtual bus v_{789}, the system partitioning result is as shown in Figure 9.9. The electrical distances from buses v_{789} to their neighboring buses are

$$ed_{v_{12345},v_{789}} = (0.3111 + 0.4966)/2 = 0.4039$$
$$ed_{6,v_{789}} = (0.2435 + 0.3335)/2 = 0.2885$$

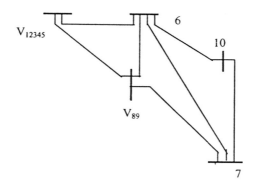

Figure 9.8 System Partition Results after Bus v_{123} and Bus v_{45} Are Merged

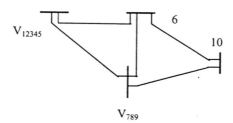

Figure 9.9 System Partitioning Result after Bus7 and Bus v_{89} Are Merged

When $r_d = 0.2885$, bus 6 and bus v_{789} are merged into a new virtual bus v_{6789}. The reference bus 10 is not supposed to participate in the system partition, as it will be automatically merged into the area where its neighboring buses are located. In this example, bus 10 will be merged into bus v_{6789} as shown in Figure 9.10.

At this time the system has been partitioned into two areas that are respectively represented by virtual buses v_{12345}, and v_{78910}, which are two sets of buses. If the number of areas is set at 2, the system partitioning will stop here with the results shown in Figure 9.11. The electrical distance between these two areas is

$$ed_{v_{12345}, v_{678910}} = (0.3111 + 0.4966)/2 = 0.4039$$

Figure 9.10 System Partition Results after Buses 6,10, and Bus V_{789} Are Merged

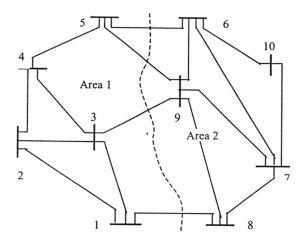

Figure 9.11 System Partitioning Results When the Number of Areas is Set to 2

When r_d is equal or greater than 0.4039, the entire system will be merged into one area. It is obvious that with the increase of r_d, the number of areas will decrease. The relationship between the number of control areas and the control area radius r_d for the test system is illustrated in Figure 9.12.

When the system is partitioned into two control areas as shown in Figure 9.11, the pilot buses of these two control areas are determined according to (9.8). The computation results for the selection of pilots in these two areas are listed in Tables 9.3 and 9.4, respectively. In this case, bus 3 is chosen as the pilot bus for area 1 and bus 9 is chosen as the pilot bus for area 2. These results are consistent with the decision of the system operator, because a bus with the maximum number of lines in an area is usually chosen as the pilot bus of that area.

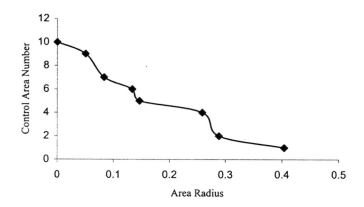

Figure 9.12 Relationship between the Number of Areas and Area Radius

Table 9.3 Pilot Bus Selection for Control Area 1

k	1	2	3	4	5
$\sum_{i,k \in S_e} ed_{i,k}$	0.5020	0.4443	0.3571	0.4278	0.6439

Table 9.4 Pilot Bus Selection for Control Area 2

k	6	7	8	9
$\sum_{i,k \in S_e} ed_{i,k}$	0.8424	0.7823	0.7564	0.6053

The partitioning of a system can be implemented by a computer program in which the increment for r_d can be built into each step. If in the example system above, where the minimum and maximum electrical distances are 0.0504 and 0.4966 respectively, we choose to conduct the system partitioning in 20 steps, then we would set the incremental r_d in each step would be set at (0.4966 - 0.0504)/20 = 0.0223. We notice, however, from the example that the system partitioning will be completed before r_d reaches its maximum value of 0.4966. The results of this case study have shown that with the help of a system operator, the proposed system partitioning algorithm can obtain satisfactory results.

9.4 DECENTRALIZED CLOSED-LOOP PRIMARY CONTROL

As we emphasized earlier, the primary voltage/VAR control is responsible for abrupt and major voltage changes that can seriously impact the power system security. Take the example of the bus voltage limits [0.85, 1.15], which means that the system will risk its security if voltages at certain buses fall out of this range. If such events occur, the primary control will take immediate action to restore bus voltages to their normal range. The bus voltages within the given secure range may not correspond to the optimal value of the objective function. In such cases, the system will operate normally, though away from its optimal solution.

It is important to distinguish between the voltage change that affects the economic solution and the voltage change that affects the system's security, as the system operator has different ways of handling these two situations. The system operator will try to eliminate violations that threaten the system's security as quickly as possible, but if the voltage change will only affect the economic solution, the system operator will leave it to secondary and tertiary voltage/VAR optimization.

The primary control will take effect in response to disturbances, regardless of how the disturbances may affect economical system operation. For example, generator voltage controllers could respond to small disturbances at the corresponding generator bus. For large disturbances such as short-circuit faults or large load fluctuations, which can cause large changes in bus voltages, the primary voltage control devices such as AVC will respond instantly to voltage changes. The response will be based on sampling signals through a closed-loop control scheme as shown in Figure 9.13.

AVC performance is crucial to the primary control for it has to respond quickly to voltage changes. As these disturbances are random, the primary control must be activated in real time with a closed-loop control.

The distributed primary control should restore a voltage change within several seconds up to several minutes. In some cases a few devices might respond to a common voltage changes at the same time. For instance, AVCs of generators connected to a common bus might all respond at the same time as the pilot bus voltage drops suddenly. So an optimal scheme needs to be designed to coordinate AVC responses to that work to restore bus voltage to its normal level.

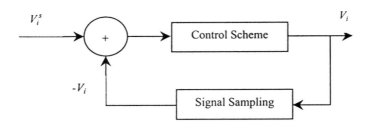

Figure 9.13 Scheme for Primary Control

9.5 DISTRIBUTED SECONDARY VOLTAGE/VAR CONTROL

9.5.1 Problem Description

As we saw earlier, the secondary voltage control tries to keep a pilot bus voltage at its optimal level. For the purpose of voltage/VAR optimization and control, pilot bus voltages are monitored in real time. If the deviation of a pilot bus voltage exceeds the preset threshold, the secondary voltage/VAR control will take remedial actions immediately. Ideally secondary control can be performed very quickly. However, backed up by the primary control, the secondary control does not have to be adjusted as quickly as the primary control. It is imperative to make an analysis of the secondary voltage control and identify reactive power resources that can be utilized for the control, and those that are most effective for the control. Sometimes, in order to maintain a smooth control process, we need to apply a piecewise control process based on several steps.

Certainly secondary voltage control can be implemented in a centralized manner. However, it may be unnecessary to perform an optimization for the entire system as voltages at a few pilot buses deviate from their preset optimal values. In fact we may prefer to realize the secondary voltage/VAR optimization and control in a distributed manner because of the local property of voltage/VAR control. Below we introduce a distributed optimization method for secondary voltage control. The method is based on system partitioning proposed in Section 9.3 which provides an online closed-loop secondary voltage/VAR optimization and control model.

9.5.2 Distributed Control Model

Regardless of the coupling between active and reactive power and P-V buses being incorporated into load flow equations, we can derive the following formulation for Q-V iterations:

$$-[B \Delta V] = [\frac{\Delta Q}{V}] \tag{9.11}$$

where B is the Jacobian matrix and V is the vector of the system voltage. If the system is partitioned into na control areas, then we have

$$\begin{bmatrix} B_{11} & B_{12} & \cdots & B_{1na} \\ B_{21} & B_{22} & \cdots & B_{2na} \\ \vdots & \vdots & \vdots & \vdots \\ B_{na1} & B_{na2} & \cdots & B_{nana} \end{bmatrix} \begin{bmatrix} \Delta V_1 \\ \Delta V_2 \\ \vdots \\ \Delta V_{na} \end{bmatrix} = \begin{bmatrix} \Delta Q_1 / V_1 \\ \Delta Q_2 / V_2 \\ \vdots \\ \Delta Q_{na} / V_{na} \end{bmatrix} \tag{9.12}$$

The system can be further partitioned based on the method described in Section 9.3 to create sufficient reactive power sources in each control area. The interactions among neighboring control areas, however, are assumed to be negligible when the secondary voltage control is applied to each control area. Figures 9.14 and 9.15 represent the reactive power flows on the tie lines as equivalent reactive power loads or injections that remain fixed during the control process.

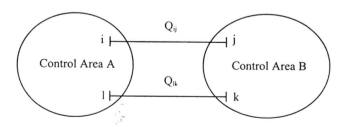

Figure 9.14 Neighboring Control Areas

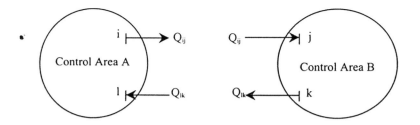

Figure 9.15 Equivalent Tie Line Reactive Power

Based on the assumptions presented above, we have

$$-\begin{bmatrix} \mathbf{B}_{11} & & & \\ & \mathbf{B}_{22} & & \\ & & \ddots & \\ & & & \mathbf{B}_{na,na} \end{bmatrix} \begin{bmatrix} \Delta \mathbf{V}_1 \\ \Delta \mathbf{V}_2 \\ \vdots \\ \Delta \mathbf{V}_{na} \end{bmatrix} = \begin{bmatrix} \Delta \mathbf{Q}_1 / \mathbf{V}_1 \\ \Delta \mathbf{Q}_2 / \mathbf{V}_2 \\ \vdots \\ \Delta \mathbf{Q}_{na} / \mathbf{V}_{na} \end{bmatrix} \quad (9.13)$$

We can rewrite (9.13) as

$$-\mathbf{B}_{ii} \Delta \mathbf{V}_i = [\frac{\Delta \mathbf{Q}_i}{\mathbf{V}_i}] \quad i = 1, 2, \cdots, na \quad (9.14)$$

Consider the *i*th control area in which load and generation buses are denoted with subscripts L and G, respectively, and the pilot bus is denoted with subscript p, which can be a generation or a load bus. Then we have

$$-\begin{bmatrix} \mathbf{B}_{LL} & \mathbf{B}_{LP} & \mathbf{B}_{LG} \\ \mathbf{B}_{PL} & \mathbf{B}_{PP} & \mathbf{B}_{PG} \\ \mathbf{B}_{GL} & \mathbf{B}_{GP} & \mathbf{B}_{GG} \end{bmatrix} \begin{bmatrix} \Delta \mathbf{V}_L \\ \Delta \mathbf{V}_P \\ \Delta \mathbf{V}_G \end{bmatrix} = \begin{bmatrix} \Delta \mathbf{Q}_L / \mathbf{V}_L \\ \Delta \mathbf{Q}_P / \mathbf{V}_P \\ \Delta \mathbf{Q}_G / \mathbf{V}_G \end{bmatrix} \quad (9.15)$$

If we assume that reactive power demands do not change during the secondary voltage control process, $\Delta \mathbf{Q}_L = 0$, then we can eliminate the load buses from (9.15) and retain generator buses and the pilot bus p. Accordingly we have

$$-\begin{bmatrix} \tilde{\mathbf{B}}_{pp} & \tilde{\mathbf{B}}_{PG} \\ \tilde{\mathbf{B}}_{GP} & \tilde{\mathbf{B}}_{GG} \end{bmatrix} \begin{bmatrix} \Delta \mathbf{V}_P \\ \Delta \mathbf{V}_G \end{bmatrix} = \begin{bmatrix} \Delta \mathbf{Q}_P / \mathbf{V}_P \\ \Delta \mathbf{Q}_G / \mathbf{V}_G \end{bmatrix} \quad (9.16)$$

Let us further assume, for the purpose of voltage control, that there is no reactive power available at the pilot bus, which means $\Delta \mathbf{Q}_P = 0$. Then we have

$$\Delta \mathbf{V}_p = -\left[\tilde{\mathbf{B}}_{PP}^{-1} \tilde{\mathbf{B}}_{PG}\right] \Delta \mathbf{V}_G \tag{9.17}$$

where (9.17) is the voltage control function for the pilot bus of the ith control area. If we choose the minimum adjustment of control variables as the optimization criterion, the optimal adjustments of generator terminal voltages can be calculated according to the following formulation

$$\min \quad J = \frac{1}{2} \Delta \mathbf{V}_G \Delta \mathbf{V}_G \tag{9.18}$$

$$\text{subject to} \quad \Delta \mathbf{V}_p = \mathbf{A} \mathbf{V}_G \tag{9.19}$$

where $\mathbf{A} = -\tilde{\mathbf{B}}_{pp}^{-1} \tilde{\mathbf{B}}_{PG}$. From (9.19), we obtain

$$\Delta \mathbf{V}_G = \mathbf{A}^T (\mathbf{A}\mathbf{A}^T)^{-1} \Delta \mathbf{V}_p \tag{9.20}$$

Accordingly (9.19) can be written more explicitly as

$$\Delta \mathbf{V}_G = -\tilde{\mathbf{B}}_{GP} (\tilde{\mathbf{B}}_{PG} \tilde{\mathbf{B}}_{GP})^{-1} \tilde{\mathbf{B}}_{PP} \Delta \mathbf{V}_p \tag{9.21}$$

Hence, each control area will perform its own secondary voltage/VAR control in a distributed manner. If we choose $\Delta \mathbf{Q}_G$ as a control variable, we derive the following formulation:

$$\Delta \mathbf{Q}_G / \mathbf{V}_G = \left[\tilde{\mathbf{B}}_{GP} - \tilde{\mathbf{B}}_{GG} \tilde{\mathbf{B}}_{GP} (\tilde{\mathbf{B}}_{PG} \tilde{\mathbf{B}}_{GP})^{-1} \tilde{\mathbf{B}}_{PP}\right] \Delta \mathbf{V}_P \tag{9.22}$$

9.5.3 Closed-Loop Secondary Voltage/VAR Control

The real-time closed-loop secondary voltage/VAR control process is shown in Figure 9.16, where an SVC is located at each area control center. The typical response period of the secondary voltage/VAR control is about three minutes.

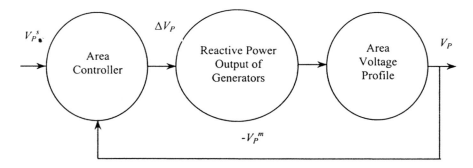

Figure 9.16 Closed-Loop Secondary Voltage/VAR Control

In Figure 9.16, the function of the area controller is to compare the real-time metered voltage V_P^m with the preset optimal V_P^s, which is the tertiary voltage/VAR optimization result, and calculate ΔV_P which will be used to calculate ΔV_G or ΔQ_G according to (9.21) and (9.22). ΔV_G or ΔQ_G will then be immediately sent to corresponding generators for reactive power adjustments. In case where ΔV_P is large, the secondary voltage control will be completed in several stages in order to get a smooth control process so that the system can move from one steady state to another. For instance, if we consider M stages, then in the kth stage we have

$$\Delta V_P^{(k)} = \Delta V_p / M \tag{9.23}$$

$$\Delta \mathbf{V}_G^{(k)} = \mathbf{A}^T (\mathbf{A}\mathbf{A}^T)^{-1} \Delta \mathbf{V}_P^{(k)} \tag{9.24}$$

All control areas should synchronize their staged control, and all adjustments should be completed within the given time limit in order to obtain a better performance for the stage control.

9.5.4 Case Studies

Recall the 10-bus system of Section 9.3.8, which was divided into two control areas as illustrated in Figure 9.11. When the interaction of these two control areas is disregarded, the voltage control in these two control areas will be implemented separately in a distributed manner. In the following, we perform the control function in area 1 in order to show how secondary voltage control works.

DISTRIBUTED CONTROL OF VOLTAGE/VAR

Suppose that the tertiary optimal value of the voltage at the pilot bus 3 in the control area is 0.995. However, the real-time load flow results show that the voltage at bus 3 is 0.920, and so it will need to be adjusted. We assume generators G_1 and G_5 (the subscript is the bus number) will be rescheduled for this voltage control and buses 2 and 4 are the load buses. The distributed secondary voltage control in control area 1 will be implemented as follows.

i) Calculate the deviation of pilot bus voltage as

$$\Delta V_3 = V_3^{op} - V_3^{me} = 0.995 - 0.920 = 0.075$$

ii) Calculate the control variable V_G

The calculation result based on (9.16) for control area 1 is

$$-\begin{bmatrix} -8.093 & 16.2384 & -8.1454 \\ 7.5691 & -16.0145 & 8.1451 \\ 0.0000 & 38.2136 & -38.2136 \end{bmatrix} \begin{bmatrix} \Delta V_P \\ \Delta V_G \end{bmatrix} = \begin{bmatrix} \Delta Q_P / V_P \\ \Delta Q_G / V_G \end{bmatrix}$$

Accordingly, we find that $\mathbf{A} = [2.0064 \; 1.0064]$. Here (9.19) shows that

$$\Delta V_G = \begin{bmatrix} \Delta V_1 \\ \Delta V_5 \end{bmatrix} = \mathbf{A}^T (\mathbf{A}\mathbf{A}^T)^{-1} \Delta V_P$$

$$= \begin{bmatrix} 0.3982 \\ 0.1997 \end{bmatrix} * 0.075$$

$$= \begin{bmatrix} 0.030 \\ 0.015 \end{bmatrix}$$

iii) Calculate reactive power adjustments of generators according to (9.22)

$$\Delta Q_G / V_G = \begin{bmatrix} \Delta Q_1 / V_1 \\ \Delta Q_5 / V_5 \end{bmatrix} = \begin{bmatrix} 12.3194 \\ 7.5854 \end{bmatrix} * 0.075 = \begin{bmatrix} 0.9240 \\ 0.5689 \end{bmatrix}$$

Suppose before the realization of voltage control, we have

$$\begin{bmatrix} V_1 \\ V_5 \end{bmatrix} = \begin{bmatrix} 0.900 \\ 0.9202 \end{bmatrix}$$

Then

$$\begin{bmatrix} \Delta Q_1 \\ \Delta Q_5 \end{bmatrix} = \begin{bmatrix} 0.8316 \\ 0.5235 \end{bmatrix}$$

iv) Let G_1 and G_5 adjust their terminal voltages or reactive power outputs within a given time limit such that the voltage at the pilot bus 3 in control area can be raised to the optimal value.

9.6 DISTRIBUTED SECONDARY VOLTAGE/VAR CONTROL BASED ON REACTIVE POWER BIDS

9.6.1 Introduction

Unlike the traditional vertically integrated power system where all reactive power resources can be used by the system operator for voltage/VAR optimization and control, the voltage/VAR control in a restructured power system is mainly managed by ancillary services. The ISO obtains the required reactive power by participating in an auction to buy the cheapest reactive power supply for voltage/VAR control. The reactive power supply in certain control areas may even be constrained by insufficient voluntary offerings during certain periods. In such instances, the ISO mandates the participants to adjust their reactive power generation for voltage/VAR control.

It is assumed that there is a sufficient level of reactive power generation available for voltage/VAR control in individual control areas. Hence, in most cases, the voltage of the pilot bus at each control area could be restored to its optimal value by just redispatching the local reactive power generation in that area. If voltages at several pilot buses deviate simultaneously from their preset values, they can be controlled independently at the respective control areas through a distributed optimization and control process.

9.6.2 Optimization Based on Reactive Power Bidding

In a restructured power system, the objective of the secondary voltage/VAR control would be to regulate bus voltages to a certain level at minimum cost. The area control center would first use cost-free facilities, including FACTS devices, tap transformers and phase shifters. If the cost-free facilities cannot regulate pilot bus voltages to their optimal levels, the area control center will then use reactive power adjustments to regulate the voltages. The area control center will choose reactive power adjustments with lowest bids in the ancillary services market. The optimization model is as follows:

$$\text{Min} \sum_{i=1}^{N_s} C_i Q_i \tag{9.25}$$

$$\text{Subject to} \quad \sum Q_i \geq Q^{req} \tag{9.26}$$

$$Q_i^{min} \leq Q_i \leq Q_i^{max} \tag{9.27}$$

where C_i is the bidding price of the ith reactive power supplier.

9.6.3 Optimization Using Sensitivities

The effect of reactive power adjustments on the pilot bus could vary with the location of the bus, which is due to the local property of voltage/VAR. The adjustments near the pilot bus are more effective than those of remote facilities. It would be more cost-effective to utilize these resources though the reactive power generation near the pilot bus may be more expensive. Therefore, we consider the sensitivity of reactive power adjustments for the secondary voltage/VAR control.

When the area control center considers the sensitivity of reactive power adjustments, the optimization function for the secondary voltage/VAR optimization and control becomes

$$\text{Min} \sum_{i=1}^{N_s} C_i W_i Q_i \tag{9.28}$$

$$\text{Subject to} \quad \sum Q_i \geq Q^{req} \tag{9.29}$$

$$Q_i^{min} \leq Q_i \leq Q_i^{max} \tag{9.30}$$

where W_i is the weighting factor for adjusting the ith reactive power resource. W_i is represented along with sensitivities for adjusting the ith reactive power source to modify the pilot bus voltage. The calculation of the weighting factor is described in Appendix III.

9.6.4 Optimization and Control Considering Area Interactions

If the voltage at a certain pilot bus is changed abruptly, additional reactive power support will be acquired from neighboring control areas. Thus, the area control center will form a large virtual control area which includes its neighboring control areas, and the optimization and control will be performed based on this newly formed virtual control area. The cheapest and the most effective reactive power adjustments will be utilized first. The mathematical model is almost the same as (9.28-9.30) except that the constraints of neighboring areas and tie lines must also be taken into consideration.

Theoretically, the newly formed virtual control area will have a weak interaction with its neighboring control areas due to the locality of voltage/VAR. Therefore, distributed voltage/VAR optimization and control will be implemented in the new virtual control area. But, if the newly formed virtual control area has significant interaction with its neighboring control areas, further merging will be required. This is due to the fact that the adjustment of reactive power generation in neighboring control areas will not only affect the pilot bus voltage of the internal control area but also the voltage value of the pilot buses of the neighboring control areas. However, when the voltage/VAR optimization and control is performed in the newly formed virtual control area, all pilot bus voltages will be kept at their individual permissible ranges. The scheme for distributed secondary voltage/VAR optimization and control with area interaction is depicted in Figure 9.17.

9.7 CENTRALIZED TERTIARY VOLTAGE/VAR OPTIMIZATION

9.7.1 Optimization Approaches

In the daily operation of a large-scale power system, the power loads draw a significant amount of reactive power from the system. Hence, a power system must provide a reliable supply of reactive power to keep bus voltages within their permissible limits. In this regard the voltage/VAR must be optimized in such a way that it will ensure the reliable and economical operation of the system. The tertiary voltage/VAR optimization

DISTRIBUTED CONTROL OF VOLTAGE/VAR 335

generally considers the power loss minimization of the entire network as its optimization objective.

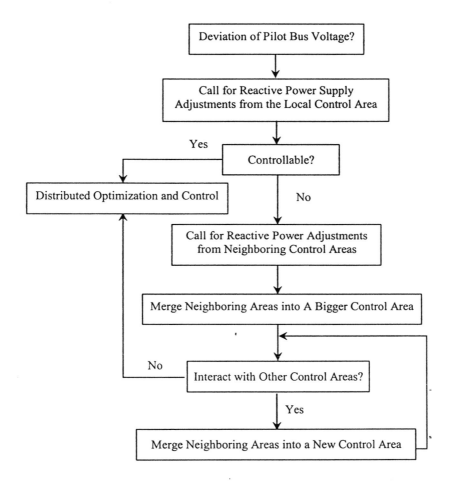

Figure 9.17 Distributed Secondary Voltage/VAR Optimization and Control with Area Interaction

As discussed earlier, the function of tertiary voltage/VAR control in the three-level hierarchical voltage/VAR control scheme is to optimize the voltage profile of the entire system with respect to some given optimization criteria. The optimization results provide a control reference for the secondary voltage/VAR control.

In a traditional vertically integrated power system, the tertiary optimization is a function of CCC. However, in a restructured power system, the optimization function is the responsibility of the ISO. The ISO does not possess any reactive power resources and will acquire reactive power from the ancillary services market. So the ISO's initial intention will be to use static resources such as transformer taps, reactors, capacitor banks, and phase shifters to optimize the system voltage level. When there is still a possibility of further improving the voltage profile, the ISO will purchase reactive power from reactive power providers. At the CCC or ISO, the tertiary voltage/VAR optimization will be performed periodically, and, if a substantial load or generation change is detected, the tertiary voltage/VAR optimization will be performed.

9.7.2 Linear Optimization Approaches

Linear programming (LP) has been recognized as an effective and reliable optimization approach that directly enforces limits on variables and constraints that are linear functions of the variables. The enforcement of such limits presents difficulties in classical nonlinear gradient or Newton techniques. The prominent drawbacks of LP lie in the requirement that all relations must be linear or formulated by linear functions. In the past, all linear techniques formulated dependent variables in terms of independent variables and eliminated dependent variables from the problem by means of a sensitivity analysis based on the power flow equation. This could reduce the number of variables of the optimization problem, but it was an inefficient and time-consuming process to form the sensitivity matrix by inversing the Jacobian matrix of large-scale power systems.

To avoid the computation of the inversion of the Jacobian matrix and to apply decomposition techniques, we incorporate all system variables for reactive power in the optimization function. Although this formulation could increase the number of variables in the optimization problem, it decreases the number of constraints, which is realized by adding load and tap-changing transformer effects to the Jacobian matrix. So there will be no separate constraints imposed on the tap settings of transformers that need to be updated implicitly. Equality constraints are applied to load and junction buses by forcing a reactive power equal to the initial value of the reactive injection into the bus. Moreover dependent and independent variables are transformed into non-negative variables suitable for LP without increasing the number of variables of the optimization problem.

The power loss of the entire system is a nonlinear function of the system variables. The objective function and constraints of the tertiary voltage/VAR optimization are linearized in the neighborhood of the current

DISTRIBUTED CONTROL OF VOLTAGE/VAR

operating state of the system. On the other hand, the variation of Jacobian coefficients is restricted to a small range by introducing a limited step size for voltage and transformer tap setting changes. The solution of the problem is based on the Dantzig-Wolfe decomposition technique in which the primal and dual solutions of the revised simplex method are utilized. The applications of LP and sparsity have reduced the required CPU time and memory space.

P-Q decoupling is usually used to simplify the optimization process. The real power is assumed to be distributed optimally before the voltage/VAR optimization is performed, and bus voltage angles are assumed to remain constant during the voltage/VAR optimization process. This assumption is required so that the coupling between phase angles and reactive variables can be assumed to be small.

9.7.3 Mathematical Models

The objective of the tertiary voltage/VAR optimization is to minimize the real power loss P_L, which can be formulated as

$$P_L = \sum_{k=1}^{N_L} g_k [V_i^2 + V_j^2 - 2V_i V_j \cos(\delta_i - \delta_j)] \tag{9.31}$$

where g_k is the conductance of line k connected between bus i and j. In order to use LP, the objective function is linearized as follows:

$$\frac{\partial P_L}{\partial V_i} = 2g_k [V_i - V_j \cos(\delta_i - \delta_j)] \tag{9.32}$$

$$\frac{\partial P_L}{\partial V_j} = 2g_k [V_j - V_i \cos(\delta_i - \delta_j)] \tag{9.33}$$

The preceding formulation is used to calculate the partial derivative of P_L with respect to voltages at buses i and j for every transmission line. Then partial derivatives associated with a certain bus are summed up to form the loss sensitivities with respect to all voltages in the system. Loss sensitivities are re-evaluated at the current operating state of the system at the end of every iteration of the optimization process. Hence, the real power loss increment ΔP_L is related to bus voltages changes as follows:

$$\Delta P_L = \left[\frac{\partial P_L}{\partial V_1} \frac{\partial P_L}{\partial V_2} \cdots \frac{\partial P_L}{\partial V_{N_B}} \right] \begin{bmatrix} \Delta V_1 \\ \Delta V_2 \\ \vdots \\ \Delta V_{N_B} \end{bmatrix} \quad (9.34)$$

or in matrix form:

$$\Delta P_L = M \cdot \Delta V \quad (9.35)$$

The control variables of the system are the reactive power generation at various buses and the tap positions of the control transformers. As noted in the formulation of the objective function, the voltage magnitude changes are the optimization problem variables. The tap positions of transformers are implicitly incorporated in the constraints and upper and lower limits imposed on the bus voltages of the system.

Generally, variations in reactive power injections are represented as a function of voltage fluctuations at the same bus. So sensitivities with respect to the control variables are calculated. These elements are modified to take into account the effects of load variations and tap-changing transformers on the modified Jacobian elements. Once the elements of the modified Jacobian matrix are determined, the variations of reactive power injections are written as follows:

$$\Delta Q = J'' \Delta V \quad (9.36)$$

We categorize the system buses as one of two types:

- Buses with reactive power generation and/or connection to tap-changing transformer terminals.

- Load and junction buses that are neither connected to reactive power generation nor to tap transformer terminals.

Let us number the buses of the first type from 1 to nc, and the buses of the second type from $nc + 1$ to nb. Then, using (9.36), we can write the following two separate equations:

$$\Delta Q_1 = J_1'' \Delta V \quad (9.37)$$

DISTRIBUTED CONTROL OF VOLTAGE/VAR

$$\Delta Q_2 = J_2'' \Delta V \tag{9.38}$$

where the dimension of the system Jacobian matrix J'' is $(nb)(nb)$, the dimension of J_1'' is $(nc)(nb)$, and that of J_2'' is $(nb-nc)(nb)$. By modeling the effect of loads and tap-changing transformers as reactive power injections that are formulated as functions of bus voltage magnitudes in J'', we can formulate the tertiary voltage/VAR operation problem as follows:

$$\text{Min } \Delta P_L = M \cdot \Delta V \tag{9.39}$$

$$\text{Subject to } \Delta Q^{min} \leq \Delta Q = J_1'' \cdot \Delta V \leq \Delta Q^{max} \tag{9.40}$$

$$J_2'' \cdot \Delta V = 0 \tag{9.41}$$

$$\Delta V^{min} \leq \Delta V \leq \Delta V^{max} \tag{9.42}$$

The control variables for this optimization are the outputs of reactive power sources in the system such as the reactive power generation of generators, the reactors, the capacitor banks, and transformer taps and phase shifters. The inequality constraints (9.40) are for reactive power source buses and for buses connected to transformer terminals. The equality constraints (9.42) are for load and junction buses.

The linearized objective function above is used for real power loss minimization by controlling transformer tap settings, generators, and switchable capacitors reactive power generation, which are implicitly considered in the proposed formulation. The detailed derivation of objective function is described in [Dee91]. Here we need to point out that the following assumptions are taken into consideration:

- Bus voltage angles are assumed to remain constant during the optimization process.

- Real power injection at each bus is constant except at the slack bus.

9.7.4 Dantzig-Wolfe Decomposition Method

Suppose that the entire system is partitioned into na control areas; each control area has one control center which is responsible for the voltage/VAR optimization and control of that area. Then the solution procedure is based on the decomposition of system equations according to the available number of areas in a power system. Buses that link different areas will formulate the linking constraints. We consider the schematic diagram of Figure 9.18 in which several areas are linked together to supply the system load. On the formulation of the dispatch problem, areas 1 through *na* will form the subproblems in the decomposition procedure.

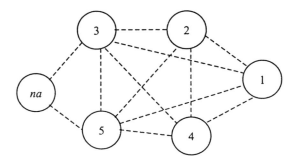

Figure 9.18 Schematic Presentation of Multi-area System

The voltage/VAR optimization problem is solved using the Dantzig-Wolfe decomposition method, which can reduce, in a very efficient way, the dimension of the problem by separating it into several subproblems; these subproblems are solved individually employing the simplex method. The master problem, which consists of the coupling constraints, will be solved by the revised simplex method.

In the Dantzig-Wolfe decomposition method, system equations have to be decomposed according to individual control areas in the system. The constraints that represent links between bus voltages within a specific area will form a subproblem [Deeb91]. Any bus within an area which has a link to buses outside that area forms linking constraints. Say the system has nb buses in total and is partitioned into NA control areas. For area number i, let b represent the number of internal buses and let c represent the number of boundary buses that link neighboring control areas. In addition, for all control areas, internal buses will be numbered first and then the boundary buses will be numbers in the Jacobian matrix. Through this arrangement, the voltage/VAR optimization problem can be formulated in the following block-angular form:

DISTRIBUTED CONTROL OF VOLTAGE/VAR

$$\text{Min} \quad \frac{\partial P_L}{\partial V_1} \Delta V_1 + \frac{\partial P_L}{\partial V_2} \Delta V_2 + \cdots + \frac{\partial P_L}{\partial V_{na}} \Delta V_{na} \quad (9.43)$$

Subject to

$$\mathbf{A}_1 \Delta \mathbf{V}_1 + \mathbf{A}_2 \Delta \mathbf{V}_2 + \cdots + \mathbf{A}_{na} \Delta \mathbf{V}_{na} = b_0 \quad (9.44)$$

$$\left.\begin{array}{r} \mathbf{D}_1 \Delta \mathbf{V}_1 = b_1 \\ \mathbf{D}_2 \Delta \mathbf{V}_2 = b_2 \\ \mathbf{D}_{na} \Delta \mathbf{V}_{na} = b_{na} \end{array}\right\} \quad (9.45)$$

$$\Delta V_i^{\min} \le \Delta V_i \le \Delta V_i^{\max}, \quad i = 1, \cdots, na \quad (9.46)$$

where $\mathbf{A}_i$ is the coefficient matrix that corresponds to linking constraints of area i and $\mathbf{D}_i$ is a coefficient matrix that corresponds to the subproblem i of area i. The equations above can be rewritten in a more general form

$$\left.\begin{array}{l} \min \sum_{i}^{na} M_i \Delta \mathbf{V}_i \\[1ex] \text{subject to} \quad \sum_{i=1}^{na} \mathbf{A}_i \Delta \mathbf{V}_i \\[1ex] \qquad \qquad \mathbf{D}_i \Delta \mathbf{V}_i = b_i \\[1ex] \qquad \qquad \Delta V_i^{\min} \le \Delta V_i \le \Delta V_i^{mzx} \\[1ex] \qquad \qquad i = 1, \cdots, na \end{array}\right\} \quad (9.47)$$

The LP formulation described by (9.47) allows us to apply the Dantzig-Wolfe decomposition technique. This way a large linear optimization problem can be converted into *na smaller* ones of the following type:

$$\left.\begin{array}{l} \min (\Delta P_{L,i} - mA_i)\Delta V_i \\ \text{subject to} \quad \mathbf{D}_i \Delta \mathbf{V}_i = b_i \\ \Delta V_i^{\min} \leq \Delta V_i \leq \Delta V_i^{mzx} \\ i = 1, \cdots, na \end{array}\right\} \quad (9.48)$$

where $\Delta P_{L,i}$ is the real power loss of area i.

These subproblems are coordinated by a master problem, which provides them with a simplex multiplier m from the dual solution. The master problem includes the overall objective function and the linking constraints. Figure 9.19 depicts the solution procedure of the Dantzig-Wolfe decomposition method that is an iterative process between the master problem and subproblems.

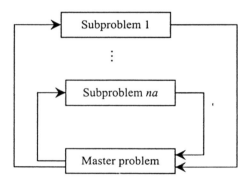

Figure 9.19 Iterative Procedure of the Dantzig-Wolfe Method

As noted in (9.48), upper and lower limits are imposed on the increments of bus voltages. Thus the size of each subproblem can be further reduced through the application of the upper bound technique so that upper and lower constraints may be dealt with implicitly by the simplex method.

In the proposed Dantzig-Wolfe method, a revised simplex method is used to solve the master problem, which is to update the basis according to the solutions of the subproblems. In each iteration, the subproblem's objective is revised according to the solution of the master problem. To further reduce the computation time and the iterations required for the

DISTRIBUTED CONTROL OF VOLTAGE/VAR

solution, the following techniques are incorporated in the proposed method:

- The number of constraints in (9.47) representing upper and lower limits of bus voltages is equal to the number of the buses of the system. By applying the upper bounding technique where variable limits are dealt with implicitly, the computation time will be reduced and thus the solution process of the proposed Dantzig-Wolfe method will be improved;

- The solution of the master problem depends on dual solutions provided by the subproblems. Depending on the criterion of the objective function, one column is introduced to the basis of the master problem. Therefore, if more information is derived from the subproblems at each iteration, the iterations required for the solution will be greatly reduced. This is implemented by adding more than one column to the basis of the master problem.

- More information about the subproblems will be used if more than one column is introduced into the basis of the master problem. The selection of these columns depends on the outcomes of the following steps:

- Update the objective function of the subproblems according to the solution of the master problem.

- Except for the subproblem that has already entered the basis, substitute the solutions of all subproblems in this new objective function, and solve the problem to determine which will enter the basis next; use the new information after introducing any vector in the basis.

- Repeat the process until the number of these inner iterations equals the number of subproblems, or the criterion of selecting the vector to enter the basis is accomplished.

9.7.5 Case Study

We consider the IEEE 30-bus system as an example to show how the proposed Dantzig-Wolfe method is used for the power loss optimization. The system is divided into three areas as shown in Figure 9.20 (see Appendix C for the system data). The loss optimization results by using the Dantzig-Wolfe method are listed in Table 9.5. The power loss convergence for the 30-bus system based on different algorithms used for solving the LP problem is shown in Table 9.5.

Table 9.5 Results of Different Methods Applied to the 30-Bus System

Iteration number	A (MW)	B (MW)
0	10.24	10.24
1	10.01	9.98
2	9.66	9.52
3	9.38	9.18
4	9.08	8.85
5	8.69	8.34
6	8.24	7.94
7	8.01	7.38
8	7.72	6.97
9	7.43	6.50
10	7.09	
11	6.63	
12	6.85	
13	6.64	
14	6.51	

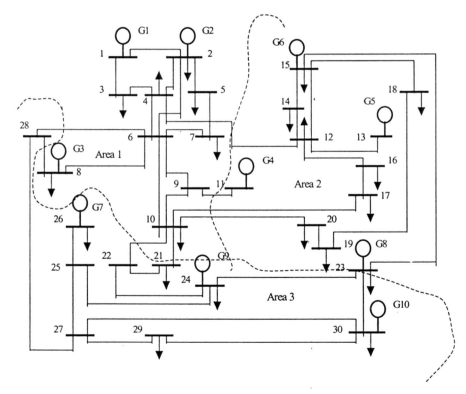

Figure 9.20 Area Partition of IEEE 30-buş System

DISTRIBUTED CONTROL OF VOLTAGE/VAR

In this table, column A is the power loss at every iteration when the simplex method is used, and column B is the power loss at every iteration when the Dantzig-Wolfe decomposition method is used.

9.7.6 Parallel Implementation of Dantzig-Wolfe Decomposition

The proposed Dantzig-Wolfe decomposition method can be implemented in a computer with a single processor. Initially the master problem and then the na subproblems can be solved sequentially, as there is no dependency among subproblems. This serial solution process is depicted in Figure 9.21 in which the framed solution process is a serial process. When na is large, this tertiary voltage/VAR optimization process will be slow.

To expedite the solution process, the serial solution method can be executed in a parallel manner, with the master problem and each subproblem solved by different processors. This parallel solution process is depicted in Figure 9.22 in which the frame solution process represents a parallel solution process. The parallel solution process can be implemented either in a parallel machine with multiple processors or on a LAN with multiple computers.

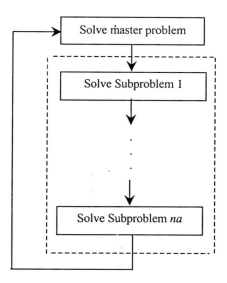

Figure 9.21 Serial Solution of the Dantzig-Wolfe Method

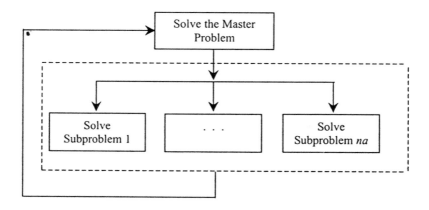

Figure 9.22 Parallel Solution of the Dantzig-Wolfe Method

9.8 DISTRIBUTED TERTIARY VOLTAGE/VAR OPTIMIZATION

9.8.1 Negligible Interactions among Neighboring Control Areas

Suppose that the entire system is partitioned into *na* control areas, and each control area has a control center which is responsible for the voltage/VAR optimization and control of its area. Assume that the system is well partitioned, which means that the adjustments of the reactive power resources in a control area will mainly affect the voltage profile of that control area with little effect on the voltage profile of its neighboring control areas. Accordingly, the tertiary voltage/VAR optimization will be accomplished in a distributed manner, where each control area will only optimize and control the voltage of its own control area.

This scheme can be utilized when the coupling among control areas is weak and power loss increments on tie lines are negligible during the optimization process. In this case we assume that tie flows are fixed and represented by a constant load or generation injection as depicted by Figures 9.23 and 9.24.

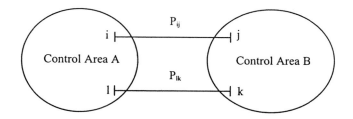

Figure 9.23 Two Neighboring Control Areas

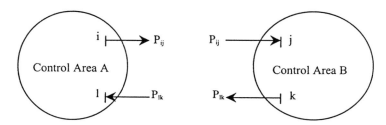

Figure 9.24 Effect of Tie Line Power on Control Area

The Dantzig-Wolfe decomposition method discussed in Section 9.7 can be applied to individual control areas according to (9.47). Hence, the size of the problem will be reduced dramatically and a much faster solution will be obtained by utilizing distributed processing.

9.8.2 Interactions among Neighboring Control Areas

When the coupling among a few control areas is strong, interactions among these control areas have to be taken into consideration in the tertiary voltage/VAR optimization and control. In this case, the tertiary voltage/VAR optimization is implemented in a distributed manner based on the WAN which is composed of computers at each control center as illustrated in Figure 9.25.

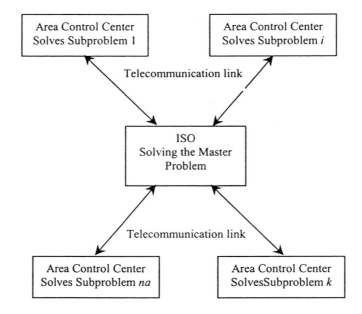

Figure 9.25 Distributed Tertiary Voltage/VAR Optimization Based on WAN

Chapter 10

Transmission Congestion Management Based on Multi-Agent Theory

10.1 INTRODUCTION

Major components of an electric power system are geographically dispersed and can exhibit global changes instantaneously as a result of local disturbances just like large telecommunications, transportation, and computer networks. An issue that the restructured power industries will face is controlling such a heterogeneous, widely dispersed, yet globally interconnected system. It is thought that it will be particularly difficult to control the power system to achieve optimal efficiency and provide maximum benefits to the consumers while allowing all market participants to compete fairly and freely.

Several alternatives for the control of electric power networks call for the ISO's intervention, as it has the responsibility and authority to facilitate less expensive power for market participants and at the same time to ensure the security and reliability for the system operation. Figure 10.1 depicts a completely centralized control scheme for the power system operation and control that will require all measured data to be transmitted to the system control center, and detailed control command and signals to be sent to controllers dispersed in the system. For most purposes, especially in emergencies, this centralized control scheme requires high-speed and two-way communications to meet real-time control requirements.

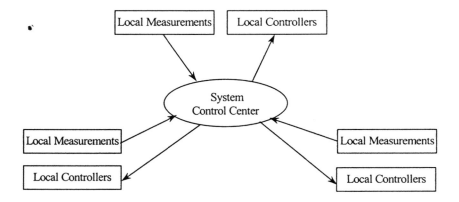

Figure 10.1 Centralized Control Scheme

The availability of various advanced control devices has rendered the power system a prime candidate to benefit from a distributed control system. A distributed control scheme may not only be useful but also necessary for free competition in an electric power market, for centralized control by the system operator will be displaced by the coordination of individual market participants.

The simplest distributed control scheme is to use distributed regulators. The performance can be improved by displacing the regulators with feedback or PID controllers. In times, however, as the distributed controllers increase in number, it may become necessary to supervise and coordinate their performance by adjusting their gains and switching points. Massive parallel processing will be required to handle the heightened complexity, the decision-making process approaches to real time, and the increased computational burden on the central computer. This way the computational load and the decision-making details can be distributed among local controllers.

As shown in Figure 10.2, the idea for implementing this approach is to model relevant power system components as independent adaptive agents. These agents are partly cooperating and partly competing with each other while pursuing a common objective set by a minimal supervisory function.

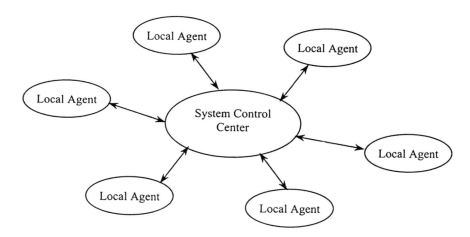

Figure 10. 2 Distributed Control with Autonomous Agents

In a general sense, an agent is a proactive, personalized, adaptive, and autonomous entity. An agent can represent certain physical entities or constitute a piece of software without any visible appearance. The use of agents in modeling the engineering systems is not new. In the last two decades, multiple quasi-autonomous agents, which were usually called actors or demons, have already been used to assist human decision-making in many areas such as transportation, logistics, and batch manufacturing. Agents are used to represent certain functions, operations, or physical entities, and agent-based modeling has been applied successfully in many fields including the commodity markets, automobile traffic control simulation, field combat simulation and other military applications, robotics and manufacturing applications, ecological simulations, videogames, and many more.

An intelligent agent is an agent that possesses certain human intelligence. An intelligent agent, as illustrated in Figure 10.3, is composed of four components: input interface, output interface, communication system, and decision-making system. For instance, in the context of a distributed control system, individual agents represent a local controller, comprising of one or more sensors and/or actuators, a microprocessor with some memory, and two-way communications. An intelligent agent can think and function like a human being, although it may not necessarily have the physical form of a human being. A host is the entity that an agent represents. An agent communicates not only with its host but also with other agents.

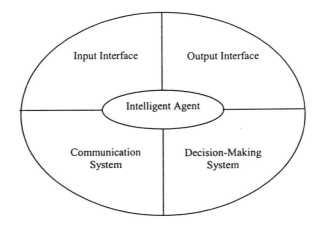

Figure 10.3 Agent Model

10.2 AGENT-BASED MODELING

From a programming point of view, agent-based modeling and simulation is largely a natural extension of the currently prevailing object-oriented methodology. In other words, agents are simply active objects that are used to model parts of a real-world system. In many computer simulations, agents have been employed to represent separate processes, which operate independently and interact with each other by exchanging the pertinent information. In these simulations each agent has a unique ID tag for addressing messages, internal data storage, and functions that respond to reactive and proactive behavior of objects and agents.

The agent-based modeling usually starts with a group of active objects of a system. Various parameters, which usually describe the inter-relational behavior of agents, are assigned to each agent. These parameters depend on the particular system being modeled, and may include attributes, capabilities or strategies; behaviors such as cooperation, competition, or conflict; sources of strength and weakness, which may be physical, technological, intellectual, social, or political. Further the modeling requires a fitness function for each agent that represents required goals and constraints for the system. The modeling includes a sufficient variety of resources and strategies for improving the agents' joint "fitness" for controlling the power system.

One revolutionary idea in the agent theory is the adaptation of agents. Adaptation is the capacity for the modification of goal-oriented individuals or the collective behavior in response to changes in the environment. There are two kinds of adaptation including active and passive adaptation. A passive adaptive agent responds to changes in its environment without modifying the environment. An active adaptive agent exerts some control or influence on its environment in order to improve its adaptive power. In effect, it conducts experiments and learns from them. Individual agents must exhibit enough flexibility to respond to environmental conditions and to other agents in a way that enhances their survival or meets other goals.

In order to learn a strategy that increases its fitness, the agent has to gather and store enough information to adequately forecast and deal with changes that occur within a single generation. In this sense, a major issue in agent design is the interplay of short- and long-term evolution. In the longer time frame, agents (softbots) must learn new strategies and/or modify old ones in order to handle the real-time control. Their long-term fitness should be based mainly on the evolution of strategies for cooperation, methods for raising the sum of the game for all players [Axel94]. However, their strategies for a short-term operation must evolve in real time, so they may create temporary conflicts in the real-time control, which must also be resolved in that same time frame. To reduce conflicts in the short term, a "look ahead" for the real-time solution is required. The strategies for where and how to "look" are evolved over the long term.

In addition to offering more flexibility and reliability, compared to that of aggregated modeling methodologies, agent-based modeling has the following advantages:

- Easier implementation of rule-based and qualitative specifications
- Easier implementation of complex boundary conditions
- Easier implementation on parallel machines
- Easier evaluation of the effect of initial conditions.

It is believed that the application of autonomous, adaptive, and intelligent agents not only can improve the performance of power system optimization and control but also can provide new business strategies such as internal reorganization, external partnerships, and market penetration for participants.

10.3 POWER SYSTEM MODELING BASED ON MULTI-AGENTS

The restructuring of power systems may ultimately be limited by the physics of electricity and the topology of the power grid. For this reason, it is of great value to model a power system in a control theory context. A power system can be taken as a complex adaptive system (CAS) for which the intelligent agent-based distributed control is the only practical way to achieve real-time optimization and control. The multiple agents-based CAS model developed by EPRI is based on a conceptual design for distributed control of the power system by intelligent agents which operate locally with minimal supervisory control [Wild97].

The design of agents for power systems attempts to integrate modeling, computation, sensing, and control to meet certain goals such as efficiency and security in a geographically distributed system, subject to unavoidable natural disasters, and operated by partly competing and partly cooperating business organizations. The CAS simulation is used in the modeling of the computational intelligence, which is required to automate the distributed control of a geographically dispersed but globally interconnected power network. This model is also used to test whether any central authority is required, or even desirable, and whether free economical cooperation and competition can, by itself, optimize the efficiency and security of the power network operation for everyone's benefit.

For the preliminary implementation of CAS, a generic model of a complete electric power system was developed by EPRI. This model employed multiple adaptive, intelligent agents to represent individual components such as generators, transformers, transmission lines, loads, and buses. These agents were taken as intelligent robots that can cooperate to ensure global optimal operation or act independently to ensure an adequate individual performance. For instance, a single bus will strive to stay within its voltage limits but still operate in the context of voltages and flows imposed on it by other agents. Similarly, other components have safety and capacity restrictions, such as maximum thermal limits, which may not be exceeded for more than specified time periods. The agents that represent these components will strive to stay in a state that is optimal both locally and globally.

More complex components, such as a power plant or a substation, are modeled as a class and hierarchy of simpler components using the object-oriented method. Agents and the subagents are represented as autonomous "active objects" that adapt gradually to their environment and improve performance even as conditions change. This model will be used to illustrate distributed sensing and control, as well as the evolution of business strategies in a power system.

The multiple intelligent agents will act in a parallel and distributed manner; they can communicate via microwaves, optical cables, or power lines the information that is necessary to both global and local optimization. If a failure occurs in a certain local area, agents within that area will immediately respond to minimize the impact on the overall network. If the failure causes a system to breakup, agents will help isolated areas operate independently, and meanwhile help them re-join the network without creating unacceptable local conditions.

When applied to the contingency planning, the agent-based approach will focus on potential failure modes and areas under the greatest stress. On the other hand, since individual agents need to collaborate with each other, it will be helpful for agents to have the complete information about others. However, this requirement may not be practical because detailed communications would require much time and broad bandwidth communication links. A feasible way to solve this shortcoming is to allow each agent a very simple representation of the entire system, and only the information about exceptions will be communicated intentionally and explicitly. Before a completely automated control system becomes practically available, intelligent agents are expected to provide power system operators with a means for the real-time optimization and control.

This complete multiple-agent model can also serve as a convenient test-bed, without requiring any a priori assumptions about global scenarios for many other studies. The model can provide an insight into the evolution of power systems in response to various economical pressures and technological advances. It can also serve as a practical way to evaluate the impact of hypothetical changes to operating practices on the power system. The power market participants may use the model to examine the potential for embarking on new partnerships.

The CAS research has provided mathematical and computational tools for the modeling and the design of power systems based on multiple, autonomous, intelligent agents, which are competing and cooperating in a distributed control environment. When combined with intelligent and fast sensors and control devices, such a distributed control scheme could ultimately provide the foundation for the real-time operation and management of a fully automated power network, which would exhibit the following features:

- Measured by local autonomous intelligent sensors
- Modeled as a hierarchy of competing and cooperating adaptive agents
- Computing in a parallel and distributed environment
- Controlling local operations automatically in conformity with global optimization criteria
- Communicating merely the essential information, possibly over power transmission lines
- Robust enough to operate sub-optimally either individually or in groups when separated by disturbances.

Since intelligent agents have already been used successfully for modeling traders in commodity markets, artificial agents representing buyers and sellers are employed to build an economic model of a bulk power market. In this application, agents observe changes in strategies used by others and adjust their bids accordingly before the market-clearing price is calculated. Unlike the games solved by von Neuman, this application signifies the repeated games with non-zero sum payoffs [Nash53].

A multiple agent-based approach is also used to perform risk-based contingency analysis [Kris98a,b]. This simulation involves two groups of intelligent agents. In one group, each agent represents a contract between a seller and a buyer; in the other, each agent represents a potential equipment failure, such as the failure of a transformer. Based on the power trading contracts and the probability of failures, each agent will estimate the probable cost of the additional power required to maintain all contracts after certain failures. The result of this simulation is a series of Pareto curves, which represent the network configurations for providing the security against contingencies at various risk levels. When the second group of agents is complete, the resulting Pareto surfaces will represent the trade-off between the losses of denying various combinations of contracts versus the loss of accepting the contracts but failing to fulfill them due to a certain contingency.

A comprehensive, high-fidelity, scenario-free modeling and optimization tool for power systems is being developed using agent-based modeling [Wild97]. This prototype tool includes four basic agent classes including generation unit agents, transmission system agents, load agents, and corporate agents. Considering that users may need to extend the model by specializing these agent classes or defining new classes to allow for different kinds of agents, the design and implementation of these agents are intended to be sufficiently generic. This project could gain strategic insights into the power market for certain applications such as real-time pricing, co-generation, and retail wheeling. Possible results of this model will be the development of conditions for attaining equilibriums in a power market, strategies, or regulations that can destabilize the market, mutually beneficial strategies, implication of differential incomplete information, and conditions under which chaotic behavior might develop. Further enhancements of this model may emphasize greater fidelity for each transaction corresponding to the network flows. In the next section, we provide an example of the application of agents in mitigating the network flow congestion in power systems.

10.4 MULTI-AGENT BASED CONGESTION MANAGEMENT

10.4.1 Congestion Management

Congestion occurs when the scheduled energy exceeds the available transmission capacity (ATC) in either the day-ahead or the hour-ahead markets. If congestion happens on a transmission line, the ISO will call for scheduling adjustments to eliminate the congestion according to the congestion management protocols. To manage the congestion, we divide the power system into zones. The zonal boundaries are determined by the corresponding values of locational marginal price (LMP) as LMP values will be quite close within each zone.

There are two types of transmission flow congestion: inter-zonal and intra-zonal. The inter-zonal congestion occurs between contiguous zones or adjacent control areas, while intra-zonal congestion is constrained within a zone. The congestion is assessed on all-wheeling transactions when applicable, and participants will be subject to congestion charges. However, charges will not be allocated to firm transmission right (FTR) holders who schedule the flows within their existing rights. If an FTR holder schedules the flows above the allocated

FTR, the holder will be subject to congestion charges. Costs associated with congestion mitigations, including losses and wheeling, will be paid by the ISO and recovered from transmission users.

Most GENCOs and DISTCOs in a power system are scheduled by scheduling coordinators (SCs), through which a pair of GENCO and DISTCO signs a transaction contract. However, there are still a number of independent GENCOs and DISTCOs, which are not associated with any SCs. An independent DISTCO buys energy directly from the power market instead of a specified GENCO, and accordingly an independent GENCO sells energy directly to the power market instead of a specified DISTCO. The impact of a pair of GENCO and DISTCO scheduled by a SC on the congestion can be determined by the source and sink of their contract, while the impact of independent GENCOs or DISTCOs on congestion will be determined by tracing techniques such as their PDFs (power distribution factors) [Sha02].

The ISO resolves congestion problems on a zonal basis. A congestion zone is a predetermined area where congestion is most likely to happen between zones. After the initial preferred schedules (IPSs) are submitted by market participants, the ISO runs the first round of congestion check. If there is no congestion, the IPS will be published as the final schedules. If congestion occurs, however; all market participants including SCs and independent participants will be informed and given an opportunity to revise their IPSs. If congestion still exists after these voluntary schedule adjustments, the ISO will use adjustment bids to alleviate the congestion. However, if adjustment bids submitted to the ISO are insufficient, the ISO will try to adjust all schedules on a pro-rata basis. A default usage charge, which is the last accepted insufficient bid, will be set by the ISO. It is up to each individual SC to determine whether to submit adjustment bids and at what price. If no adjustment bid is submitted, SCs and other independent market participants will be considered as the "price taker" and will have to pay whatever the cost of congestion ends up to be. In case where the ISO is forced to resolve the congestion, each SC's portfolio will have to be balanced by its adjustments.

The ISO employs a market-first policy for the real-time management of intra-zonal congestion. Intra-zonal congestion is managed by utilizing energy adjustment bids based on their effectiveness and in merit order. Resources are to be increased or decreased on both sides of the intra-zonal congestion interface to relieve the congestion. Intra-zonal

CONGESTION MANAGEMENT BASED ON MULTI-AGENTS

congestion management is implemented so as not to inadvertently create inter-zonal congestion. The difference between an incremental bid and decremental bid is the grid operation charge associated with the congestion. In the event that there are inadequate bids to solve the congestion, reliability must-run resources can be utilized.

Generally, the objectives of congestion management are described as follows:

- Allocate transmission capacity according to economic principles
- Create and present price signals to the market to guide generation, transmission and load management operating decisions and investments
- Adjust scheduled generation and load levels to prevent overloads of transmission facilities
- Minimize the ISO's intervention in real time by allowing the forward market to manage the congestion

The present congestion management is implemented through a linear program, which has defects in several aspects. The most distinguishing one is that the adjustments of relevant participants are limited by their initial adjustment bids, which were submitted at the same time with their initial preferred schedules. These adjustments bids are largely "blind", because as participants submit bids they do not know:

- Where the congestion could occur
- How severe the congestion could be

After they get the congestion information from the ISO, the market participants may find their previous adjustment bids were not the best. If they do not have an opportunity to revise their adjustment bids, some participants could get a severe charge. Since they cause the congestion unconsciously, the participants should be given the right to "repent," which means they would alleviate the congestion voluntarily and at their own cost. On the other hand, the present method also lacks flexibility.

New trends in congestion management emphasize decentralized decision-making and operations, diversity, and choice in accomplishing tasks. New approaches should relate the individual's benefits to the system's operations, since restructuring to some extent means substituting price signals for commands as a way to meet operational requirements. The price signal persuades market participants to behave consistently with operational requirements: aligning market interests with ISO interests. The effectiveness of congestion mitigation depends on how participants react based on the congestion information that they receive from the ISO.

Generally, congestion charges are calculated based on the amount of energy transmitted over the congested transmission lines. The congestion charges collected by the ISO are first used to pay participants with counter schedules who helped alleviate congestion, and any remainder would be allocated to transmission owners or FTR holders.

10.4.2 Application of Agents

The agent theory has been proved suitable for solving problems where many entities compete and at the same time cooperate. Hence, agent theory can be applied to the congestion management when congestion occurs as GENCOs, DISTCOs, TRANSCOs, and SCs, compete for their own profits but also have to cooperate to eliminate the congestion most effectively. In this chapter, we propose a new congestion management protocol to eliminate the congestion more effectively and flexibly. Market participants are represented by their individual intelligent agents, which can seek useful information for their hosts, and discuss and negotiate trades with other agents. These agents can make their decisions on behalf of their hosts. By employing intelligent agents, market participants can make use of much useful information for the congestion mitigation.

Compared with the existing congestion management methods, the proposed approach presents the following advantages:

- In traditional adjustment bidding methods, the level of adjustments are determined by the ISO. However, in the agent method, agents determine their adjustments according to the information they have obtained.

- In traditional adjustment bidding methods, market participants are passive. However, in the agent method, agents are active since they hold the rights on final decision.

- In the agent method, each agent can communicate with other agents for the information, which is usually regarded as trade secrets, such as the available adjustment range and bidding price. This information can help for agents make their optimum schedules and adjustments.

- Any participant can participate in the congestion mitigation if its behavior is helpful to the congestion mitigation.

Next, we use a multi-agent based model to solve the congestion mitigation problem in a distributed manner.

10.4.3 Agent Models of Market Participants

Based on the framework of Figure 10.3, agent models of GENCO, DISTCO, TRANSCO, SC, and ISO are depicted in Figures 10.4 through 10.8 respectively. The difference between the agent models of an SC-scheduled GENCO and that of an independent GENCO is that the former may communicate with its associated SC. Communication of an agent with its host is necessary since the host may need to inform its agent frequently of important instructions.

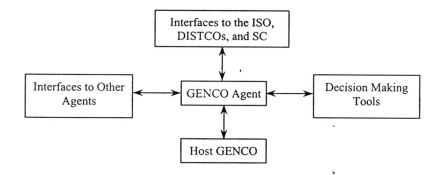

Figure 10.4 Agent Model of GENCO

The intelligent agents of GENCO, DISTCO, TRANSCO, SC, and the ISO are different, and their hosts are responsible for their design according to their particular requirements and functions. The participants who are very concerned with congestion may select a high-performance agent, while others may opt for a simple agent, or merely to submit an adjustment range with their IPSs and not utilize an agent.

One reason we want to use intelligent agents in congestion management is that market participants have difficulty in maintaining their trade secrets when they cooperate for the effective mitigation of transmission congestion. When intelligent agents are used, market participants can get any information they need to re-adjust their generation or demand, which can help them make optimal decisions. This is because intelligent agents are represented by codes instead of their real names when they negotiate with other participants. This way, trade secrets are not discharged if the negotiation fails. Another advantage of using intelligent agents is that the agents' hosts can delegate certain jobs to the agents while the host is dealing with other issues. At any time the host can deal directly with its agent.

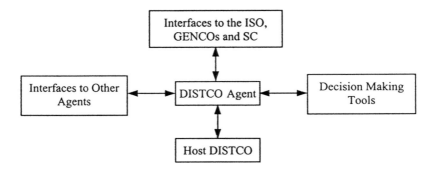

Figure 10.5 Agent Model of DISTCO

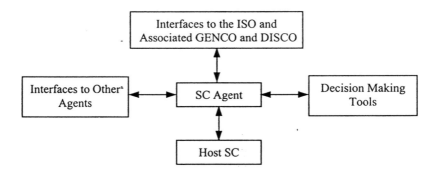

Figure 10.6 Agent Model of SC

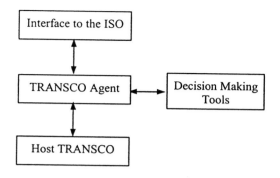

Figure 10.7 Agent Model of TRANSCO

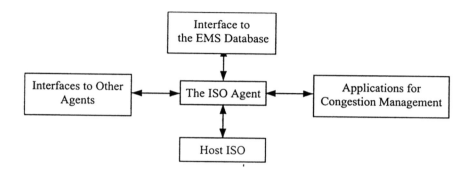

Figure 10.8 Agent Model of ISO

10.5 MULTI-AGENT SCHEME FOR CONGESTION MITIGATION

Consider that market participants send their intelligent agents to the ISO. These intelligent agents will be merged into a particularly organized congestion management package as a software module of the ISO agent. The package provides a mechanism, similar to a "meeting room" or "bulletin board" where intelligent agents exchange information and negotiate with each other to seek the best way to mitigate transmission congestion. Each agent can get information about the congestion from, or publish its comments for the congestion solution methods on, this public bulletin board, and contact other agents directly. Such an information exchange mechanism is demonstrated on Figure 10.9.

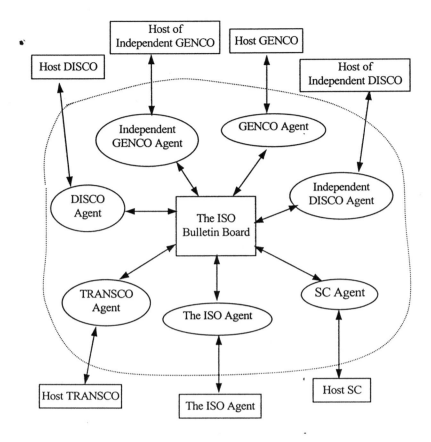

Figure 10.9 Information Exchange among Agents

To let agents exchange information freely, the ISO assigns each agent a particular code that will be used instead of the agent's true identity during the whole congestion management process. Only the ISO agent is aware of the true identity of agents.

When congestion occurs, the ISO agent broadcasts the information to all participants able to make decisions and to offer voluntary adjustments for congestion management. Not all adjustments are helpful to congestion mitigation, and not all agents may be available to take part in scheduling adjustments. Therefore, the ISO divides GENCO and DISTCO agents into two groups according to their contributions to the congested line. The first group represents agents whose decremental adjustments could mitigate the congestion; the second group represents agents whose incremental adjustments could mitigate the congestion.

Figure 10.10 gives an example representing SC_c: GENCO k and DISTCO L, SC_d: GENCO H and DISTCO N, SC_f: GENCO Z and DISTCO Y, as well as independent participants GENCO E and DISTCO M. Suppose that the transmission line between buses i and j is congested and the system is thus divided into two congestion zones. Then, GENCOs in zone A and DISTCOs in zone B belong to the first group, and DISTCOs in zone A and the GENCOs in zone B belong to the second group.

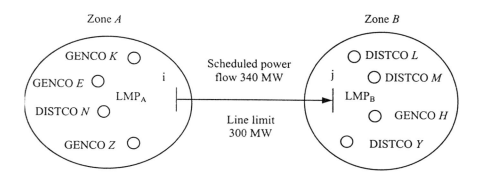

Figure 10.10 Transmission Congestion

To facilitate the negotiation, the agents in the first group are referred to as "red agents" and the agents in the second group are referred to as "green agents." The red agents are subject to congestion charges if the congestion is not mitigated, while the green agents can get congestion credits if they provide adjustments for congestion mitigation. Suppose that GENCO K supplies 100 MW power to DISTCO L, and GENCO H supplies 80MW power to DISTCO N, all through the congested transmission line ij. The initial LMPs of the two congestion zones are $LMP^{(0)}{}_A$ = 350\$/MW and $LMP^{(0)}{}_B$ = 450\$/MW, which are calculated based on the bidding prices submitted by market participants. Once the ISO publishes this information, SC_c pays a congestion charge of 100 * $(LMP^{(0)}{}_B - LMP^{(0)}{}_A)$ = 100 MW * (450 - 350) \$/MW = \$10,000 and SC_d gets the congestion credit of 80 * $(LMP^{(0)}{}_B - LMP^{(0)}{}_A)$ = 80 MW * (450 - 350) \$/MW = \$8,000. On the other hand, only the cooperation of a pair of GENCO and DISTCO agents in the same group will be accepted by the ISO which could otherwise violate the energy balance.

When congestion occurs, the ISO agent publishes the information about the congestion to all agents as follows:

- Load flow information, especially on congested transmission lines

- Initial zonal LMPs interfacing the congested lines
- Adjustment resources and their bids
- Nodal PDFs
- External transfer schedules
- Zonal congestion costs
- Total SC interchange in each congestion zone.

Based on this information and the negotiation results with other agents, individual agents make their final decision on voluntary adjustments for congestion mitigation. The process of congestion mitigation comprises three stages:

1. Red agents decrease their generation and demand voluntarily at their own cost, since the congestion is theoretically caused by the red agents. If the congestion can be removed by voluntary adjustments, the ISO will not impose any congestion charges on agents.

2. If voluntary adjustments of red agents are insufficient to mitigate the congestion, the ISO calls for voluntary adjustment from the green agents, which have nothing to do with the congestion but can make a profit when they help mitigate the congestion. At this stage, congestion charges and credits are used to persuade green agents to offer incremental adjustments.

3. If the adjustments offered by green agents are not enough to mitigate the congestion, the ISO agent takes mandatory adjustments.

Figure 10.11 depicts the scheme for congestion mitigation.

10.6 APPLICATION OF PDF TO CONGESTION MANAGEMENT

The LP method used currently by the ISO for congestion mitigation is based on a merit order of bidding prices. Although the ISO tries to minimize the cost for congestion mitigation, in some cases this method cannot guarantee that the cost is minimized because it only takes the bidding price into account and does not consider the power distribution factors (PDFs).

CONGESTION MANAGEMENT BASED ON MULTI-AGENTS

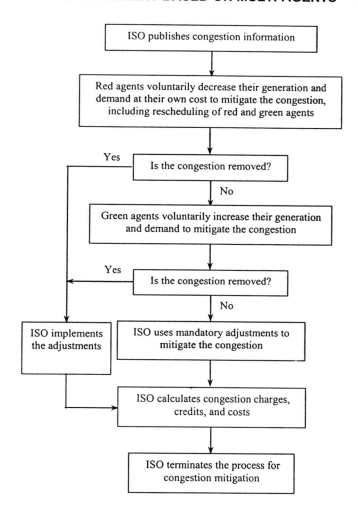

Figure 10.11 Congestion Mitigation Based on Multi-Agent Participation

The PDF method, which is a common method for the load flow computation, is detailed in Appendix VI. The following example illustrates this point. Suppose that GENCOs L, M, and N, which are in the red agents group, offer the adjustment bids listed in Table 10.1.

If GENCOs L and M with lower bidding prices adjust their generation to mitigate the congestion, and disregard the fact that their adjustments are less effective on the congestion, the cost of congestion mitigation is calculated as follows.

Table 10.1 Congestion Adjustment Bids

Agent	Bid Price	PDF	Adjustment Range
GENCO L	100 $/MW	0.10	[0, 250] MW
GENCO M	150 $/MW	0.15	[0, 150] MW
GENCO N	200 $/MW	0.30	[0, 150] MW

GENCO L first decreases its generation by 250MW:

Congestion relief: 250MW * 0.10 = 25MW

Adjustment cost: 250MW * 100$/MW = $25,000

GENCO M decreases 100 MW of generation to mitigate 15MW of congestion:

Congestion relief: 100 MW * 0.15 = 15 MW

Adjustment cost: 100 MW * 150$/MW = $15,000

The total adjustment is 250 MW + 100 MW = 350 MW

The total cost to mitigate the congestion is $25,000 + $15,000 = $40,000.

If power distribution factors are taken into consideration, GENCO N will decrease 133.4 MW of its generation to mitigate the congestion

Congestion relief: 133.4 MW * 0.3 = 40 MW

Total adjustment is 133.4 MW

Total congestion cost is: 133.4 MW * 200 $/MW = $26,680

The difference between the two schemes is $40,000 - $26,680 = $13,320.

Although the price of GENCO N is higher than that of GENCO L and GENCO M, the adjustment of GENCO N is more effective because GENCO N is closer to the congestion line and has a larger PDF. This example demonstrates the necessity of considering PDFs in congestion mitigation.

10.7 OBJECTIVES OF MARKET PARTICIPANTS

The participants in a power market have different objectives for their trading activities. Usually bilateral contracts involve a preset energy price upon which both sides of the contract reach an agreement, whereas for most independent participants, the energy transactions are settled based on MCP. In either case, congestion charges and credits are taken as an important factor in agent design. It is perceived that both congestion charges and credits will be due to supply and demand since in congestion mitigation, adjustments in demand and generation must be kept in balance.

10.7.1 Objective of a GENCO

The objective of a GENCO is to make the maximum profit from its trading activity. Suppose the initial preferred schedule of the ith GENCO $P_{G,i}^{(0)}$ in the red agent group (r) is denoted as $P_{G,r,i}^{(0)}$, and the corresponding congestion flow (CP) that would be subject to congestion charge is denoted as $CP_{G,r,i}^{(0)}$. This charge is determined in terms of the FTR that the GENCO owns on the congested line. For example, if $LMP_A^{(0)}$ and $LMP_B^{(0)}$ are the initial LPMs of zones A and B when the congestion happens, then the congestion charge (CC) of the ith red GENCO agent will be calculated as

$$CC_{G,r,i} = CP_{G,r,i}^{(0)}(LMP_B^{(0)} - LMP_A^{(0)})/2 \\ = \alpha_{G,r,i} P_{G,r,i}^{(0)}(LMP_B^{(0)} - LMP_A^{(0)})/2 \qquad (10.1)$$

where $\alpha_{G,r,i}$ is the PDF of the ith red GENCO on the congested line.

Figure 10.12 shows that $\Delta LMP = (LMP_B^{(0)} - LMP_A^{(0)})$ when there is no adjustment for congestion mitigation, which would result in the ith independent red GENCO's price (GP) as follows:

$$GP = MCP - \alpha_{G,r,i}(LMP_B^{(0)} - LMP_A^{(0)})/2 \qquad (10.2)$$

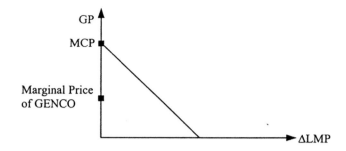

Figure 10.12 Actual Price of a Red GENCO

Now suppose that the initial preferred schedule of the ith green GENCO is $P_{G,g,i}^{(0)}$, and $CP_{G,g,i}^{(0)}$ is subject to congestion credit. Similarly, congestion credit (CD) of the ith green GENCO agent would be calculated as

$$CD_{G,g,i} = CP_{G,g,i}^{(0)}(LMP_B^{(0)} - LMP_A^{(0)})/2 \\ = \alpha_{G,g,i} G_{G,g,i}^{(0)}(LMP_B^{(0)} - LMP_A^{(0)})/2 \quad (10.3)$$

Suppose that the ith red GENCO agent offers voluntary adjustment for the congestion mitigation. Then its objective function is given as

$$Max\ Revenue = P_{G,r,i} * MCP - CP_{G,r,i}(LMP_B - LMP_A)/2 \quad (10.4)$$

where $P_{G,r,i}$ is the final adjusted schedule, and LMP_A and LMP_B are the LMPs of the two congestion zones. The two LMP values could be different in case where congestion could not be mitigated by voluntary adjustments.

The objective function of the ith bilaterally contracted red GENCO through an SC is

$$Max\ Revenue = P_{G,r,i} * P_c - CP_{G,r,i}(LMP_B - LMP_A)/2 \quad (10.5)$$

where $P_{G,r,i}$ is the final schedule for the red GENCO and P_c is the energy price set by the contract.

CONGESTION MANAGEMENT BASED ON MULTI-AGENTS

Besides the congestion credits, green agents will also be rewarded by the ISO for their voluntary adjustments. The objective function of the ith independent green GENCO is

$$\text{Max Revenue} = P_{G,g,i} * MCP + CP_{G,g,i}(LMP_B - LMP_A)/2 + \Delta P_{G,g,i}^{adj} * C_{G,g,i}^{adj} \quad (10.6)$$

where $\Delta P_{G,g,i}^{adj}$ is the adjustment for congestion mitigation and $C_{G,g,i}^{adj}$ is the adjustment bid.

The objective function of the ith bilaterally contracted green GENCO through a green SC agent is

$$\text{Max Revenue} = P_{G,g,i} * P_c + CP_{G,g,i}(LMP_B - LMP_A)/2 + \Delta P_{G,g,i}^{adj} * C_{G,g,i}^{adj} \quad (10.7)$$

where $P_{G,g,i} = P_{G,g,i}^{(0)} + \Delta P_{G,g,i}^{adj}$, $CP_{G,g,i} = \alpha_{G,g,i} P_{G,g,i}$. The decremental production usually means the decrement in revenue. If the congestion can be mitigated by voluntary adjustments, the LMPs of the entire system would be the same and participants would not be subject to congestion charges.

10.7.2 Objective of a DISTCO Agent

The objective of a DISTCO agent is to minimize the payment for its energy transaction. Suppose that the initial preferred schedule of the ith red DISTCO is $P_{D,r,i}^{(0)}$ MW, and $CP_{D,r,i}^{(0)}$ is the fraction $P_{D,r,i}^{(0)}$ subject to a congestion charge. Suppose that $LMP_B^{(0)}$ and $LMP_A^{(0)}$ are the initial LMPs of zone A and B as congestion occurs. Then the congestion charge of the ith red DISTCO agent will be calculated as

$$CC_{D,r,i} = CP_{D,r,i}^{(0)}(LMP_B^{(0)} - LMP_A^{(0)})/2 \\ = \alpha_{D,r,i} P_{D,r,i}^{(0)}(LMP_B^{(0)} - LMP_A^{(0)})/2 \quad (10.8)$$

where $\alpha_{D,r,i}$ is the PDF of the ith red DISTCO to the congested line.

Suppose that the ith DISTCO offers voluntary adjustment for congestion mitigation, and $P_{D,r,i}^{adj}$ denotes the amount of this adjustment. A positive $P_{D,r,i}^{adj}$ means an incremental adjustment, while a negative $P_{D,r,i}^{adj}$ means a decrement adjustment. The objective function of the ith DISTCO can be formulated as

$$\text{Min Payment} = P_{D,r,i} * MCP + CP_{D,r,i}(LMP_B - LMP_A)/2 \quad (10.9)$$

The objective function of the ith bilaterally contracted red DISTCO agent is

$$\text{Min Payment} = P_{D,r,i} * P_c + CP_{D,r,i}(LMP_B - LMP_A)/2 \quad (10.10)$$

where in both (10.9) and (10.10), $P_{D,r,i} = P_{D,r,i}^{(0)} - P_{D,r,i}^{adj}$, $CP_{D,r,i} = \alpha_{D,r,i} P_{D,r,i}$. Here $CP_{D,r,i}$ is the part of power that would be subject to congestion charge if the congestion could not be mitigated. The objective function of the ith independent green DISTCO agent is

$$\text{Min Payment} = P_{D,g,i} * MCP - CP_{D,g,i}(LMP_B' - LMP_A)/2 \\ - \Delta P_{D,g,i}^{adj} * C_{D,g,i}^{adj} \quad (10.11)$$

The objective function of the ith bilaterally contracted green DISTCO is

$$\text{Min Payment} = P_{D,g,i} * P_c - CP_{D,g,i}(LMP_B - LMP_A)/2 \\ - \Delta P_{D,g,i}^{adj} * C_{D,g,i}^{adj} \quad (10.12)$$

The decremental demand usually means the loss in revenue. However, if $P_{D,g,i}^{adj}$ is higher than MCP, the adjustment could bear a profit for the DISTCO. However, $\Delta P_{D,g,i}^{adj}$ should be within a certain range so that the DISTCO can guarantee energy supply to its customers.

10.7.3 Objective of a TRANSCO Agent

In a restructured electricity environment, participants (agents) would need to pay for use of transmission lines. A TRANSCO gets revenue by providing transmission capacity to other market participants. The revenue of a TRANSCO is constrained by its transmission capacity. The objective function of a TRANSCO can be formulated as

$$\max \sum_i \sum_j E_i * T_{j,i} \qquad (10.13)$$

where E_i is the price for transmitting one unit of power over the ith transmission line, and $T_{j,i}$ is the transmission capacity used by the jth agent. A TRANSCO intends to provide as much transmission capacity as possible. In some emergency cases, the ISO could ask the TRANSCO agent to put additional transmission lines into service for temporary use.

10.7.4 Objective of the ISO Agent

The ISO will not intend to profit from congestion mitigation because of its role as a non-profit organization. Instead, since it is responsible for the optimal congestion mitigation, the ISO has the objective of minimizing the cost of congestion mitigation. When PDFs are taken into account for congestion mitigation, the optimization function of the ISO can be stated as

$$\min \sum_i (\Delta P_{G,i} * C_{G,i} / \alpha_{G,i} + \Delta P_{D,i} * C_{D,i} / \alpha_{D,i}) \qquad (10.14)$$

where $\Delta P_{G,i}, C_{G,i}$, and $\alpha_{G,i}$ are the adjustment, the bidding price, and the PDF of the ith GENCO, respectively, for the congested transmission line; similarly, $\Delta P_{G,i}, C_{D,i}$ and $\alpha_{D,i}$ are the adjustment, the bidding price, and the PDF of the ith DISTCO, respectively, for the congested transmission line. This function provides proper scheduling adjustments for all market participants. As mandatory adjustments are practiced, the ISO will change participants' supply and demand.

The SC agent is responsible for coordinating its associated GENCO and DISTCO agents as it negotiates with other agents on behalf of its associated GENCO and DISTCO agents. In addition, an SC agent can help collect useful information on behalf of its associated GENCO and DISTCO agents through communications with other agents

To maintain the energy balance and to respect the market separation rules, GENCO and DISTCO that belong to one SC would need to cooperate for implementing the adjustments. To entice participants to actively participate in the schedule adjustment, it is usually assumed that congestion charges and adjustment revenues are shared equally by contracting agents. However, independent GENCOs and DISTCOs will not be allowed to make adjustments prior to identifying their partners' adjustments unless the ISO becomes responsible for maintaining the energy balance.

10.8 DECISION-MAKING PROCESS OF AGENTS

10.8.1 TRANSCO Agent

A TRANSCO agent could possibly mitigate transmission congestion without any adjustments in the preferred schedules of GENCOs and DISTCOs. For simplicity we do not consider any competition among TRANSCO agents. A TRANSCO agent submits bids to ISO for the utilization of additional lines by the ISO. The ISO will negotiate the submitted bid with the corresponding TRANSCO for buying the right to use the line, and selling that right to other market participants.

The ISO will then publish the price and the allocated cost according to the participants' PDFs. If certain participants, who utilize more than half of the capacity of the congested line, would agree to accept the published price, the ISO will then utilize the new line. Otherwise, the TRANSCO agent will be required to re-bid its price. If no compromise can be reached, the ISO agent will reject the usage of the new transmission line and call for voluntary adjustment bids.

Let us consider a simple example for demonstrating the process of a TRANSCO agent's participation in congestion mitigation. In the example, the TRANSCO has a standby transmission line ik that is not put into service for certain reasons such as preventive maintenance. Here, X/Y in Figures 10.13 and 10.14 represents the ratio of the actual power X to the upper limit of transmission capacity Y.

CONGESTION MANAGEMENT BASED ON MULTI-AGENTS

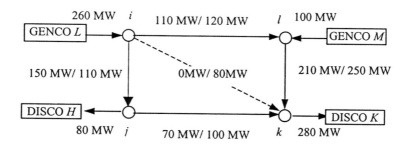

Figure 10.13 Transmission Congestion without Standby Transmission Lines

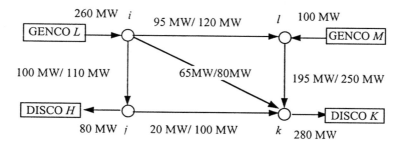

Figure 10.14 Congestion Mitigation with a Standby Transmission Line

Figure 10.13 shows the case where there is transmission congestion on line ij, and the TRANSCO agent submits a bid of C_t for the use of line ik by the ISO to mitigate the congestion. If this bid is accepted by the ISO, line ik is put into service as shown in Figure 10.14 where the congestion on line ij is mitigated. GENCOs L and M, and DISTCOs H and K will have to pay for this transmission service, and the cost of each agent is calculated according to

$$S_i^t = C_t * \alpha_i / \sum_{i=1}^{i=4} \alpha_i \qquad (10.15)$$

where α_i is the power of the ith participant that flows on line ik, which is calculated based on PDFs.

If GENCOs L and M, and DISTCOs H and K plan to mitigate the congestion by voluntary decremental adjustments, then the cost of each agent is

$$S_i^b = \Delta P_i * (C_i^{(0)} - C_i) \qquad (10.16)$$

where ΔP_i and C_i are voluntary decremental adjustments and the price of the ith agent for the congestion mitigation $C_i^{(0)}$ is the price of IPS.

If $S_i^t < S_i^b$, participants will agree to use the standby transmission line; otherwise, participants will resort to adjustment bids. All agents will declare their choices to the ISO, which will make the final decision accordingly.

10.8.2 GENCO and DISTCO Agents

An important issue in a GENCO's or a DISTCO's decision is the estimated congestion charge. When $(LMP_B^{(0)} - LMP_A^{(0)})$ is high, a GENCO agent and its partner DISTCO agent might consider using voluntary adjustments for congestion mitigation. The adjustments of GENCO and DISTCO would serve two functions. First, they would reduce the usage of congested lines; second, they would decrease the difference of $(LMP_B - LMP_A)$ as the congestion is mitigated. However, the realization of these two issues will depend on the competition of adjustment bids.

Suppose that the total number of red agents is N_{red}, and the overload on the congested transmission line is P^{over}, which is theoretically caused by red agents. To mitigate the congestion, we would have

$$P^{over} = \sum_{i=1}^{N_{redg}} (\alpha_{G,i} \Delta P_{G,i}) + \sum_{i=1}^{N_{redd}} (\alpha_{D,i} \Delta P_{D,i}) \qquad (10.17)$$

where N_{redg} and N_{redd} are the numbers of red GENCO and DISTCO agents respectively, and $N_{red} = N_{redg} + N_{redd}$.

Assuming that every red agent will have the same opportunity to participate in congestion mitigation, which means they offer similar adjustment bids, the *k*th red GENCO and DISTCO would decrease their respective generation and demand by

$$\Delta p_{G,k} = \frac{1}{\alpha_{G,k}} \left(p^{over} - \left(\sum_{\substack{i=1 \\ i \neq k}}^{N_{redg}} (\alpha_{G,i} \Delta P_{G,i}) + \sum_{i=1}^{N_{redd}} (\alpha_{D,i} \Delta P_{D,i}) \right) \right) \quad (10.18)$$

$$\Delta p_{D,k} = \frac{1}{\alpha_{D,k}} \left(p^{over} - \left(\sum_{i=1}^{N_{redg}} (\alpha_{G,i} \Delta P_{G,i}) + \sum_{\substack{i=1 \\ i \neq k}}^{N_{redd}} (\alpha_{D,i} \Delta P_{D,i}) \right) \right) \quad (10.19)$$

Because of the energy balance requirement, each SC should keep its generation and demand adjustments in equilibrium. However, this does not include any SCs' rescheduling.

Most often, agents will compete for congestion adjustment bids; they will re-adjust their power generation or demand according to the severity of the congestion. The common goal would be to eliminate the congestion while maximizing their profits. However, not every red agent may want to participate in congestion mitigation. If a red agent is not subject to a large congestion charge, it will not offer adjustment bids for congestion mitigation. Some other red agents might be constrained by their contracts and thus prohibited from participating in congestion mitigation. Therefore, the actual adjustment of a red agent is expected to be higher than the value calculated from (10.18) and (10.19). Nevertheless, as some agents offer to make larger adjustment, other agents will get fewer opportunities to offer adjustments than might be expected.

10.9 FIRST-STAGE ADJUSTMENTS

The first stage is to let the relevant red agents adjust their schedules voluntarily at their own cost. The ISO publishes congestion information including the place, the amount of overload, the color of each participating agent on a blackboard which can be viewed by all agents as shown in Table 10.2. An agent wanting to modify its IPS would consider the adjustment range and bids. To prevent some agents from making a profit by raising bid prices, we could impose a constraint that bid prices cannot be higher than that of initial bids.

Table 10.2 Congestion Information

Location of Congestion	Overload (MW)	LMP_A ($/MWh)	LMP_B ($/MWh)
j-k	50	50	76

10.9.1 Rescheduling of GENCOs

An effective way to reduce congestion charges is to reschedule generation and demand among SCs; this especially benefits SCs whose associated DISTCOs do not have much capacity for re-adjustment. In the example, shown in Figure 10.15, SC_c is associated with GENCO K in zone A and DISTCO L in zone B. According to the contract, GENCO K must supply 250 MW to DISTCO L on the following day, but there is congestion on transmission line ij. If DISTCO L does not provide any decrements, then GENCO K and DISTCO L could be fined 250 MW * (LMP_B - LMP_A) $/MW = 250MW * (450 - 350) $/MW = $25,000. But if SC_c, GENCO K and DISTCO L agents can identify GENCO agents, for instance GENCO H, who can supply an additional 200MW, then GENCO K and DISTCO L will receive a reduced congestion charge by 200 * 100 = $20,000 per hour.

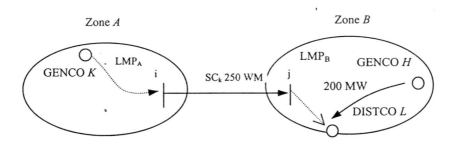

Figure 10.15 Rescheduling of SCs

In addition to the transfer of payment by DISTCO K from GENCO K to GENCO H, GENCO E could share the saved congestion charge of GENCO K and DISTCO L. Since GENCO K and DISTCO K are not in a position to make the deal feasible, most of the reduction in the congestion charge will be transferred to GENCO H. Suppose that GENCO H bids its marginal price $C_{H\_marginal}$ in the power market. When $C_{H\_marginal}$ is higher then MCP, GENCO H will not be accepted by the market. But, if MCP + (LMP_B - LMP_A) are equal to or greater than $C_{H\_marginal}$, GENCO

CONGESTION MANAGEMENT BASED ON MULTI-AGENTS

H will possibly sign the contract with GENCO K and DISTCO L, to supply the generation.

The problem here is to find such a GENCO E for SC_k. The SC_c, GENCO K and DISTCO L agents will seek any useful information and negotiate with any potential green agent candidate. To facilitate the search for a partner, SC_c, GENCO K and DISTCO L, and GENCO H publish the messages seen in Figures 10.16 and 10.17 on the bulletin board, in which ΔC is the incremental price offer above MCP.

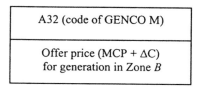

Figure 10.16 Message Used in Asking Generation

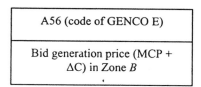

Figure 10.17 Message Used for Offering Generation

After viewing the information published on the bulletin board, the SC_k, GENCO K, and DISTCO L and GENCO H will negotiate on the price. To maintain their proprietary data, agents use their codes instead of their real identity during negation. Whether GENCO H is associated with a SC or is an independent GENCO, it will be offering additional generation with an appropriate price. Independent GENCO agents of zone A will also compete for rescheduling when it is subject to congestion charges.

Successful agents will inform the ISO of their negotiation results for the rescheduling of SCs before certain deadlines, and the ISO will publish this information as shown in Table 10.3 on the bulletin board, for all agents.

Table 10.3 Rescheduling of Agents

Agent code	Zone	PDF of previous GENCO	PDF of DISTCO	PDF of new GENCO	Amount of energy to be rescheduled (MW)
45	A	0.32	0.28	0.22	30

According to Table 10.3, the ISO agent calculates the mitigation of congestion by the rescheduling of GENCOs as

$$P_{left}^{over} = P^{over} - \sum_{i}^{N_{res}} \left(\alpha_{G_{pre},i} \Delta P_i + \alpha_{G_{new},i} \Delta P_i \right) \quad (10.20)$$

Where N_{res} is the total number of rescheduling of GENCOs. $\alpha_{G_{pre},i}$ and $\alpha_{G_{new},i}$ are PDFs of the previous and the new GENCOs of the ith rescheduling in Table 10.3. ΔP_i is the amount of negotiated power to be rescheduled.

If P_{left}^{over} 0, the congestion would be successfully mitigated through the rescheduling of GENCOs. The rescheduling of SCs is at additional cost for red agents. If green agents obtain appropriate offers, they will accept the offers. For a meshed network, the congestion mitigation would involve PDFs that make the situation more complicated.

10.9.2 Adjustment Process of Red Agents

An agent can submit to the ISO a new decremental adjustment scheme that consists of an adjustment range and a bidding price. An SC agent submits this adjustment request on behalf of its associated GENCO and DISTCO agents. A pair of independent GENCO and DISTCO agents can also submit their adjustment requests to the ISO as one unit after they have reached an agreement on decremental adjustments for their respective generation and demand. The ISO then publishes on the bulletin the available information for voluntary adjustments as shown in Table 10.4.

Table 10.4 Voluntary Adjustment Bids of the Red Agents

Code of SC Agent	Adjustment range [P_{min}, P_{max}] (MW)	Bidding Price P ($/MWh)	Minimum PDF	Maximum PDF
45	[0, 30]	50	0.08	0.28

The ISO agent calculates the following values:

$$\Delta p_{min} = \sum_{i=1}^{N} \alpha_{min,i} \Delta p_{min,i} \qquad (10.21)$$

$$\Delta p_{max} = \sum_{i=1}^{N} \alpha_{max,i} \Delta p_{max,i} \qquad (10.22)$$

where $\alpha_{min,i}$ and $\alpha_{max,i}$ are the minimum and maximum PDFs of GENCO and DISTCO agents associated with the ith SC, $\Delta p_{min,i}$ and $\Delta p_{max,i}$ are the minimum and maximum adjustments of GENCO and DISTCO agents associated with the ith SC.

If $\Delta p_{min} \geq p^{over}$, which means that sufficient adjustments are available, the ISO agent will inform participants that each agent needs to adjust its minimum amount to mitigate the congestion. However, each pair of agents has the right to adjust their generation and demand as long as their adjustments are within the submitted range to the ISO. In the case where $\Delta p_{max} < p^{over}$, voluntary adjustments would be insufficient and mandatory adjustments are required. In this case most agents would want to implement their maximum adjustments to reduce congestion charges.

In the case where $\Delta p_{max} > p^{over} > \Delta p_{min}$, there would be space for agents to adjust their generation and demand to eliminate the congestion. The ISO agent will publish a list based on the merit order of bidding price/APDF as shown in Table 10.5, where M is the total number of participating agents, and the second column is given in an ascending order. APDF is the average PDF of paired GENCO and DISTCO agents.

Table 10.5 Order for Decremental Adjustment

Order	Agent Code	Price ($/MWh)/APDF
1	A12	38.23
⋮		
M	A57	48.12

The kth pair of agents would adjust its generation and demand according to the following formulation:

$$P^{over}_{left-new} = P^{over}_{left} - \sum_{i}^{k-1}\left(\alpha_{G,i}\Delta P_{G,i} + \alpha_{D,i}\Delta P_{D,i}\right) \qquad (10.23)$$

where $\Delta P_{G,i} = \Delta P_{D,i}$ to maintain a power balance, and $\Delta P_{G,i} = \Delta P_{D,i} \in \left[P_i^{min}, P_i^{max}\right]$ is the optimization result for the ith agent.

An agent can estimate its loss by assuming that the congestion is entirely mitigated; that is $LMP_B - LMP_A = 0$. All agents must check if their losses are within acceptable range. When congestion is mitigated, the financial loss to an independent GENCO is estimated as

$$L_{G,i} = (P^{(0)}_{G,i} - P_{G,i}) * MCP \qquad (10.24)$$

where $P^{(0)}_{G,i}$ is the initial preferred scheduling submitted to the ISO, $P_{G,i}$ is the generation after the voluntary adjustment for congestion mitigation. MCP is the market-clearing price set by the power exchange.

Similarly, the financial loss to a GENCO or a DISTCO, scheduled through an SC, is estimated as

$$L_{G,i} = (P^{(0)}_{G,i} - P_{G,i}) * C_{G,i} / 2 \qquad (10.25)$$

where $C_{G,i}$ is the energy price set in the bilateral contract.

The ISO agent supervises the entire adjustment process. It calculates $P^{over}_{left\_new}$ after each agent has completed its adjustment, when $P^{over}_{left\_new}$ becomes zero. When the congestion is mitigated, the ISO agent terminates the adjustment process. However, if the congestion is not mitigated after the first round of adjustments, the ISO may possibly allow agents to perform a second round of adjustments. The second round of adjustments would be broader, and the bidding price would be lower than that in the first round, and this implicitly keeps the first round of adjustments valid. If the congestion would not be eliminated after the second round of adjustments, additional voluntary adjustments of green agents would be sought.

10.10 SECOND STAGE ADJUSTMENTS

At the second stage, the ISO uses congestion credits to entice participants to provide more voluntary adjustments. The red agents who would be subject to congestion charges are supposed to have done their best to adjust their generation and demand at the first stage. Hence, only green agents would make voluntary adjustments for congestion credits at the secondary stage. Theoretically, green agents have nothing to do with the congestion and thus have no obligation to participate in the congestion mitigation. In a competitive environment, certain factors might inspire these agents to participate in congestion mitigation. Green agents would profit by acquiring more congestion credits than if they offered voluntary adjustments for congestion mitigation. The only requirement is that their adjustments accommodate the congestion mitigation; otherwise, their adjustments will be rejected by the ISO.

Green SCs, which are associated with DISTCO and GENCO agents in zones A and B, respectively, would be able to increase their generation and demand when congestion credits are made because the DISTCOs in zone A can buy energy from GENCOs in zone A at a price not higher than LMP_A, while GENCOs in zone B can sell energy at a price above LMP_B. However, since LMP_B is greater than LMP_A, and there are no extra conditions, green SCs cannot accommodate both sides. So congestion credit must be considered.

When a green SC agent wants to participate in congestion mitigation, it must seek relevant information from both congestion zones, pass it to its associated GENCO and DISTCO agents, and let them negotiate a bidding price. If independent green GENCO and DISTCO agents participate in the congestion mitigation, they will seek a partner at this secondary stage, and then negotiate with that partner on a bidding price to be submitted and accepted by both sides.

A pair of GENCO and DISTCO agents will submit one price to the ISO. The ISO publishes the adjustment information as shown in Table 10.6 on the bulletin board after it receives all the adjustment bids.

Table 10.6 Available Incremental Adjustments of Green Agents

Agent Code	Adjustment Amount (MW)	Bidding price ($MWh)	Minimum PDF	Maximum PDF
23	45	56	0.11	0.34

According to the information in Table 10.6, the ISO calculates

$$\Delta p_{green} = \sum_{i=1}^{N_{green}} (\alpha_{G,g,i} \Delta P_{G,g,i} + \alpha_{D,g,i} \Delta P_{D,g,i}) \quad (10.26)$$

where N_{green} is the total number of green agents participating in the congestion mitigation. $\Delta P_{G,g,i}$ and $\Delta P_{D,g,i}$ are incremental generation and demand of the ith green SC. $\Delta P_{G,g,i}$ is equal to $\Delta P_{D,g,i}$, $\alpha_{G,g,i}$, and $\alpha_{D,g,i}$ are PDFs of the associated GENCO and DISTCO agents. If $\Delta p_{green} < p_{left\_new}^{over}$, the voluntary adjustments are insufficient; however, for $\Delta p_{green} \geq p_{left\_new}^{over}$, the ISO agent will inform the participating green agents of a possibility effective congestion mitigation. In either case, the ISO agent will publish a list as shown in Table 10.7 in which the third column is given in an ascending order.

Table 10.7 Order for Incremental Adjustments

Order	Agent Code	Price ($/MWh)/PDF
1	A44	35.78
⋮		
M	A27	29.45

The first pair of agents on the list would adjust their generation and demand first, and remaining agents provide their adjustments in the orders given on the table. The ISO agent monitors the extent to which the congestion is mitigated, which is similar to that of the first stage. The adjustment process will continue until the congestion is mitigated or all agents have exhausted their adjustment ranges. This process is performed once, and if the adjustment is insufficient, the ISO will use mandatory adjustments to mitigate the congestion.

10.11 CASE STUDIES

10.11.1 Congestion Information

The ISO locates a 40 MW overflow on the line between nodes i and j in Figure 10.10 after clearing the market. The congestion mitigation process is discussed as follows.

The ISO assigns a code to each agent to conceal itsr identity as shown in Table 10.8. Then, the ISO publishes the PDF and agent property information given in Tables 10.9 and 10.10 on the bulletin board. A red agent can only be decremented and a green agent can only be incremented. Each agent submits its adjustment range and bidding price to the ISO along with its IPS.

Table 10.8 Codes of Agents

Agent code	Represented Participant
A01	GENCO K
A05	DISTCO M
A10	DISTCO N
A12	DISTCO L
A15	GENCO E
A20	GENCO H
A22	SC_c
A25	SC_d
A30	GENCO Z
A32	DISTCO Y
A33	SC_f

Table 10.9 Congestion Information

Congestion Place	Overloaded Energy	LMP_A	LMP_B
Line ij	40 MW	220 $/MW	370 $/MW

Table 10.10 PDF and Property of Agents

Agent Code	PDF	Property	Adjustment Range (MW)	Bidding Price ($/MW)
A01	0.21	R	[-50, 150]	350
A05	0.15	R	[0,0]	350
A10	-0.09	G	[0,100]	340
A12	0.17	R	[-50, 150]	350
A15	-0.25	G	[0,100]	340
A20	-0.18	G	[0,0]	500
A22	0.19	R	[-50, 150]	350
A25	-0.17	G	[0,100]	340
A30	0.23	R	[-100,100]	390
A32	0.27	R	[-100,100]	390
A33	0.25	R	[-100,100]	390

10.11.2 The First-Stage Adjustment of Red Agents

SC_c is associated with the GENCO K and DISTCO L, and 100 MW is scheduled between these two participants. Accordingly SC_k, GENCO K, and DISTCO L, are subject to following congestion charge:

$$100 \text{MW} * (370 - 220) \text{ \$/MW} = \$15,000$$

Suppose that the contract price of SC_c is 250 $/MW and MCP is 280 $/MW. To avoid congestion charges, SC_c tries to locate other SCs for rescheduling. So it publishes the information given in Figure 10.18 on the bulletin board. Suppose that GENCO H cannot supply energy to the market, since its bid was 300 $/MW. This GENCO will now get a chance to re-submit by publishing the information in Figure 10.19 on the bulletin board.

A01 offers a price $(250 + 70) for generation in zone B

Figure 10.18 Message Asking for Generation

CONGESTION MANAGEMENT BASED ON MULTI-AGENTS

> A20 bids a price $(250 + 60)$ for generation in zone B

Figure 10. 19 Message Offering Generation

Then SC_c would select A20 as its business partner. Suppose that the partners make an agreement to exchange 20 MW and SC_c reduces the congestion charge by 20 MW $\times$ (370 - 220) \$/MW = \$3000. If the congestion cannot be mitigated, A20 will make 20 MW * 320 \$/MW = \$3000. According to the PDF information in Table 10.10, congestion is mitigated by 0.21 * 20 + 0.18 * 20 = 7.8 MW after the rescheduling of SC. Hence, line *ij* would still have (40 - 7.8) MW = 32.2 MW of overloaded flow. Suppose that these two SCs are the only ones that will provide rescheduling. The ISO agent calculates new LMPs as LMP_A = 200 \$/MW and LMP_B = 310 \$/MW. Then, the congestion credit for A20 is 20MW * 0.18 * (310 - 200) \$/MW = \$396. So, SC_k is subject to a congestion charge of 80 MW * 0.21 * (310 - 200) \$/MW = \$1848, and A05 is subject to congestion charge of 50 MW * 0.15 * (310 - 200) \$/MW = \$825.

Suppose that the SCs' congestion charges are calculated according to the scheduled energy between zones A and B rather than by using PDFs. A33 has scheduled 40 MW for A30 from zone A to zone B, so it is subject to the congestion charge of 40 MW * (310 - 200) \$/MW = \$4400; A30 and A40 will each bear half of this congestion charge. The agents' congestion charges based on new LPMs are listed in Table 10.11.

Table 10.11 Congestion Charges of Red Agents

Agent code	Property	Congestion Charge ($)
A01	R	462
A05	R	412.5
A12	R	462
A22	R	924
A30	R	2200
A32	R	2200
A33	R	4400

These agents will offer voluntary adjustments to mitigate the congestion by submitting bid prices and adjustment ranges. The bid price cannot be higher than the adjustment bidding price in the IPS in order to keep participants from making excessive profit. Suppose that A05 cannot find a partner for its adjustment. Then it can neither meet the market requirement of balanced power nor participate in the adjustment. A22 and A33 submit their voluntary adjustment bids to the ISO on behalf of their GENCOs and DISTCOs. The ISO will publish this information as listed in Table 10.12. From the information in Table 10.12, the ISO agent will produce and publish the information shown in Table 10.13. The APDF in the table is the average of PDFs of the associated participants.

Table 10.12 Voluntary Adjustment Bids of Red Agents

Agent code	Bidding Price	Adjustment range [$\Delta Pmin, \Delta Pmax$]	Minimum PDF	Maximum PDF
A22	250	[10,40]	0.17	0.21
A33	200	[15,50]	0.23	0.27

Table 10.13 Order for Decremental Adjustment

Order	Agent Code	Price/APDF
1	A33	800
2	A22	1316

A33 gets the right to adjust its schedule first. Suppose that it decides to make a reduction of 30 MW. Then the ISO agent calculates the remaining overloaded flow on the congested transmission line according to (10.23):

$$p_{left-new}^{over} = 32.2 - (30*0.23 + 30*0.27) = 17.2 \text{ MW}$$

Since the congestion is not completely mitigated, A22 will have to make an adjustment. Suppose that A22 decides to make a reduction of 40 MW so that the remaining overloaded flow on the congested transmission line is

$$p_{left-new}^{over} = 17.2 - (40*0.17 + 40*0.21) = 2.0 \text{ MW}$$

There is still 2 MW to be mitigated on the congested transmission line. Suppose that the adjustment could be repeated. Then, if A22 and A33 have further adjustment possibilities, they can do so in the second round. Let us suppose that A22 and A33 cannot make adjustments in a second round. Hence, we need a second stage for mitigating the congestion.

10.11.3 The Second-Stage Adjustment of Green Agents

Suppose that A25 is the only agent that can provide a voluntary adjustment for congestion mitigation. A25 submits the information in Table 10.14 to the ISO as it decides to make a 10 MW additional adjustment.

Table 10.14 Voluntary Adjustment Bids of Green Agents

Agent Code	Bidding Price	Adjustment range $[\Delta Pmin, \Delta Pmax]$	Minimum PDF	Maximum PDF
A25	240	[5, 15]	0.09	0.25

Before the additional adjustment of A25 is realized, the ISO calculates the remaining overloaded flow on the congested transmission line as

$$P_{left-new}^{over} = 2.0 - (10 * 0.09 + 10 * 0.21) = -1.0 \text{ MW}$$

A negative value means that the power flow is below the transmission limit. So the ISO agent asks A25 to make the following adjustment:

$$\Delta p_{inc} = 2.0/(0.09 + 0.21) = 6.67 \text{ MW}$$

Accordingly the transmission congestion is entirely mitigated. In case the congestion was not entirely mitigated, the ISO would use the mandatory tools to mitigate the congestion in order to maintain system reliability.

10.12 CONCLUSIONS

The congestion mitigation scheme based on the multi-agent model has more flexibility than the traditional LP method. It provides market participants with additional freedom to make reasonable adjustments. Market participants are allowed not only to make decremental adjustments based on LP but also allowed to make increments based on their impact on the transmission line just as green agents did at the second stage. In this proposed agent method, market participants should make a choice between accepting congestion charges and offering voluntary adjustments at their own costs. Once the congestion occurs, participants who have physically utilized the congested lines will have to bear the related economic losses. Although congestion loss or revenue due to limited adjustment ranges may be minute, they may be substantial based on the submitted IPS. The general goal is to mitigate the congestion effectively, so the agent with the bigger distribution factor and lower price will be adjusted initially.

The flexibility of this proposed agent method manifests itself in several ways. For instance, as intelligent agents are well designed, they attempt to solve the congestion problem on behalf of their host. In applying an ISO agent, the ISO can work on its more critical tasks while the ISO agent is working on the congestion management task.

Chapter 11

Integration, Control, and Operation of Distributed Generation

11.1 INTRODUCTION

Distributed generation (DG) encompasses any small-scale electricity generation technology that provides electric power at a site close to consumers. The size of DG could range from a few kilowatts to hundreds of megawatts. DG units, which are scattered throughout the distribution system, will be connected to a consumer's facility, the utility's distribution system, the power transmission grid, or a combination of these options.

Today there is growing interest in DG, particularly as on-site generation for businesses and homeowners, which is stimulated by better power quality, higher reliability, and fewer environmental problems [Del01, Dug93, Dug00, Web17]. This interest is strengthened by the availability of more efficient and modular electric supply technologies.

DG technology is often lumped with distributed storage, and their combination is referred to as a distributed energy resource (DER) that represents a modular electric generation or storage installed at consumer sites. In some cases, the DER will include controllable loads. Dispersed generation[1] (DG), a subsection of DER, refers to smaller generating units that are usually installed at consumer sites and isolated from a utility system. In most cases, dispersed generation, which is seldom operated in parallel with a utility system, will be used as a standby power source during power outages.

---
[1] Also referred to as Distributed Generation

Technological advances in DG have resulted in small-scale generation that is cost-competitive with large power plants. Compared with traditional large-scale generation stations, DG is a less expensive, flexible and environmentally friendly power source. These features enhance its position in market competition. It is perceived that the efficiency of most existing large generation units is in the range of 28% to 35%, indicating that they convert 28% to 35% of their input energy into electric power. By contrast, efficiencies in the range of 40% to 55% are attributed to several DG units including small fuel cells and various combined cycle and gas turbines units suitable for utilization as DG. The improved efficiency of combustion turbines, which is about 50% when operated as combined cycle plants, has transformed DG units from expensive peaking units to base-load generators. Microturbines in the tens-of-kilowatts range are offered to the market at a cost that is competitive with the delivered price of retail energy. Internal combustion engines are significantly improved to generate electricity with impressive efficiency. In addition, fuel cells, photovoltaics, and wind turbines have become a commercial viability. Cogeneration has provided much wider economic possibilities for smaller units that could accelerate their deployment in restructured electricity markets. Some of the specific attributes of DG for utilization in restructure power systems are as follows.

- **Performance.** The performance-based pricing mechanism in the restructured power systems should encourage wide utilization of DG. The electric power market is like a comb that is made up of many niches with different frameworks for energy needs. DG offers a very wide range of options and hence will appeal to consumers in many of these niches. DG is reliable and economic for use by commercial, residential, and industrial consumers. Moreover, DG can operate as a standby power source, which is especially important to some critical and sensitive loads possibly subject to curtailments or interruptions. Besides providing base load generation, peak shaving, standby power supply, and cogeneration, DG is expected to have an even greater function in a restructured electricity environment in providing spinning reserve, reactive power supply and other ancillary services for system reliability, power quality and congestion mitigation.

- **Cost of delivery.** A vital consideration for the utilization of DG is that it will not incur transmission and distribution costs. This will greatly reduce its marginal costs, enhance its competition, and reduce severe congestion penalties [Del01, Ram01]. The other advantage in competition is derived from DG's flexibility. DG can be sized appropriately to match the specific needs of consumers and can be quickly installed almost anywhere to capture the market value at key locations. DG can operate flexibly to follow hourly fluctuations in energy prices.

- **Environmental concerns.** Environmental concerns are the key driver for utilizing DG technologies, particularly renewable energy technologies. The emission level of DG is significantly lower than that of central generating stations. Though there is a debate over what constitutes renewable, wind and solar technologies are generally acceptable entities in this category. Large-scale hydropower plants are also considered renewable resources, but not necessarily environmentally friendly because of their impact on fish and wildlife species and the land that would be flooded. The most favorable renewable generation option at this time is the wind farm. Geothermal and biomass applications can also be cost-effective sources of energy but difficult to site and more difficult to implement. The construction of new renewable generation resources is partly financed through premium green energy markets.

- **Energy from Brownfield.** DG makes it possible to tap energy resources that would otherwise be wasted. There are many oil fields that are remote or produce low-quality gas that is not economical to transport, so the gas is ignited in a flare to dispose of it. DG technologies such as microturbines appear to be excellent means for recovering this kind of energy. Biomass power is electricity produced from biomass fuels. Biomass consists of plant materials and animal products. Biomass fuels include residues from the wood and paper products industries, food production and processing, trees and grasses grown specifically to be used as energy crops, and gaseous fuels produced from solid biomass, animal wastes, and landfills. Some of these resources could produce more than 1000 kW of DG. Biomass technologies convert renewable biomass fuels into electricity using modern boilers, gasifiers, turbines, generators, fuel cells, and other methods.

- **Production cost.** The success of DG in the market competition is due to its ability to provide consumers with the lowest cost solution to meet their particular needs. The energy price mostly depends on the production costs, and many factors can contribute to the overall pricing of certain generation technologies. Generation efficiency is most important in this category, because even a small increase in efficiency can yield a significant increase in profitability. Other operating costs can be reduced by enhancing the reliability. High reliability reduces the need for outages, and therefore the need for high-cost backup power; in addition, high reliability can reduce the demand for O&M personnel and replacement parts.

- **Flexibility.** DG provides significant flexibilities in power system operations, such as quick ramp-up and shutdown, and enables investors to take advantage of short-term sales opportunities. DG's high efficiency, flexibility, and low-maintenance requirements, would keep its O&M costs down. Since DG systems are installed at distribution sites, neither consumers nor energy providers would pay transmission and distribution tariffs. In circumstances where DG faces challenges imposed by capital and production costs that are higher than those of larger generating systems, these challenges can be balanced against factors such as the opportunity for waste heat utilization, increased reliability at the site, elimination of peak load constraints, reduction in transmission and distribution charges, reduction in line losses, improved power quality, and more flexible response to market changes.

- **Retail competition.** Retail competition encourages more on-site generation by energy consumers as DG becomes a means of managing the uncertainty in the energy supply market. DG investors include retail customers, electric utilities, and various ESPs (energy service providers) among which are IPPs (independent power producers). Retail customers would evaluate the need for installing a DG system as an alternative to the following operations:

    o Identifying load shedding and load shifting opportunities

    o Quantifying outage costs to determine whether a standby power is needed

    o Evaluating energy requirements to determine whether a DG alternative is more economical.

- **Deployment.** So far numerous DG units have been installed within distribution systems. The DPCA (Distributed Power Coalition of America) has estimated that, within the next 20 years, DG projects could capture 20% of new generating capacity. There are no such risks as stranded cost in utilizing DG as most of the energy will be consumed by local consumers. Modular plug-and-play interfaces and intelligent adaptive control technology, along with a regulatory and institutional environment that recognizes the benefits of DG, would result in ubiquitous deployment of DG. This deployment could represent some of the integrated components of efficient, clean, and reliable energy systems for buildings and industrial facilities.

- **Reliability.** For the reliable and efficient operation of a distribution system with DG units, the operation and control strategy must accommodate both the engineering needs to maintain collective services as well as the economic push for independent and decentralized decision-making [Bar00, Beg01, Can0, Dal01, Din01, McD02].

This chapter will address technical and economics issues associated with integrating many small DG generators into the distribution system and a competitive electricity market. The role of DG in minimizing energy costs by reducing losses, replacing expensive energy at peaks, and relieving congestion will be investigated. New system monitoring and control methods based on the SCADA/DMS will be introduced. The chapter will also discuss retail market mechanisms whereby DG participates in the competitive markets that will be established as power industries are further restructured.

11.2 DG TECHNOLOGIES

In general, DG can make use of energy derived from wind, solar, geothermal, biopower, and fossil fuels. Typical DG technologies available include wind turbines, photovoltaic panels, fuel cells, combustion turbines, gas turbines, and combustion engines. Several of these technologies could offer clean, efficient, and cost-effective electric energy

11.2.1 Wind Turbines

Wind turbine generates electricity by making use of renewable wind energy. As early as in the first half of the 20th century, small amounts of electricity were generated by windmill to light rural farmhouses. Today wind power is the fastest growing energy source in the world. By the end of 2000, total world wind capacity was about 17,000 MW, enough to generate about 34 billion kWh a year.

Wind is a clean renewable energy that is becoming increasingly popular with utility customers and policy makers. Many wind power plants consisting of wind turbines are being developed to meet the need for such a clean, sustainable power source. The working principle of wind turbine is very simple, which implies the construction of a wind power plant would be easier than that of a conventional fossil fuel power plant. Wind turbines can be coupled to a synchronous or induction generator and operated at constant or variable speed. A variable speed turbine has a simpler mechanical system and is usually the preferred solution for a new installation. Wind energy systems use an inverter to convert a dc current in to the standard ac current.

11.2.2 Photovoltaic Systems

Photovoltaic (PV) systems make use of solar energy to produce electricity. Modular photovoltaic cells can be installed at any place where the sun shines and have been commercially demonstrated in extremely sensitive environments. However, because of the intermittent nature of the PV generation, a dedicated battery needs to be integrated into the PV system for a better performance. With the availability of advanced batteries, it will be possible to store large amounts of energy during off-peak periods for use during peak hours. The current high capital costs of PV make these systems a niche technology that can compete more on the basis of environmental benefits than on economics. PV has been used extensively in space power programs. As hardware costs decline, terrestrial applications of PV would become more economical.

11.2.3 Fuel Cell

Fuel cells are unique self-contained, energy conversion devices that convert fuel, such as the natural gas, into electricity at near atmospheric pressure. The electricity-producing process of fuel cells is quite like that of a battery; but unlike a battery, which produces electricity from stored chemicals, fuel cells produce electricity when hydrogen fuel is delivered to the negative pole of the cell and oxygen in air is delivered to the positive pole. In fuel cells, hydrogen and oxygen are separated by an electrolyte to induce an electrochemical potential. This potential is converted into electricity (dc) by hydrogen protons moving through the electrolyte and electrons flowing through a separate electrical circuit. The dc voltage produced by the cell is converted into ac using a dc/ac inverter.

The hydrogen fuel used for fuel cells can be obtained from a variety of sources. However, the most economical technique is the steam reforming of natural gas, which is a chemical process that strips the hydrogen from both fuel and steam. According to the types of liquid and solid media used for creating the fuel cell's electrochemical reactions, fuel cells would be classified into the following four groups:

- Molten carbonate fuel cells (MCFC)
- Phosphoric acid fuel cells (PAFC)
- Proton exchange membrane fuel cells (PEM)
- Solid oxide fuel cells (SOFC)

Each group represents a distinct technology with its own performance characteristics. Generally, applying direct electrochemical reactions in fuels cells would be more efficient than using fuel to drive a heat engine to produce electricity. When utilizing fuel cells, unreacted fuel along with steam generated from heating the recovered water is recycled to operate fuel processors. The remaining heat can be used for water or space heating by the consumer. Therefore, fuel cells are highly efficient with efficiencies ranging from 35%-40% for PAFC up to 60% for MCFC and SOFC. Fuel cells have a high degree of reliability, are easily adaptable to new fuels, require little on-site attention, and can be operated remotely.

Fuel cells are capable of producing reliable electricity for residential, commercial, industrial, and transportation applications. Fuel cells possess high-power densities and, owing to their compact nature, are very easy to site in locations where real estate is scarce. Compared with conventional centralized power plants which usually demand a large space to site prime movers and complicated auxiliary systems, a typical fuel cell with an output of 200 kilowatts requires an area of 10 by 25 by 12 feet in size. Fuel cells are also known for high flexibility and modularity for installation. A consumer can start with a small fuel cell unit and stack additional capacity to meet varying capacity requirements. This prefabricated nature makes it easy to quickly construct fuel cells. In addition, the short lead-time for installation allows consumers to expand generation capacity in a timely manner. For instance, when the load growth is rather abrupt in a service territory, the electric utility may satisfy the demand by quickly establishing a power plant based on fuel cells. The investment in this case is less risky since fuel cells can be relocated easily.

Fuel cells also present major environmental incentives. Because fuel cells use a non-combustion process to produce electricity, they show no vibration and do not give rise to any noise. It is perceived that, due to technological innovations, fuel cell emissions would become negligible as compared to conventional means of producing electricity.

The chief obstacle to commonly accepted fuel cells is the cost, namely the equipment cost for fuel cells is currently much higher than the lowest cost types of DG (usually diesel generator sets). The present hope is for the automotive industry to adopt the fuel cell as its primary energy source, which might bring the cost down to where it would be economically competitive with other means of electricity generation. Currently, fuel cells have proved economical in certain niche applications that are highly subsidized.

11.2.4 Combustion Turbines

Combustion turbines (CTs) represent an established DG technology ranging from several hundred kilowatts to hundreds of megawatts. CTs could burn natural gas, a variety of petroleum fuels, or have a dual-fuel configuration. With the capacity of 1 to 30 MW, CTs could produce high-quality heat that would generate steam for power generation, industrial use, or district heating. CT emissions could be reduced to very low levels using dry combustion techniques, water or steam injection, or exhaust treatment. Maintenance costs of CTs per unit of output power are among the lowest for DG. Low-maintenance and high quality waste heat would make CTs an excellent choice for industrial or commercial CHP (combined heat and power) applications with larger than 5 MW capacity.

11.2.5 Microturbines

Microturbines belong to the smallest category of combustion turbines. Microturbines were originally designed for vehicular application such as aircrafts and buses but are now receiving much attention in power industries. Microturbines are simple, compact, and robust, and they are suitable for DG use. Recently, microturbine units have been used as a stationary source of DG.

Usually the capacity of microturbine units ranges from 20 to 750 kW, in the manageable size of a refrigerator. In most configurations, the microturbine shaft spins at 100,000 rpm and the output is converted to the standard power grid frequency as illustrated in Figure 11.1.The dc/ac

DISTRIBUTED GENERATION

converter is used for generating ac at a variable frequency. If additional power is needed, the microturbine shaft is revolved at a higher speed without any need for a precise speed regulation.

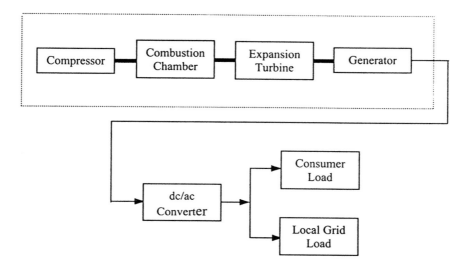

Figure 11.1 Microturbine Generation Unit

Microturbines have considerable advantages over other kinds of DG units. For instance, to further increase the efficiency of microturbine units, a special heat exchanger, called a recuperator, can be employed to capture the exhaust thermal energy for preheating the compressed air before the burner. Most microturbines are capable of producing electricity at an efficiency of 25%-30%. As a special category of CTs, microturbines share most of the advantages and disadvantages with combustion turbines. Except for low emissions, major features of microturbines in their market niche include:

- Compact design
- Multi-fuel capability
- Low maintenance rate
- Durability
- High reliability
- Quick response to load fluctuation.

11.2.6 Internal Combustion Engines

Combustion engines, especially reciprocating internal combustion (IC) engines, represent a widespread technology which is used in automobiles, construction and mining equipment, marine propulsion, and other types of power generation from small portable generator sets to large engines, powering generators with several megawatts of capacity.

Spark ignition IC engines for power generation would mostly use natural gas as fuel but could also run on propane or gasoline. Diesel cycle, compression ignition IC engines could operate on diesel fuel, heavy oil or in a dual-fuel configuration that would burn natural gas primarily with a small amount of diesel pilot fuel that could be switched to 100% diesel. Current IC engines would offer low fuel cost, easy start-up, proven reliability when properly maintained, good load-following characteristics, and heat recovery potential. IC engines with heat recovery have become a popular form of DG. Emissions of IC engines are reduced significantly by exhaust catalysts and by better design and control of the combustion process.

11.2.7 Comparison of DG Technologies

In general, economics of electric power systems will depend on capital costs, operating efficiencies, fuel costs, as well as operation and maintenance costs. The DG technologies discussed above are considered compatible with other merchant power generation options and are utilized in today's restructuring environment. Each technology has its own strengths and weaknesses for competition and so would be aimed at a specific market segment. Photovoltaic cells are less versatile than wind energy systems, which can be combined with other forms of energy sources such as natural gas to form a hybrid system that is cost-effective and, supply a continuous source of power. Environmentally-friendly renewable energy technologies such as wind turbines and photovoltaic, and clean and efficient fossil-fuel technclogies such as gas turbines and fuel cells are new generating technologies driving the utilization of DG. These renewable generators usually have a small size and can be easily connected to distribution grids [Dis99].

A side-by-side comparison of these DG technologies, as of 2001, is given in Table 11.1. All these DG systems are dispatchable momentarily except wind turbine and photovoltaic systems, which depend on climatic conditions.

Table 11.1 Performance Characteristics of DG Technologies

Technology	Fuel Cell	Wind Turbine	Photovoltaic System	Combustion Turbine
Capacity Range	500 kW – 25 MW	200 kW – 2MW	10 kW – 0.8 MW	1 kW – 0.5 MW
Efficiency	29 – 42%	40 – 57%	25%	6 – 19%
Capital Cost ($/kW)	450 – 1870	700 – 2800	600 – 2300	900 – 1900
O&M Costs ($/kW)	0.003 – 0.007	0.004 – 0.011	0.003 – 0.009	0.001 – 0.005
Dispatchable?	Yes	Yes	Yes	Yes

11.3 BENEFITS OF DG TECHNOLOGIES

DG technologies promise a great many benefits, including cheaper power supply, higher power quality, less capital investment, and improved system performance. These benefits will be transmitted to a large segment of electric power system users among them consumers, electric utilities, ESPs, transmission operation, and power marketers. Some of these issues are discussed next.

11.3.1 Consumers

Most DG units are located near customers and are operated by electric utilities, ESPs, or customers. DG technologies are becoming increasingly efficient and reliable, and will soon be competitive with conventional grid-based power supply. DG can greatly reduce consumers' expenditure for electricity while consumers with on-site DG will benefit from the additional cost-effectiveness and reliability.

In comparison with traditional centralized generation facilities, most DG units are installed at or near consumers' sites. This way, DG owners obtain a cheaper power source without needing to pay for costly transmission and distribution loss and services. This arrangement is especially important to consumers in areas with frequent transmission congestions. In addition, by retaining DG as a standby power supply, critical consumers can double their power supply reliability and guard against power outages that could otherwise pose catastrophic outcomes to their businesses. For instance, a hospital could install an on-site fuel cell system to avoid power outages that would cripple its sensitive equipment and systems. There is also an opportunity for DG owners to make a profit by selling their energy or ancillary services to the power market, especially during peak demand hours.

11.3.2 Electric Utilities and ESPs

To compete for supplying electricity, utilities and ESPs would need to develop new strategies in delivering electricity to customers. Since DG is a cheaper power source and does not require any transmission and distribution services, one strategy for competition is to provide on-site DG as a kind of value-added service to consumers. Both electric utilities and ESPs could use DG units to provide bundled services to consumers and compete outside their service territories. Besides, DG could be used by utilities as a means of retaining consumers and protecting their service territories from intrusions. Likewise, DG could facilitate a utility's invasion of other territories that were inaccessible due to transmission or siting constraints.

Some electric utilities might decide to install on-site DG units near their customer sites, rather than upgrading old distribution feeders, which could be more costly. Electric utilities could use their DG units to defer the investment in power delivery facilities such as distribution lines and transformers. For instance, electric utilities could lease diesel generator sets and mount them at customer sites for shaving peak loads. Electric utilities and ESPs could add DG with low capital investments and short lead-time for construction to supply small capacity increments.

DG could not only be used to defer the construction of new distribution lines, it could also be used to improve power quality and maintain reliability. Most DG units have a power electronic interface that can be properly configured to provide ancillary services when necessary. In this regard, DG units could improve the voltage profile of the local distribution system.

In summary, electric utilities and ESPs could benefit from the following aspects of DG:

- Reduction in operating costs
- Improvement of power quality and reliability
- Deferral of capital investments for power delivery

DISTRIBUTED GENERATION

11.3.3 Transmission System Operation

Although DG units are mostly installed in local distribution systems, they can indirectly affect the operation and planning of generation and transmission systems. As to the operation of a transmission system, following benefits could potentially be obtained from the utilization of DG:

- Deferral of new transmission investments
- Mitigation of transmission congestion
- Improvement of power reliability

11.3.4 Impact on Power Market

The integration of DG into the power grid could influence the operation of power markets. The formation of a retail power market where consumers can choose their power provider is the expected result of power industry restructuring. In the retail power market, DG will compete with traditional centralized power generation. New business models for power market need to be developed to support the realization of the full economic value of DG, and they should take into account the value of DG in delaying or avoiding transmission and distribution system upgrades, the use of DG for ancillary services and for improving system reliability, power quality, and reducing line losses.

These distinctive features of DG make it fit well into the combined PoolCo/bilateral (i.e., hybrid markets). DG units installed at consumer sites will affect certain purchasing requirements from the pool. In particular,

- General shift to CHP district heating systems would move the source of generation closer to the location of electricity consumption and tend to reduce bulk power transfers on transmission systems.

- The transmission system will likely accommodate more renewable energy generating units to cope with changes in capacity which is in response to market demand.

- There is a potential for unpredictable fluctuations in market trades as a result of large trenches of DG such as wind and solar generation.

11.3.5 Environmental Benefits

Concerns for environment have stimulated the deployment of DG. The DG's environmental benefits can be observed and shared by participants in various ways, as discussed later in this chapter. As an alternative to central generation, DG units supply energy just at or near the point of use. Therefore, transmission and distribution energy losses can be greatly reduced, which means less electricity would be generated from traditional power plants which mostly use fossil fuels and produce emission. It is believed that DG will be the best choice for obtaining pollution credits when the mandate for clean power is enacted.

11.4 BARRIERS TO DG UTILIZATION

Although, the employment of DG could provide benefits to almost all entities of power systems, DG is encountering various barriers that could hinder its connection to the utility's grid and the power market. These barriers are categorized as technical, business practices, and regulatory barriers, which are discussed next.

11.4.1 Technical Barriers

Utility requirements of the DG utilization intend to address engineering compatibility of DG with the grid and various operational requirements, which include specifications for power quality, dispatch, safety, reliability, metering, local distribution system, and control issues. Several organizations including IEEE have established working groups for exploring technical and operational barriers of DG which emphasize safety and reliability issues. For example, a utility lineman who is repairing downed power lines must be certain that the line is not carrying any power during the repair period. The technical barriers are divided into the two broad categories of transmission and distribution.

Transmission System Barriers. When they are practically integrated to utilities' grids and operated in parallel with central station generators, various DG units will bring significant changes to the system operation. In essence, the existing transmission systems were designed for the centralized control of power transmission and distribution, which may not yield to DG provisions. DG owners could be required to pay for pre-interconnection engineering studies, and this would add a significant cost to that of the DG system. The utility system operators would not have any real experience in operating numerous tiny generating units scattered

DISTRIBUTED GENERATION

across their system. Ultimately, additional protective relays, transfer switches, and net meters would have to be installed. New SCADA systems would have to be created to acquire real-time data and control DG units. Operation issues such as synchronization and system stability are to be studied carefully. Traditional load flow computation methods, which consider unidirectional electric current in the distribution system, would have to be modified. The current power market mechanisms for trading the wholesale power would need to be modified to meet DG retail power requirements.

Distribution System Barriers. Technical challenges associated with the utilization of DG in distribution systems are divided into three categories:

- DG interfaces with distribution systems
- Operation and control of DG
- Planning and design of DG

In distribution systems, frequency would not fluctuate as much. So, most of the operation effort could be focused on maintaining the voltage profile of the system. With the introduction of DG, however, additional attention is directed toward the potential destabilizing effect of DG. This effect will depend on the level of DG penetration and the characteristics of the distribution system.

When connected to the local distribution system, the effect of DG on the system operation will focus on the following factors:

- Siting of DG in a distribution system
- Operation of DG and the distribution system
- Outages of DG and the distribution system reliability

The siting and the operation of DG could affect the operation state of a distribution system as a DG outage could affect the stability and the reliability of a distribution system. However, detailed quantitative studies will be necessary in order to quantify the potential impact of DG on the operation of a specific distribution system.

11.4.2 Barriers on Business Practices

Business practice barriers usually relate to the contractual and procedural requirements for system operation and market participation. Among such barriers are contract length and complexity, contract terms and conditions, application fees, insurance and indemnification requirements, identification of an authorized utility contact, consistency of requirements, operational requirements, timely response, and delays.

The principal business impediments to DG come from the existing business practices that prevent DG not only from being integrated into the utility's grids but also from participating in the market competition. DG is also experiencing various barriers that hedge its participation in the retail power market. One major obstacle for entering DG to the power market is the unavailability of net metering. Net metering, which represents the ability to sell power back to a utility, is expected to open doors for DG. The existing market regulations for competitive electricity markets do not provide equal treatment to the centralized power and DG. Only in a perfect market would there be no significant market barriers to market participants; in this sense, market regulations are required to ensure equal treatment of all market participants. These impediments will hopefully fade away as utilities see the opportunities to reduce their own costs. Ironically, DG may begin to appear less competitive if the cost of grid-based power shrinks in the restructuring paradigm.

11.4.3 Regulatory Policies

Regulatory barriers include matters of policy that usually fall within the jurisdiction of state utility regulatory commissions or FERC. The barriers arise from or are governed by statutes, policies, tariffs, and regulatory filings by utilities, which are approved by the regulatory authority. The examples of regulatory barriers include direct utility prohibition, high exit fees, selective discounting to discourage distributed power generation, ISO procedures and costs, and tariffs including high demand charges, low buyback tariffs, and high uplift tariffs. In addition, regulatory prohibition of interconnection, unreasonable backup and standby tariffs, local distribution system access pricing, transmission and distribution tariff constraints, independent system operators (ISO) requirements, and environmental permitting are listed in this category.

Backup charges, i.e., a charge for maintaining electric supply to backup a DG unit if it fails to operate, have ranged from a few dollars to well over $200 per kilowatt hour. More appropriately designed tariffs can provide standby and backup power services without incurring prohibitive charges. Standby services include power to supplement or replace a consumer's on-site generation. Backup services include power supplied to a consumer during an unscheduled or emergency outage of the on-site generation. Regulatory barriers are beginning to be addressed on a state-by-state basis, but many state variations of rules and regulations will continue to undermine the economic advantages of many distributed power projects and hinder this developing market. In one of the strange cases, a utility required the exclusive control of a sponsor's generation unit and used it to reduce its own system peaks. In this case, the DG owner was prohibited from using the unit to reduce its own energy bill.

Zoning, air permitting, water use permits, comprehensive environmental plan approval, and other regulatory processes can delay and increase the cost of DG projects. These issues typically relate to site-specific concerns. Environmental regulations are not currently administered such that they would give credit for the overall pollution reduction effects of high-efficiency DG technologies. Permitting costs are another barrier. Until the EPA declares the microturbine to be a low-emission technology, the permitting process will be the same as that for a much larger piece of equipment. Hence, relevant energy and environmental rules and policies should be established as soon as possible, and should cover much of the following issues:

- Methodology to quantify benefits for specific DG projects
- Interconnection standards
- Integrated operation to capture benefits
- Environmental regulations
- Economic regulations
- Contractual relationships

Although regulatory barriers have been significant, the restructuring of power industry is evolving such that it might encourage the utilization of DG. Utilities are presently permitted to recover stranded assets using various types of tariffs including exit fees, competitive transition charges, access charges, and other means. Policies for the application of these charges to DG could significantly delay the benefits of these projects. However, certain social benefits such as environmental protection and regional economic development may justify special treatments. Most states intend to apply stranded costs only to power purchases from the grid. On the other hand, consumer-owned generation requires in most cases a backup source of power to meet load requirements during generation outages. Utilities now charge not only for the power used but also for the standby generation and distribution capacity (i.e., ancillary services). In addition, the separation of generation from T&D provides an equal access to competitive market participants. However, in many cases regulators have legitimized selected DG investments by power distribution companies that can provide grid support for localized areas.

11.5 DG INTEGRATION TO POWER GRID

Consumers, ESPs, and electric utilities could install DG units to supply their own energy or to make money by selling electricity to the market. The first step to realize these purposes is to connect DG units to the power grid. In 1978, PURPA (Public Utilities Regulatory Policy Act) mandated utilities to allow IPPs to be interconnected with power grids. In 1992, EPAct mandated open access to power systems by allowing generating facilities running either by IPPs or by consumers to be connected to power grids. Though DG could bring many benefits to consumers, ESPs, and electric utilities, the integration of DG to the power grid has presented many new challenges to power system restructuring, operation, market mechanisms, and regulations. As we discussed above, DG integration could face many technical and institutional barriers, which would block DG from penetrating into power grids and participating in market competition. The integration of DG into the power grid is hampered by the lack of uniform requirements for interconnection and a feasible energy management system. Several private and government sectors such as IEEE, EPRI, and DOE are developing interconnection standards for DG.

As shown in Figure 11.2, integrating DG into a power grid has two implications. One is the technical integration (i.e., hardware and software) which includes issues relevant to power system operation, control and optimization; the other is the market integration, which includes the establishment of new market mechanisms that allow DG to participate in a market competition.

DISTRIBUTED GENERATION

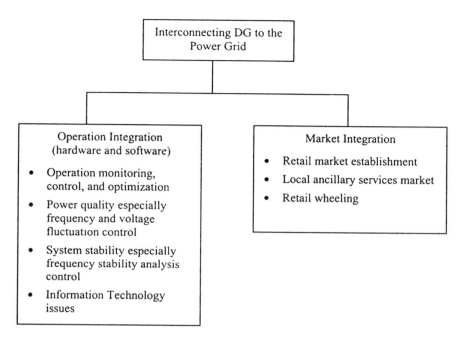

Figure 11.2 Integration of DG into Power Grids

Some of the issues for the integration of DG to power grids are as follows:

- How will a multitude of small-scale DG units influence the operation and control of the distribution system that was originally designed to operate with a unique power supply from substations?

- What kind of market mechanism is most effective and efficient for the DG participation?

- How will the restructuring proceed based on technological advances and competitive market forces created by DG?

- What kind of coordination between market forces and real-time control will be required for distribution system operation?

- Which will be in response to the new and potentially conflicting economic and technical demands of a growing number of DG?

- How will ancillary services be maintained in the new environment?

- How will the information technology and software issues will be handled?

As a large number of DG with diverse characteristics have been installed in the distribution system, most engineering and operation concerns are focused on the development of new control approaches and tools, which include new monitoring schemes, new reliability and security analysis, and new market mechanisms for DG to participate in competition. The integration of DG would cause system reliability and stability problems; in addition, it would impose new challenges on the planning of transmission and distribution systems. Most of these challenges are of a technical nature, and in-depth studies will help us understand the impact of physical behaviors of DG on the power system performance.

11.5.1 Integration

The alternatives for connecting DG systems to local distribution systems are shown in Figure 11.3. Alternatives a, b, and c are mostly used by consumers. However, alternatives a and b with double power sources provide higher reliability, while alternative a can directly sell the excess energy to the market. In addition, alternative c is used by consumers who have difficulties in making direct connections to power grids. Alternative d is mostly used by utilities and ESPs for the mere purpose of supplying energy to the market by way of advanced DG technologies.

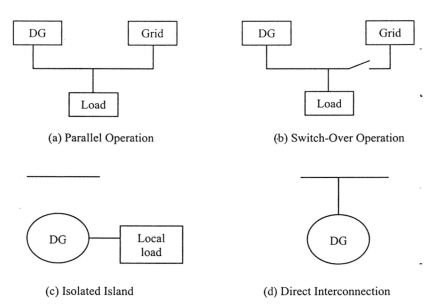

Figure 11.3. Interconnection Fashions

DISTRIBUTED GENERATION

Generally, connection charges are comprised of an amortization of DG assets, which is payable while the assets are in service. A termination charge is payable if the connection is terminated prior to the end of the DG's life.

DG units are rarely directly connected to transmission grids. This is because, compared with traditional central generation, DG has limitations for bulk power production. Though additional revenues are possible by increasing the size of DG units, economies of scale shows that the marginal improvement in efficiency would become insignificant once the size reaches certain limits, e.g., 50 MW. Other practical concerns might prevent the manufacturing of incredibly large thermal generating units as efficiency, fuel cost, and availability would be the decisive factors.

Coal power generating units are among the most competitive and large generators whose efficiency may be lower than many DG units. However, due to the abundance of cheap coal, their cost of bulk power production would be much lower. Since existing thermal units represent sunk costs and the stakeholders will try to make the best of their investments, thermal generators will remain a significant fixture in the power industry for the foreseeable future. To win an economic showdown, DG units would need to be more economical than the central generating station and its associated T&D systems. In fact, the reason DG options are so attractive is not only because of their high efficiency but also because they avoid T&D costs. For DG units, proximity to consumers is as important as efficiency since T&D components bear a significant capital and continuing O&M costs.

11.5.2 Management of DG Units

Since "fit it and forget it" is a common rule for most DG investors, the operation of DG units is usually assigned to UDCs (utility distribution companies) or ESPs. Many large consumers might further want to outsource the responsibility for managing their energy needs to a competent energy provider to maximize the reliability of the presumed services. In the past, distribution systems were operated by local electric utility companies; however, with the power market opening to DG units, it will be more appropriate for local electric utilities, with a number of DG units, to maintain fair competition by delegating the responsibility to a neutral entity, namely the DNO. Like the ISO, the DNO would be an independent and non-profit organization responsible for the operation of the distribution system with multiple DG generators. There could be more than one DNO in a distribution system. Since distribution systems are normally independent of each other, each DNO could monitor and control the distribution system that is under its jurisdiction independently. The ever-developing Internet technology can be used in conjunction with SCADA/DMS to lower the communication costs and improve the designated service. The DNO can use the Internet to remotely control and dispatch consumer-owned DG units in its service territory.

11.5.3 Concerns for Utilizing DG in Distribution Systems

In general, electric utilities have the following three basic concerns about the integration of DG to the power grid:

- **Safety.** DG units could inadvertently energize part of the distribution or transmission system when that part of the system is not supposed to be energized. This could endanger utility personnel and the general public, and damage utility or other customer equipment.

- **Power quality.** DG units could cause voltage flickers, harmonics, or other power quality problems to nearby customers.

- **Stability.** Multiple DG units operating in parallel could cause stability problems in adjacent power systems.

These concerns manifest themselves in numerous technical requirements such as the need for utility-grade breakers, dedicated isolation transformers, and feeder relay coordination. These legitimate concerns have led to considerable debate over interconnection requirements. It is conceivable that these issues would have to be resolved before a major deployment of DG units occurs in power systems. Several state commissioners have already created standards for DG integration since it is believed that DG could be an effective solution at least to generation shortage.

Distribution systems are designed fundamentally differently than transmission systems. While transmission systems normally accommodate multiple energy sources, the distribution system expects only one source in a radial configuration. DG units can bring new challenges to the operation of distribution system in following four key areas:

- **Load flow.** Distribution system equipment is designed to accommodate the designated loads plus perceived contingencies to maintain supplies under abnormal conditions and hence meet the security requirements. Inappropriately sized DG units connected to the distribution network can sufficiently alter power flows to either exceed network capabilities or adversely affect distribution network losses.

- **Voltage control.** A key element of the design of radial distribution networks is the use of tapered circuits where the size and capacity of the circuit tapers off along its longitude; this arrangement will maximize the use of allowable voltage variations within statutory regulations while minimizing the cost. Generation connected to tapered circuits tends to increase voltages, potentially above statutory limits, particularly when connected to long rural circuits.

- **Network security.** Strategies for connecting DG units must ensure the security of the overall distribution system, although these strategies could presumably change if the system is managed differently. Distribution system design standards, which aim at maintaining the security of the system at an acceptable level, are taken as the benchmark for the security assessment of new DG connections

- **Fault levels.** Distribution systems are traditionally designed to have a low number of faults consistent with switchgear ratings and operational limitations. This feature enables large and fluctuating loads to be connected economically for minimizing losses, while minimizing the effect on other consumers. Due to a small margin between operation and rating of distribution equipment, integration of DG into distribution grids is likely to increase short-circuit currents above plant capabilities. Induction motors, which form part of the distribution load, contribute to short-circuit current and also erode this margin.

11.6 OPERATION OF DISTRIBUTION SYSTEM WITH DG

Many distribution systems have utilized advanced digital devices such as protective relaying systems, transfer switches, controls, remote monitoring, and communications technologies. These new technologies promise to address the DNO's concerns while providing value-added features for new DG units. Furthermore, recent advances in communication and control, and especially the Internet revolution, have increased the feasibility of remote control and dispatch of DG units while decreasing the cost of operation significantly.

11.6.1 Transition to a More Active System

The distribution system was initially designed to operate with minimal real-time intervention except for responding to faults, allowing routine maintenance and network modifications. This philosophy of passive operation has placed a number of limitations on the integration of DG into the power grid. It is envisioned that the active operation of the distribution system could accommodate a large number of DG units while improving the performance of the entire distribution system. The active operation of a distribution system refers to its intelligent adaptation to changes in local loads and DG units. The transition from the present passive mode to a fully active mode can be accomplished in the following three stages:

- **Passive.** The DNO would only be reacting to abnormal situations and planned abnormalities or longer-term restrictions. This can be realized by using existing monitoring and control infrastructure at key nodes.

- **Intermediately active.** In addition to the functions of the passive stage, the DNO would provide real-time monitoring of the distribution system, scheduling of DG units, and load management, for parts of the distribution system.

- **Fully active.** In addition to the functions of the intermediately active stage, the DNO is equipped with real-time modeling capability for the monitoring and control of the distribution system security. The need for enhanced control and communication systems, as well as appropriate market mechanisms, could develop over time as the DNO becomes more actively familiar with a network operation.

Two key issues may arise during the process of transition from a passive operation mode to a fully active one:

- The need for enhancement of the current SCADA/DMS infrastructure to enable the DNO to fulfill real-time monitoring and control functions. The existing SCAD/DMS system can be enhanced to provide the additional functionalities. Enhancements would be needed to make additional information on the real-time performance of the system available to the DNO so that it can make the right decisions. In order to implement the DNO decisions, facilities to control key generation items such as generator active and reactive power output, transformer tap position, and circuit breakers will be needed. To provide this information and control facilities, communications among DG owners, load consumers, and the DNO would be needed. Some of the existing SCADA systems could be extended to provide these additional features. Those systems with more limited capacities may require upgrading or even replacement. The restructuring initiative in distribution systems SCADA/DMS focuses on capturing the price elasticity or the price responsiveness of consumers, since the process will not be as complex or time-consuming as some of the other initiatives.

- The establishment of new market mechanisms for electricity retailing to enable the DNO to effectively manage the power generation and load. The fully active management of the distribution system would require the development of market mechanisms that are specially tailored to meet the DG requirements. The contractual framework of the market mechanism would have to ensure that services procured from DG compare with those provided from networks. There is a key link between such market mechanisms and the development of active distribution systems. It is essential to determine how active management of a distribution system would interface with the wholesale trading mechanisms and whether there is a need for a local market mechanism that allows the distribution system operator to balance the system to accommodate load and generation needs at minimum network costs. The regulatory mechanisms associated with DG would need to be carefully examined and developed. Regulation has a role in helping establish the optimum balance among different consumer groups needs that conflict in areas such as network cost, supply quality, and security.

11.6.2 Enhanced SCADA/DMS

The passive operation mode does not meet the requirements for the operation of DG units in a complex environment. To ensure the security of distribution system operation, pertinent information must be exchanged between the DNO and DG units. Also, a distribution system SCADA/DMS infrastructure must be enhanced with new functions in order to realize the real-time monitoring and control of the distribution system.

The operation of a distribution system in an active mode is more complex than that of a passive system. There are many areas where existing planning and operation practices and procedures will need to be enhanced. The DNO is central to managing the distribution network of generators, suppliers and consumers efficiently while ensuring acceptable levels of security and quality. This mode of operation requires complete metering, data control, and management, and communications and utility operations for the control of DG units.

The SCADA system collects various kinds of real-time information in a distribution system, such as the active and reactive power provided by each DG unit, which is critical to the real-time monitoring and control. The DNO control commands will also be sent to DG units or loads through the SCADA system. The main tasks of DMS are to maintain power supply reliability and to keep costs at minimum and profits at maximum. DMS is equipped with many application programs, such as load flow and state estimation, which help the DNO analyze the reliability of its network.

DMS can monitor and control a distribution system using the SCADA/DMS. A SCADA/DMS can track and manage loads, maintain voltage profiles and maximize the efficiency of the distribution system. This function optimizes power purchase commitments and costs. With the aid of DMS real-time monitoring functions, the DNO can identify system failures immediately and dispatch repair crews quickly. The DNO can also shift power within the grid to manage the daily load requirements among consumer segments and forecast power needs for efficient purchase from the transmission grid.

Once the required functions for SCADA/DMS and market mechanisms have been established, the DNO will provide the following management services and functions:

- Real-time monitoring and control of distribution system operation
- DG units scheduling and dispatching
- Maintaining the system voltage profile
- Real-time operation studies such as online load flow and state estimation
- System frequency stability analysis

DISTRIBUTED GENERATION 417

- Power quality improvement
- Control of fault levels during abnormal system conditions
- Power supply restoration after system faults
- DG units outage coordination

Besides, the DNO will provide the following functions for market services:

- Feasibility verification of retail transactions
- System load forecasting
- Spot price forecasting.

Feasibility verification of retail transactions is a critical function of the DNO, as it is similar to the ISO's congestion check for wholesale transactions in the transmission system. Aggregators submit the information on DG supplies and load demand to the DNO. Then, together with the other information such as the supply of power from the utility' grid to distribution buses, the DNO uses the load flow computation to verify the feasibility of retail transactions. Correspondingly, services that are available in a distribution services market are provided to a transmission services market.

11.6. 3 Role of UDCs and ESPs

Similar to the transmission system, distribution systems remain a regulated utility while other products and services are open to competition. In this situation, DG investors participate in the retail market through both direct access to the market and contracting with local UDCs or ESPs. The day-to-day even minute-to-minute operation and coordination of various DG units are managed by local UDCs or ESPs, which are supervised by the DNO in an emergency. The supervision requires tariff and metering arrangements for load and generation. There are also issues related to network constraints, and scheduling of active and reactive power generation, to meet distribution network requirements, which will become increasingly important as networks become actively managed.

To enhance their competition, UDCs and ESPs provide many other services such as project development, operation cooperation, management of customer energy facilities, risk management, and even financing. Their roles will also include energy market development for DG, joint developments of DG demonstration projects with consumers, manufacturers, and utilities, and reduction in the costs associated with the DG project development. On the other hand, customers face additional challenges in managing their energy consumption in a competitive environment.

11.6. 4 Distributed Monitoring and Control

A distribution network is usually composed of a number of distribution systems. Most of these distribution systems operate independently based on a radial network structure. Some of them may cooperate for the distribution network reconfiguration if they are under the control of the same DNO and can be physically interconnected with switches. Such an operation structure is depicted in Figure 11.4.

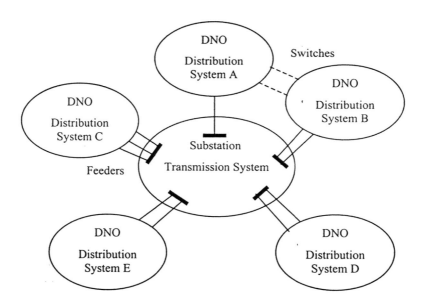

Figure 11.4 Distributed Processing of Distribution System

The real-time data for the monitoring and control of the distribution systems will be collected by the SCADA system at the DNO. When there are many DG units operating in parallel, a large distribution system could be divided into a number of control areas for the effectiveness of monitoring and control, which is depicted in Figure 11.5.

DISTRIBUTED GENERATION

In each control area, DG units will be controlled by associated UDCs, ESPs, or the DNO. A control area could set up a secondary control center when there are many DG units to monitor and control. The scheme for the real-time monitoring and control of distribution systems with several DG units is depicted with Figure 11.6.

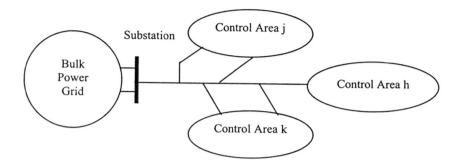

Figure 11.5 Control Areas of Distribution System

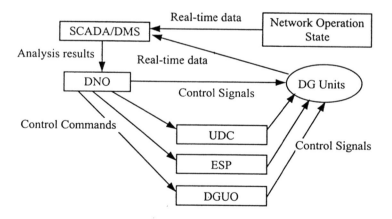

Figure 11.6 Real-Time Monitoring and Control of Distribution System

11.7 LOAD FLOW COMPUTATION

With the introduction of DG to distribution systems, DNO would need effective tools, possibly both hardware and software, to monitor and control the operation of the distribution system. The load flow computation is the most fundamental tool for the real-time monitoring and control of the distribution system operation in a DMS. Similar to the load flow computation for transmission systems, the Newton method was initially applied to the load flow computation of distribution systems. However, compared with transmission systems, feeder line resistance of distribution lines is much higher, and this could result in a high ratio of line resistance to impedance. In many cases the Newton method will face convergence problems.

Many researchers have developed new solution methods of which DistFlow is the one that has received a wide attention. However, DistFlow is based on the assumption that power will have a unidirectional flow on feeders and their lateral branches of distribution systems, which was only correct before DG systems were introduced to distribution systems. This difference is demonstrated in Figures 11.7 and 11.8. As these figures show, the power flow on some branches and main feeders could change direction when DG units inject power into the grid. When DG units inject power to a remote bus away from the substation, the DG system can cause stability and security problems in the distribution system.

In this new situation, a new power flow method will be developed. Because the load flow direction in each feeder and branch is unknown before the load flow results are computed, DistFlow cannot be applied in this situation. Figure 11.8 points out that a distribution system with multiple DG resembles a transmission system.'

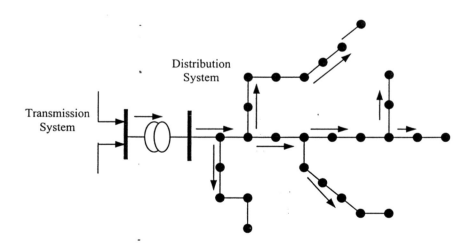

Figure 11.7 Unidirectional Power Flow of Distribution System

DISTRIBUTED GENERATION

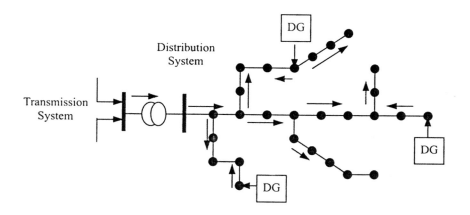

Figure 11.8 Multi-Directional Power Flow of Distribution System with DG

As DNO may oversee several independent distribution systems, the load flow computation can be processed in a distributed manner. In Figure 11.9, distribution systems A, B, and C are only related by the status of buses k, l, and m respectively, and the load flow computation for these distribution systems could be processed independently in a distributed manner as discussed in Chapter 6.

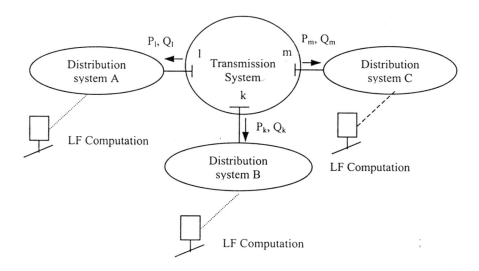

Figure 11.9 Distributed Computation of Load Flow for Distribution System

UDCs may use the load flow computation for their wholesale transactions with the PX. DNO will use the load flow results to monitor and control distribution systems. Besides, load flow results will be sent to the ISO for the computation of transmission system load flow.

There are several reasons for the separation of load flow computation in transmission and distribution systems. First, transmission and distribution systems have different missions and physical characteristics, are different in design and operation, and have different requirements for load flow computation. Second, transmission and distribution systems are monitored and controlled in real time by two different entities, i.e., ISO and DNO, which collect different real-time information through their respective SCADA systems with different physical territories in power systems. Third, power market mechanisms for transmission and distribution systems would be quite different as DNO will be responsible for the retail power market while the ISO is responsible for the wholesale market.

11.8 STATE ESTIMATION

Traditionally, distribution systems are designed to operate in a passive mode where there is a need for human interference. Because of such design and operation philosophy, few utilities have used state estimation for the real-time monitoring and control of their distribution systems. As additional DG units appear in distribution systems, the operation of distribution systems will become increasingly complicated. In order to implement a fully active operation of distribution systems, the DNO has to enhance the existing DMS infrastructure by further utilization of state estimation and load flow.

State estimation was applied rarely to distribution systems, and is fairly new to distribution systems with DG. State estimation is an indispensable tool for the DNO to monitor and control the operation of a distribution system in real time. In this respect, state estimation for several distribution systems can be processed in a distributed manner, while the state estimation for individual distribution systems is processed in a distributed manner.

11.9 FREQUENCY STABILITY ANALYSIS

In the past, there was a single substation that supplied power to a distribution system, and it was assumed that the distribution system had

DISTRIBUTED GENERATION

no stability problem. Stability analyses were the responsibility of central generating stations connected to the transmission grid. However, in the DG era, stability concerns arise with the penetration of DG in a distribution system. This concern is mounting as the number of DG units that supply multiple distribution consumers is increasing. In such cases, if corrective actions especially secondary controls are not taken in time, the system frequency may drift from the nominal value and finally cause the loss of synchronism.

Different types of DG are likely to cause instability and the larger the number of DG units, the more likely is the chance of instability. Stability analysis relates the characteristics and operational status of DG in distribution systems, and it should be performed by the DNO as an indispensable function of DMS. The first step to study the distribution system stability is to define the system model.

11.9.1 System Models

The stability analysis here is focused on frequency dynamics within the distribution system, as shown in Figure 11.10. In this design, the transmission system behind the substation is modeled as an infinite bus for the distribution system. This way, individual DNOs perform their own stability analysis for their distribution systems.

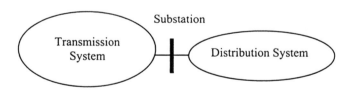

Figure 11.10 Model of a Distribution System

A distribution system can be defined by specifying its location, topology, and sizes and types of loads and DG units. The load flow equations given below are the mathematical representation of the distribution system:

$$P_i = \sum_{j=1}^{n} V_i V_j \left(g_{ij} \cos(\theta_i - \theta_j) + b_{ij} \sin(\theta_i - \theta_j) \right) \qquad (11.1)$$

$$Q_i = \sum_{j=1}^{n} V_i V_j \left(g_{ij} \sin(\theta_i - \theta_j) - b_{ij} \cos(\theta_i - \theta_j) \right) \qquad (11.2)$$

where P_i and Q_i are the real and reactive power at bus i, V_i and V_j are voltage magnitudes of buses i and j, g_{ij} and b_{ij} are the line admittance parameters, and θ_i and θ_j, are the rotor angles of buses i and j, respectively.

Since different DG units have different dynamic characteristics, state space models for individual units need to be developed. Suppose that the disturbances are small enough, and thus small-signal linear models are used for the stability analysis of distribution systems [Nwa91a]. Various DG units in the distribution system are modeled as follows:

$$x_g = A_g \dot{x}_g + B_g P_g \qquad (11.3)$$

where x_g is the state vector of DG units, and $\dot{x}_g$ is the time derivative of x_g; A_g is the local system matrix which consists of linear coefficients of DG units parameters; P_g is the vector of the real power output of DG units, and B_g is the coefficient matrix. (11.1) and (11.3) describe two different parts of the distribution system, and to integrate these two models, the power output of each DG unit is chosen as the coupling variable. The state equation of P_g is written as:

$$\dot{P}_g = K_p \omega_g + D_p \dot{P}_D \qquad (11.4)$$

The two coefficient matrices K_p and D_p are derived from the Jacobian matrix; $\dot{P}_g$ represents the random load fluctuation. Then, the complete system model is built by augmenting variable $\dot{P}_g$ with the local state spaces of individual DG units, and the corresponding system state equation is formulated as

$$x_e = A x_e + D_p \dot{P}_D \qquad (11.5)$$

where x_e is the vector of the augmented state space variables and A is the system matrix [Nwa91b].

DISTRIBUTED GENERATION

The inputs to (11.5) are system disturbances, which are specified by the location and the timing of disturbances. The outputs of (11.5) are the dynamic behaviors of state variables, including the local and the coupling variables. This model is used in this chapter for analyzing the system frequency stability due to random disturbances or to implement control measures. In addition, dynamic interactions among DG units are analyzed as different control strategies are introduced for enhancing the frequency stability.

11.9.2 Stability Analysis

Maintaining the system frequency has not traditionally been a concern because there was no generator in a distribution system. It is now a very important issue in distribution systems because multiple DG units operating simultaneously could cause a loss of synchronism.

In several ways, the frequency stability criterion of distribution systems with DG units is different from that of a high voltage transmission system with large central stations. First, compared with central station generators, small-scale DG units have smaller inertias, which leads to a stronger coupling between the local state space and the system variables. Second, the distribution system has large line impedances that could strengthen the coupling between the distribution system and the frequency. In some cases, instability occurs as the number of DG units increases, which is explained by the fact that smaller inertias create stronger coupling between the frequency and the system dynamics, while the line resistance is not large enough to provide a sufficient dampening. Third, governors of local DG units can be used to maintain frequency stability, which may not occur much in transmission systems as the governors of large-scale generators react very slowly.

To maintain frequency stability of a distribution system with a multitude of DG units, sufficient attention needs to be paid to the control parameters of local DG units such as time constants and gains. Accordingly, a more effective procedure for the real-time control of state variables is to seek proper ranges for DG control parameters by calculating the sensitivity of eigenvalues to parameters. It may also be desirable to extend secondary controls to DG units so that the frequency can be restored to its nominal level as quickly as possible.

The eigenvalue analysis of the system matrices, A_g and A in (11.3) and (11.5), is used to identify the cause of the frequency instability. The eigenvalues for individual DG units and for the system are easily calculated. In many scenarios, even if each unit is individually stable in terms of their eigenvalues, the whole system can be unstable. If the unstable modes cold be uniquely associated with certain state variables, frequency stability might be easily regained by identifying and implementing proper control to those state variables. Unfortunately, state variables associated with unstable modes vary with different system configurations. To solve this problem, participation factors are used to identify the association of an individual state variable with a specific instability mode. A participation factor provides a measure of the contribution of a state variable to an eigenvalue and thus can be used to guide the controls to state variables [Nwa92].

The frequency drift or instability can be corrected in time through a closed-loop frequency control infrastructure as shown in Figure 11.11. Based on the frequency analysis results, the DNO will send control signals to UDCs and ESPs, who operate DG units, to adjust the power output of DG units that are committed to the DNO for operation control. This control infrastructure will be the same one as that for secondary frequency control.

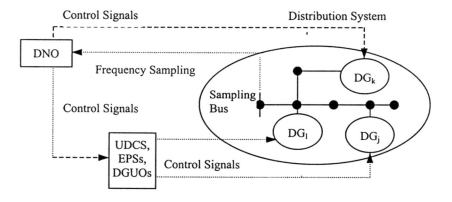

Figure 11.11 Closed-Loop Frequency Control

11.10 DISTRIBUTED VOLTAGE SUPPORT

From the DNO's viewpoint, a very attractive feature of DG is the potential for providing ancillary services, especially the spinning reserve and voltage support to power systems. A distribution system supplies power directly to consumers where the voltage control has become the most important concern in the distribution system operation. In the traditional distribution system, utilities could not easily maintain the voltage profile of a system because the substation was the only reactive power source. The length of the main feeder had to be limited within certain ranges, and in many cases, expensive capacitor banks had to be installed near the load to maintain the desired voltage level at certain buses. This capacitor option incurred a lot of extra capital investment and operation maintaining charges.

Unlike the system frequency, which is the same at any bus across the entire system, voltages at different buses can be different as long as they fall within a desired voltage profile. Bus voltages are regulated at regional levels rather than at the system level, where one or more pilot buses are chosen from the system to provide regional reference points. So the voltages of pilot buses must be maintained within the desired voltage levels to maintain a feasible voltage profile for the system. The voltage level of a pilot bus is maintained by making use of reactive power sources near the pilot bus. Due to this local property of voltage control, DG units that are sited within the distribution system and close to consumers are more suitable to provide voltage support than the central station generators connected to the transmission network.

Most DG units have certain types of power conditioning equipment that can be set to generate power with a leading or lagging power factor and hence can be dispatched to compensate the local reactive power mismatch. By controlling power conditioning equipment, DG units can increase or decrease their reactive power supply, which not only maintains the DG units' terminal voltage within given limits but also provides VAR support to the system. In some extreme cases, DG units are operated as static VAR compensators to supply reactive power to the system. In addition, DG units improve the voltage profile simply by supplying both power and reactive power to consumers, which reduces not only the power losses but also voltage drops caused by power delivery along transmission lines and feeders.

A better voltage profile of a distribution system can be obtained by realizing the following two measures:

- **Real-time monitoring and control of DG units.** A pilot bus represents the voltage level at a large number of adjacent buses. Previously, only a very few buses were chosen as pilot buses in the distribution system, where each pilot bus had a difficult task of maintaining the voltage level of a great many buses in large distribution areas. Besides pilot buses, terminal buses of DG units could often be taken as quasi pilot buses for the purpose of voltage control. Then the pilot bus could more accurately represent the voltage level of a relatively smaller area. Furthermore, most DG units had the ability to maintain their terminal voltages within specific limits, which helped maintain the voltage level of their adjacent buses.

- **Distributed voltage control by DG units.** With numerous DG units operating in the distribution system, the entire system can be divided into a number of control areas, as shown in Figure 11.12. For distributed voltage control, each control area will have one pilot bus. If voltage control interactions among adjacent control areas are negligible, the distributed voltage control in each area will be implemented independently. However, when more than one DG unit is participating in the voltage control of the same pilot bus, as depicted in Figure 11.13, the DNO coordination effort will be required. The prerequisite for such a distributed control scheme is that the DNO will have the right to operate the DG units in the system. Real-time closed-loop voltage control will be applied to DG units that are committed to the DNO by their investors, as their real-time sampling data of voltage will be sent to the DNO via the SCADA system.

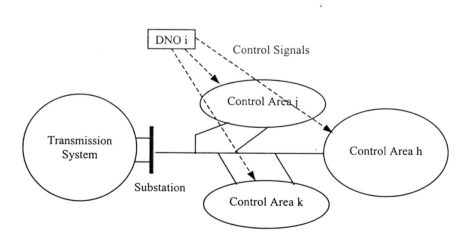

Figure 11.12 Distributed Voltage Control of Distribution System

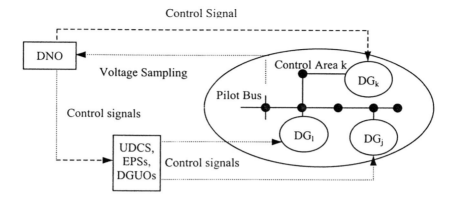

Figure 11.13 Distributed Closed-loop Voltage Control

11.11 DG IN POWER MARKET COMPETITION

11.11.1 Retail Wheeling

The restructuring of the power industry has implications for the corporate structure of electric utilities, the pricing of energy products and services, the quality of service delivered to consumers, and the selection of energy generating technologies. The concept of applying the ISO to the wholesale power market at the transmission level is reasonably defined and accepted by the industry. The next stage of restructuring could extend the competitive market to the retail level for creating a competitive retail market which allows direct access to consumers. Unlike the wholesale market where electric energy is traded at the bulk level, the retail market will allow consumers to choose energy providers. Hence, after the establishment of the retail power market, consumers will not be restricted to buying energy from their local electric utilities. The establishment of the retail market could lead to further technological and institutional changes, among which the concept of retail wheeling is critical.

The purpose of retail wheeling is more than lowering the energy costs; it is an entirely new paradigm for marketing the transmission, distribution, and consumption of electricity. In a time of retail wheeling, electric utilities could sell energy to consumers that are far from their home territories; similarly, consumers can purchase energy from electric companies that are remotely located. To avoid the transmission and distribution tariffs of intermediaries, which will drive the prices up, energy providers prefer to utilize DG technologies. Conceptually, retail wheeling is the principle of opening up to competition for energy sales to retail consumers, which is intended to allow consumers to purchase cheaper electricity from providers other than their local UDCs. Such a retail competition is critical to determining the DG's role in the power industry. DG can obtain opportunities to sell power as well as necessary ancillary services to consumers, through the establishment of a retail power market.

The first step in developing a retail competition is to open services in the revenue cycle, such as metering and billing, to competition. Proponents of the retail competition seek access to the wealth of consumer information that would become available through direct access to consumer. This step is only the beginning, however. From the perspective of DG, the retail competition brings about more changes. To ensure that technical changes are made to the way distribution systems are designed and operated, an appropriate commercial and regulatory framework would need to be created that defines the criteria based on which DG participates in the power market.

As to DG, retail wheeling will happen when consumers in a distribution system are interested in purchasing DG units' energy that is produced in other distribution systems. In such cases, the total energy produced by DG units in a distribution system will exceed the local network' demand; however, the excess energy will be represented as an equivalent generator that injects power into the transmission grid. The retail wheeling of the excess energy will depend on the operating condition of the substation transformer. If any injection to the transmission is prohibited, the excess energy will not be sold. In either case, T&D costs and associated ancillary services will be imposed on the energy sellers and purchasers, in addition to the possible congestion penalties.

The energy injected by DG into a transmission system could flow freely from any source to any sink in the power system. For instance, the DG energy in Figure 11.14 is wheeled to a different distribution system than where it is intended (i.e., the load in the figure). This condition is referred to here as virtual wheeling since the intended flow is not physically realized. In other words, the contract path (dotted line) does not conform to physical path of power flow. A more realizable retail wheeling of DG is depicted in Figure 11.15.

DISTRIBUTED GENERATION

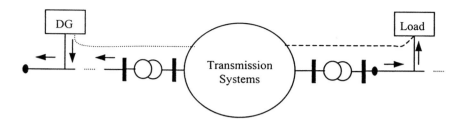

Figure 11.14 Virtual Retail Wheeling

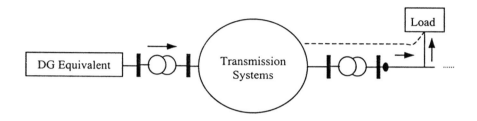

Figure 11.15 DG Retail Wheeling

11.11.2 Ancillary Services

As we mentioned earlier, DG units can provide ancillary services including spinning reserve and reactive power to the system. Flexibility in the DG operation, especially quick ramping up and shutting down, will not only enable DG unit operators to take advantage of short-term sales opportunities but also enables them to take DG units as a choice for spinning reserve. In addition, DG can provide reactive power for the local voltage support, because most DG units are equipped with power electronic converters which employ line-commutated switching devices, such as thyristors, or certain self commutated devices to convert the generated dc power to the standard ac power. These power electronic interfaces are configured to provide reactive power for the purpose of ancillary services. Therefore, DG units can provide support for the system frequency stability and help maintain voltage profiles which are incurred by load fluctuations or other abnormal events in distribution systems. When supporting voltages, DG units function like a synchronous condenser, a static VAR compensator, or FACTS devices in transmission grids.

There has already been an ancillary services market in the transmission level operated by the ISO. The ancillary services from various DG units can be used to improve the performance of the system operation. However, the lack of a local distribution system market for ancillary services may have adverse effects on the retail market competition. The lack of market may create a discrepancy in ancillary services as there is no mechanism to treat services that DG units provide equitably with those provided by the distribution system. Hence, with the establishment of a power retail market, there is an imperative need for establishing an ancillary services market with DG, so that DG units can participate in the ancillary services competition.

Ancillary services of DG units, shown in Figures 11.16 and 11.17, could greatly improve the system's performance. Before DG could provide ancillary services, the generation system was the only source of ancillary services. The delivery of ancillary services by traditional means can cause extra power losses and incur the additional cost of power transmission and distribution. DG is located closer to consumers and more prompt to response in emergencies. With ancillary services available from DG in local distribution networks, consumers can buy ancillary services along with DG energy, which reduces power losses on transmission and distribution systems and improves the system performance.

DG units can compete for ancillary services only when ancillary services are unbundled from the energy supply. In this sense, policies and market mechanisms for ancillary services will greatly influence DG planning and investment. In other words, strong policy incentives and effective market mechanisms for ancillary services are prerequisites for making a full use of DG potentials.

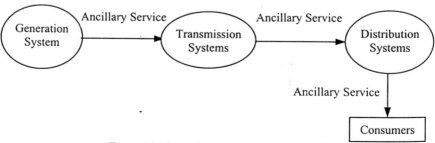

Figure 11.16 Ancillary Services without DG

DISTRIBUTED GENERATION

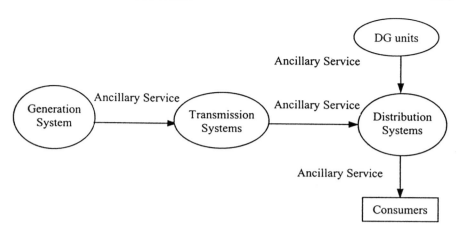

Figure 11.17 Ancillary Services with DG

11.11.3 Role of Aggregators

A key issue about the direct participation of DGs in a retail market is that the ISO may not be able to deal with a large number of individual resources. When the DG integration is fully implemented, the information to be processed by the ISO would be huge and incur very high communications requirements. For this reason, aggregators can provide a substantial help for managing DG units. In general, the ISO may not be particularly interested in learning the details and concerns of each DG customer; likewise, customers have neither the time nor the interest in learning about the load control and management issues. Aggregators can bridge this gap. By handling communications with a large number of DG units, aggregators present the ISO with a single point of contact for a reasonable amount of capacity, quite similar to the ISO's interface with generating resources.

Wholesale and retail markets have different functions, yet they interact extensively. In such a complex market environment, where there could be many participants, aggregators will play an indispensable role in the normal operation of power markets. Aggregators would collect energy supply and demand information and then trade energy in the power market. There are aggregators for the wholesale market, but they play a more important role in the retail market. This is because there are a much larger number of participants in the retail market than in the wholesale market. Besides, most DG investors would prefer to commit their energy trades and management to a third partner who is more familiar with the power market and knows more about the functionality of their DG units.

The role of aggregators in retail market is depicted in Figure 11.18. Aggregators can greatly reduce the burden of the ISO and the local DNO when there is a tremendously large number of participants in retail market. Once transactions are approved, the aggregated DG capacity becomes immediately dispatchable by the DNO, especially when required to provide peak power and spinning reserve capacities.

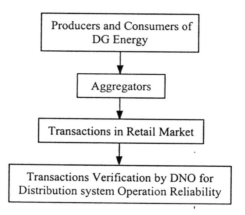

Figure 11.18 DG Power Trading in Retail Market

The aggregation of DG will be mainly focused on the following three issues:

- Energy supplied by DG units
- Energy demand for DG units
- Ancillary services supplied by DG units

Aggregators look into the local distribution systems as a market for DG because major advantage of DG lie in the elimination of T&D costs. However, if some DG energy has to be wheeled to other regions, associated T&D cost and possible congestion penalties have to be taken into account.

Bulk power producers will mainly participate in the wholesale electricity market, but with the aid of aggregators, they can also take part in the retail market for competition. The ISO will be responsible for the

DISTRIBUTED GENERATION

verification of the reliability of transactions that need to be transmitted across the transmission network, no matter if they are made in the wholesale market or in the retail market. The DNO is only responsible for the verification of the reliability of retail transactions that are to be transmitted within the distribution system. The comprehensive power market paradigm for both wholesale and retail markets is shown in Figure 11.19.

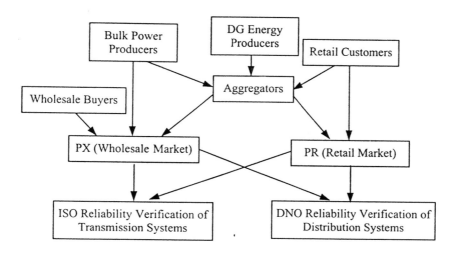

Figure 11.19 Comprehensive Power Market

When they help DG investors participate in the ancillary service market of the power system, aggregators first perform the commercial function of identifying which DG investors are interested in, and capable of participating in, ancillary service markets. Aggregators determine their collective capability to provide each ancillary service and negotiate with each DG to determine what a DG would be willing to provide at any given price. With this information, aggregators negotiate with the ISO for the supply of ancillary services. Negotiations between aggregators, system operators, and DG owners are presumed to iterative.

Aggregators would need to establish physical communications with the ISO so that the ISO is able to treat the aggregated DG as a single resource. However, aggregators establish a communications network to DG units that can be used to convey both control and price signals.

Aggregators also perform important real-time decision functions. They negotiate the price and committed quantity for aggregation by interacting with the ancillary service market and DG owners in real time. The aggregator will assemble the collection of DG capabilities into a coherent ancillary service bid for day-ahead and hour-ahead markets. If the ISO called for less than full deployment, the aggregator must be capable of exercising control over the collection of resources. For example, the aggregator may have sold 200 MW of supplemental operating reserve. The ISO may only call for 100 MW to be deployed. The aggregator could tell each DG to provide only 50% of what it offered, or it could tell half of the DG units to deploy. Or a combination of the two methods will be used based on bids to deploy DG units. Finally, the aggregator must notify DG owners of commitments that resulted from each round of market clearing.

An aggregator will also be involved in the performance evaluation and compensation of each DG unit. Compensation for the aggregated response will be based on the market rules established for ancillary service providers. The existing metering at each DG site will record individual data and report to the aggregator at the end of the billing cycle. Individual meters may record deployment requests as well as DG response to ensure that the communications system functions properly. Market provision of ancillary services, as opposed to the vertically integrated utility's command-and-control system, requires that prices be negotiated through a bidding scheme before each market closes. Faster communications will be required when deployment of reserves is required. However, this type of signal can be broadcast to corresponding resources and need not be specific to each entity. Certification and after-the-fact metering can replace the real-time monitoring of each DG unit. Aggregators will provide a valuable service by assembling DG collections to present large blocks of controllable power to the ISO.

11.12 CONGESTION ELIMINATION

The wide applications of DG technologies may be utilized in congestion elimination. Congestion pricing should be an incentive for DG investors, since DG units can be easily installed in high price areas of the system. Customers in these areas will find the employment of a compact DG unit more attractive than that of a central station. When many DG investors compete in these areas, the market will decide which competitors will win. On the other hand, DG owners who are counting on serving a high-price congested area run the risk that a TRANSCO might decide to build a new

DISTRIBUTED GENERATION

transmission line to alleviate the congestion. If this happens, even if DG technologies are economically more viable, the DG owners could lose because the transmission cost and local energy prices would drop dramatically. In this sense, if other benefits are disregarded, DG technologies located in low-price areas will not offer any price advantages.

Theoretically, DG units could take part in congestion bidding in competition with central generation stations. However, currently bulk power production has a cost advantage over most DG technologies. If the ISO could not distinguish among the types of power suppliers that participate in the bidding, DG would lose its competitive edge when its advantages, such as environmental impacts and high efficiency, are neglected. The problem here is that there is no great gain for the ISO to distinguish the energy types for the purpose of congestion management, since DG technologies can get special policy support in their routine applications. Even if the ISO could distinguish among the types of power suppliers participating in congestion bidding, the participation of DG in congestion elimination is further quite limited because it will be affected by the operation mode of the distribution system. Because most DG units supply energy only to local consumers, if the retail wheeling cannot be implemented, a DG investor must find a partner, that is, a consumer in the local distribution system for the purpose of congestion elimination.

This process is illustrated in Figure 11.20. GENCO *A* contracts with consumer *X* to supply YMW power, but this transaction is subject to a severe congestion charge. In this case, DG *K* would like to supply part or all of their contracted energy to consumer *X* for GENCO *A*.

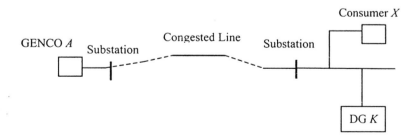

Figure 11.20 Congestion Mitigation with DG

Therefore, the congestion can be mitigated and the congestion charge of GENCO A and consumer X will be reduced. The difficulty here is how DG K or its aggregator finds GENCO A and consumer X, and how they will negotiate the service prices of DG K. The agent theory can be used to resolve such kinds of problems.

Chapter 12

Special Topics in Power System Information System

Presently, the Internet based e-commerce makes it possible to trade electricity in a power market in real time. The geographic information system (GIS) and global positioning system (GPS) are being applied to the power system's real-time monitoring and control, and this has greatly improved the operation and management of power systems. In this last chapter of the book, we discuss several advanced tools for the power system information processing. These include:

- E-commerce of electricity
- Geographic information system
- Global positioning system

12.1 E-COMMERCE OF ELECTRICITY

The e-commerce, which is the abbreviation for electronic commerce, refers to any type of sales or purchases conducted in real time and via an electronic network; the infrastructure of such an electronic network can be on the Internet, leased communication lines, or even wireless connections. The online trading is an alternative term for e-commerce, though it mainly refers to the Internet application. The main purpose of e-commerce is to make transaction processes faster, more effective, more dependable, and more responsive to changing conditions. The Internet provides a wide arena for applying e-commerce to almost all kinds of commodities. In recent years, e-commerce has expanded significantly as many companies embrace worldwide online trading, and now most businesses have learned that their market is just a mouse away, if not at their fingertips. It is estimated that the global e-commerce will reach a multi-trillion dollar level of business-to-business and business-to-consumer sales in the near future.

Electric power is viewed as a perfect commodity for trading by e-commerce. With the restructuring of power industries, electric power becomes a commodity that consumers can acquire over the Internet. Under such a condition, any power company is able to trade electricity in an electronic power market, and buyers can choose their business partners in terms of their preference. But unlike other commodities such as food and oil that can vary in quality, electricity is fungible, and the simplicity of dealing with such a fungible commodity has greatly motivated the application of e-commerce to electricity transaction. On the other hand, e-commerce can meet the rigid real-time requirement of. electricity transaction, by matching supplies with the demand in a very quick manner. Following the e-commerce development in other industries, the e-commerce for power trading grows along two possible tracks. On one track, power suppliers set up shops on the Internet to sell power directly to consumers. On the second track, consumers and suppliers submit bids to a power exchange market for trading power. Although most e-commerce cases in other industries started on the first track, the second track is widely adopted in power industries.

Like other activities that involve the exchange of information among entities, electricity trading has been completely transformed by the utilization of the Internet, and the development of the Internet has prompted the restructuring of power industries and the establishment of electric power markets. As a public facility, the Internet allows power market participants to trade not only electric power but also various services, including the transmission rights for power delivery and ancillary services that support the reliable operation of a power grid.

Because of the special characteristics of power system operation that require the real-time balancing of supply and demand, a central entity like a power exchange is generally designated to be responsible for the electricity trading activities of the whole system, including interchanges with neighboring control areas. This independent entity could be separated from the ISO so that the ISO can concentrate on the system operation, but an alternative choice is to set up a separate department within the ISO to perform market functions; for example, the PJM ISO has such kind of organizational structure. No matter what structure a power system adopts, a power trading system should comprise an energy auction market and an ancillary services auction market that have different trading contents and

SPECIAL TOPICS IN INFORMATION SYSTEM

purpose, even though from the system operation perspective, energy, and auxiliary services are closely related and their transactions need to be optimally coordinated. Transaction activities of these two markets are mainly based on e-commerce through which market participants dispersed in the entire system can participate by using the Internet.

All power exchanges are established on a certain online trading system, which is based on the Internet or some other kind of private computer network. Online trading systems now evolve in the direction that will improve price transparency, lower transaction costs, and make traders more productive. A trader equipped with only a Web browser and some necessary Web-enabled applications can make transactions almost anywhere and at any time. On the other hand, power exchanges provide but a marketplace to facilitate power trading based on a bid mechanism; they are not a party to any transactions and thus do not bear any credit risks. Besides electronic exchanges, some hybrid trading systems can also use voice brokering. Although trading efficiency is critical to online trading, customers are more concerned whether a trading system can provide them with a competitive edge to maximize their revenue rather than merely saving time. A market participant who offers a bid must take into account a number of associated factors to win the bid.

In an intensely competitive power market, especially in a wholesale power market, power trading is a risky activity just because electricity is a special kind of commodity and is different from other commodities in the following aspects:

- Electricity is non-storable.
- Demand and supply should be kept in balance on a moment-to-moment basis.
- Electricity price is closely correlated with that of many other volatile commodities.
- Trading activities are related to the reliability of power grid operation.

These particular characteristics of electricity cause volatility of the market price.

12.2 ADVANTAGES OF E-COMMERCE FOR ELECTRICITY

The e-commerce for electricity has demonstrated the following advantages over conventional trading approaches:

- **Easy accessibility.** In using online trading, market participants do not need to be present on a trading floor, and face-to-face negotiation will no longer be necessary. An online trader can obtain even more complete information from the Web.

- **Real-time pricing information.** Online trading provides real-time pricing information as a value-added service, which is crucial for decision making in an intensely competitive real-time market. In today's "e" era, traditional news media like newspapers and TV are perceived to be too slow to be practical.

- **Minimizing risk.** By instantly locking transactions at the point when they are optimal, power exchanges help market participants get their best trading prices while minimizing risks, trading time, and administrative gridlocks of traditional trading schemes.

- **Saving time and money.** In an increasingly competitive power market, it is difficult for market participants to effectively monitor the moment-to-moment fluctuations of the market situation. Online trading can transact electricity in real time and at the right price. By automatically setting customized procurement goals for customers, power exchanges can let their customers concentrate on their core businesses. Power exchanges post the customers' energy requirements on the real-time power market where qualified energy suppliers will compete. There are no proposals for the customers to write, no negotiations, and no complex analyses of contract terms. In addition, the customers do not need to spend time seeking partners, because an online trading system will maintain continual access to a comprehensive roster of power suppliers and consumers. Some power exchanges provide their services free of charge to consumers as they get a transaction fee from power suppliers.

12.3 POWER TRADING SYSTEM

To promote and stimulate effective competition, restructured power systems capitalize on the Internet for conducting the e-commerce which can almost automate all core processes of business. Many power companies have already embarked on deploying online trading systems for e-commerce. For instance, by making use of the online trading system, transmission companies can develop an information bus that connects their business critical applications to enable information from these disparate systems to be shared companywide; power marketers can also integrate systems for credit management, power trading, and risk management; ESPs can integrate applications for contract management, billing, and so on, and to exchange data with generation companies, distribution companies, and power exchanges.

12.3.1 Classifications

The trading system or platform is a workspace composed of one or more computers and software programs for e-commerce. According to their utilization of the Internet, trading systems can be divided into the following two categories:

- Internet-based platform
- Web-based platform

Though both categories are based on the Internet and in many cases are used interchangeably, there are some differences in terms of the terminologies for the Internet and the Web. The Internet is a public network that connects computers around the world. Through the use of TCP/IP, the Internet allows a customer to access resources on another computer at any place in the world; hence, information can be exchanged on the Internet. The Web is an Internet-based service that utilizes the Internet. By using the Web, one can browse documents stored on the Web servers located anywhere in the world. Obviously a Web-based platform is by nature Internet-based, but the reverse is not necessarily true. A purely Internet-based platform does not require application-specific customer software or browser plug-ins; other advantages of the Internet-based platform include simpler installation and integration, easier upgrading, greater reliability, higher security, and more functionality. However, with the advent of multi-media functions, the Web provides more attractive and convenient interfaces, and Web-based platforms have become more popular.

Before the power market becomes open to power retail, there is to consider a unique wholesale power exchange (WPX) in the system which is usually operated by a department of the ISO and by an entity independent of the ISO. When the power system restructuring reaches the stage of power retail, there will be a number of retail power exchanges (RPXs) that will be operated and managed by SCs, brokers, marketers, and ESPs. In this circumstance, it will be the level of liquidity that determines the success of an RPX. The market liquidity is a measure of whether or not buyers and sellers are able to find each other and transact quickly and satisfactorily. The number of active participants and the daily trading volume are the indicators of the market liquidity. Alternatively, the smaller the bid-offer spread, the greater will be the liquidity. Increased liquidity represents the ability for an RPX to perform e-commerce. Theoretically, the level of liquidity can be demonstrated by an RPX's ability to attract a critical number of participants, which can gradually generate the necessary inertia to increase the market liquidity and the trading velocity; in other words, liquidity begets liquidity. Because traders usually spend half of their trading time in looking for useful information, RPXs could provide price discovery services to allow traders to directly reach an appropriate community of partners.

Market participants make power transactions by using the Internet interface provided by the power trading system. As multiple power exchanges occur in the restructured power system, power trading will be processed in the distributed manner illustrated in Figure 12.1. In such circumstances, different power exchanges will have different market-clearing price. The MCP of the WPX will be determined based on a bidding mechanism, while the MCP of an RPX will be determined by the bilateral contracts of power suppliers and consumers, which could be affected by the liquidity of the RPX. After the market is cleared, all power exchanges will submit their balanced power transactions to the ISO for reliability verification.

12.3.2 Configuration Requirements

The very basic requirement for a simple power trading system is that it should provide a mechanism for posting bids and offers. An important issue in power trading is the certainty factor that could motivate a trader's confidence. Certainty could be implemented by the rules, penalties, failure terms, and dispute judgment tools that govern the trading system and define proper trading practices and responsibilities of both trading sides. Certainty is guaranteed if trades are completed and delivered as agreed.

SPECIAL TOPICS IN INFORMATION SYSTEM

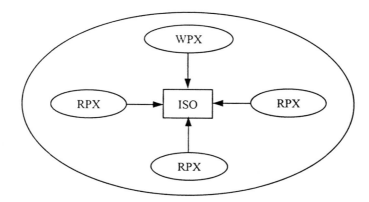

Figure 12.1   Power Market with Multiple Power Exhanges

A trading system needs to treat all participants in a fair manner; otherwise, customers will no longer trust the trading system and turn away to look for other alternatives. Moreover, all market participants should comply with market rules that include credit verifications and legal enforcement procedures. All traders would be required to show that they are able to satisfy financial obligations in with high degree of certainty. The trading system's governance should be independent of market participants. In all, a power trading system should have following characteristics:

- Independent administration
- Effective rules
- User-friendly interfaces and powerful tools
- Adequate market information
- Comprehensive trading opportunities

Because power trading is not a stand-alone activity, most power trading systems offer end-to-end software solutions which integrate front-, middle-, and back-office applications. Actually, some critical matters for a trading system, such as bandwidth, speed, scalability, security, reliability, flexibility, compatibility, and ease of integration with other systems, are just behind the scene. Besides, they provide value-added services; for instance, some power companies are trying to facilitate transactions by providing their customers with additional functions such as online news, powerful market data analysis tools, and decision support systems [Dav02,

Lee02, Sko02, Vis02]. In this regard, a good power trading system should provide appropriate functions such as information function, analysis function, risk management function, and decision-making function that enable their customers in a number of ways:

- In forecasting supply and demand
- In forecasting electricity price
- In hedging possible risks to a maximum extent

However, to meet the challenge of increasingly intense market competition, a power trading system needs to automate the business processes for interactions with customers, employees, and trading partners including energy seller and buyers, PXs, and ancillary service providers. A power trading system would need to exchange business information with partners, customers, and employees reliably and securely across corporate networks and over the Internet. In all, a power trading system should have following four kinds of functions [Voj01]:

- **Business process management.** This function controls and coordinates the exchange of information and transactions with trading partners.

- **Business-to-business communications.** This function enables secure and reliable exchange of information and transactions with trading partners over the Internet to support collaborative business processes.

- **Enterprise application integration.** This function enables secure and reliable movement of information and transactions in and out of internal business applications.

- **Real-time analysis.** This function continuously monitors and analyzes business processes to proactively identify and respond to problems as they occur.

Further, a power trading system should be able to scale well in both complexity and performance as the system expands its business.

12.3.3 Intelligent Power Trading System

Power trading activities on the Internet involve a number of important functions. These include understanding the seller/buyer's behavior, dealing with incomplete player information, and negotiating deals like a human being. Although most power trading systems have installed advanced software programs to address complicated issues involved in power trading, present power trading systems have weaknesses in many aspects of their functionality. In fact, advanced tools are critical to the success of consumers in their market competition; therefore, power trading systems are expected to be capable of automatically seeking cheaper energy sources on the Internet and bid them in the auction market. Hence, new application tools must be developed to improve the performance of present power trading systems. In this regard, artificial intelligence techniques such as intelligent agents and game theoretic modeling can be used in implementing intelligent power trading systems [Sri99].

An intelligent trading system will have many new functions for handling incomplete information about the market, forming bidding strategies, learning partner's behaviors, and negotiating deals. The wide application of intelligent software agents will eventually lead to the appearance of an intelligent power trading system. Intelligent software agents can communicate and cooperate with each other to perform many functions for power trading, which will help explore the dynamics of power market and provide a useful vehicle for investigating the trading strategies of market participants. An intelligent software agent can help power traders analyze the behavior patterns of other traders, negotiate a deal with other traders, bid successfully in auctions, and even collect time-relevant information from various Internet sources.

When an intelligent power trading system is implemented, concerns should be focused on the following aspects:

- Where should intelligent software agents be used?
- What tasks can an intelligent software agent perform?
- What knowledge must a competent intelligent software agent possess to solve problems efficiently and effectively?
- What scheme can be used for coordinating a group of agent?

12.4 TRANSACTION SECURITY

Secure exchange of business information is critical to all market participants because their profits could be purposely reduced if their transaction information is made known to their competitors. Although the

Internet makes the communication and sharing of information more convenient and efficient than ever before, its open accessibility also significantly increases risks for information security at the same time. Certain security measures must be provided to market participants. Many secure procedures such as transaction registration, password, and security key have effectively reduced the risks that market participants could otherwise meet. Some standardization measures can help not only lower the transaction security risk level but also improve price transparency, lower transaction costs, make traders more productive, and even provide a means to quantify risks.

With the rapid development of e-commerce, the credit risk has become another important concern. By credit risk we mean here a chance that the counterparty defaults on the transaction, consequently leaving its partner with the physical product outlay. Although most participants have their own trading limits and restrictions, the risk of trading out of limits increases when their trading activity increases. Since credit risk is now as real time as power market, it is necessary for the power trading system to provide certain virtual credit services for the customers. For instance, to avoid credit risk, some PXs will provide a credit preference feature to allow traders to specify the counterparties with whom they are willing to trade, while some other power exchanges can automatically prevent customers from further trading once credit limits have been exceeded. In providing a credit risk management component, they ensure that customers will stay within their credit limits

12.5 POWER AUCTION MARKETS

12.5.1 Day-ahead Market

The day-ahead market determines the basic energy price for the following day on an hourly basis. Different power markets have different schedules for the following day, and the opening time of the day-ahead power market is different for individual power systems. In systems like that of California, the day-ahead market opens at 6:00 am and closes at 1:00 pm of the day ahead of the scheduling day. When the day-ahead market opens, market participants, either buyers or sellers, submit their bids electronically to the PX prior to the deadline set by the PX. Some bids may be sent by audit e-mail. Besides the amount and price of energy, the content of a bid for the day-ahead power market should include the adjustment bid and the ancillary services bid. Adjustment bids are used for congestion management and the ancillary services are used for enhancing the

SPECIAL TOPICS IN INFORMATION SYSTEM

reliability of power transmission. Once a bid is sent out, it cannot be changed unless modification is required by the PX.

For each hour of the following day, sellers bid a supply schedule at various prices and buyers bid a demand schedule at various prices using respective electronic datasheets. The PX then verifies the submitted bids and whether their formats and contents meet specific requirements. Incomplete or invalid bids will be returned to traders by e-mail and each associated trader will have one additional chance to modify its bid. After the verification of all submitted bids, the PX aggregates all the supply bids in the ascending price order, aggregates all the demand bids in the descending price order, and determines the MCP at each hour based on participants' supply/demand bids. The 24 hourly MPCs are determined for the following scheduling day, and each MPC is enforced at the beginning of the corresponding hour.

The bids initially submitted for the day-ahead market are actually portfolio bids. In other words, power supply bids are not required to refer to any specific generation unit or power plant, and demands bids are not required to provide specific location for loads. The first step in the power auction concerns the amounts and prices of bids. However, once the PX has determined the MCP, winning participants in the day-ahead market are required to split their bids into specific generation unit and load schedules, respectively. The PX submits the corresponding market information along with the bilateral load and generation contracts to the ISO for system reliability verification.

Based the generation and load information received from the PX, the ISO determines whether there is a possibility of any transmission congestion as a result of these schedules. If transmission congestion appears during certain periods of the scheduling day, the ISO calls for load and generation adjustments. The winning participants submit adjustment schedules voluntarily to the PX or their individual SCs. The adjustment auction process is quite similar to that of the day-ahead market. If there are insufficient voluntary adjustments, the ISO will issue mandatory adjustments to participants to mitigate the potential transmission congestion. The ISO will then announce the final day-ahead schedules.

The power trading in the day-ahead market is depicted in Figure 12.2. In the figure, the iterative process is represented by the dashed line for utilizing adjustment bids in the day-ahead market. When the market closes, the PX declares the energy prices for the day-ahead market.

Transactions are subject to mutual payment obligation between the PX and participants, and settlements are based on schedules within three days after each trading day. The energy price determined in the adjustment auction market will be used for calculating congestion charges [Sha02].

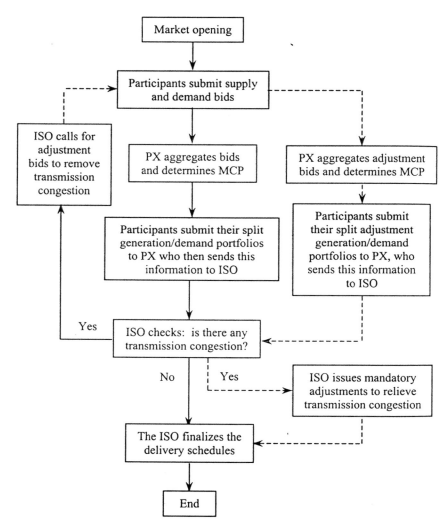

Figure 12.2 Day-Ahead Market Trading Process

12.5.2 Hour-ahead Market

The mission in the hour-ahead power market is to provide an opportunity for market participants to adjust their transactions in the previous day–

ahead power market. There are two major reasons for setting up such an hour-ahead market. First, the hour-ahead market provides an opportunity for participants to adjust their day-ahead commitments based on the newly obtained information. These adjustments will greatly increase the liquidity of the market. Since power transactions have already been set in the day-ahead market, the available transfer capacity of transmission system gets to be limited in the hour-ahead market, which means that the amount of power transactions in hour-ahead market will be limited. Second, suppliers and consumers who participated in the day-head market may be interested in making changes to their scheduled power trades.

The hour-ahead market opens two hours ahead of the practical scheduling and delivery. Participants perform a similar bidding process in the hour-ahead market as in the day-ahead market. When the hour-ahead market opens, participants submit their supply and or demand bids to the PX, and the PX aggregates the supply bids and the demand bids in order to determine the MCP. Because the processing time is short, there is not an iterative process in the hour-ahead market as in the day-ahead market. Once the PX has declared the MPC, all the winning traders will immediately provide to the PX with their split portfolio information including schedules of generation units, point of demands, adjustment bids for congestion management, and ancillary service bids. Based on this specific generation and load information, the ISO checks the system operation reliability. If there is no transmission congestion, the ISO finalizes these transactions without any modification. However, if there is transmission congestion, the ISO will remove the congestion using the existing congestion bids of suppliers and consumers. If the existing congestion bids are insufficient, the ISO will take mandatory measures and participants will have to accept these adjustments without any condition. The ISO informs the PX of its decision and when the market closes the PX declares the electricity price and individual trading quantity of each participant. The power trading procedure of the hour-ahead market is depicted in Figure 12.3. The trading process of the hour-ahead market also uses electronic datasheet.

12.5.3 Next Hour Market

FERC has designated next hour market (NHM) service as voluntary for a transmission provider to offer. The use of NHM service is limited to interchange transactions having duration of one clock-hour and requested no earlier than 60 minutes prior to the start time of the transaction. A transmission provider offering NHM service will allow an eligible transmission customer to request a NHM service reservation electronically using protocols compliant with the NERC ETAG specification.

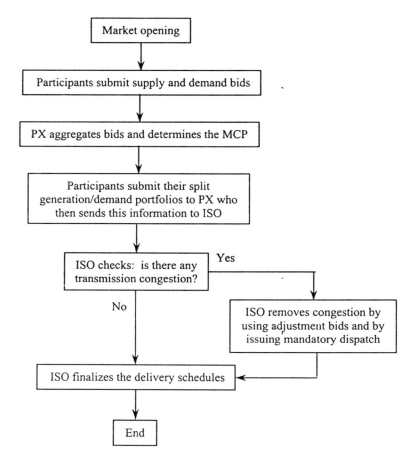

Figure 12.3 Hour-Ahead Market Trading Process

12.5.4 Real-time Operation

Generators and loads that have already been scheduled can also submit supplemental bids in the event that supply exceeds demand in real time. An incremental or decremental bid represents a price that a generator or load source would agree on for the right to reduce or increase their supply and demand. Supplemental bids must be submitted to the ISO within certain time limit, for instance 45 minutes prior to the operating hour.

The ISO first combines the bids into a system wide bid curve for incremental energy, and then determines the real-time dispatch schedule for the real-time market. This schedule is usually determined at 10-minute intervals in the hour of operation, starting from the beginning of each hour. In the case of power undersupply, the ISO will increase the lowest generation bid to restore energy balance; while in the case of power oversupply, the ISO decrease the highest generation bid to restore energy balance. The similar dispatch criteria can also be applied to available load sources. In the case of power undersupply, the ISO will reduce the load with the lowest decremental bid; while in the case of power oversupply, the ISO increase the highest incremental bid to restore energy balance. The 10-minute real time energy price is then set by the bidding price of the last generation or load source that was called on to adjust its schedule. This is the normal procedure for the ISO to operate the real-time power market. In most cases, the ISO will try use the cheapest generation or adjustable load resources to eliminate the real time imbalance, however, for some particular emergences, the ISO will first dispatch the most effective resources rather than the cheapest one to eliminate the system imbalance. In this case, the energy price of the real time market will be mostly determined by resources the ISO uses.

A constrained-dispatch is used for the determination of real-time operation to avoid the occurrence of transmission congestion. Because the reliability is vitally crucial to a power system, some states have made special polices to encourage participants to actively take part in the maintenance of the system reliability. For example, in California, the energy suppliers who have committed their capacity to one of ancillary services markets, except regulation, will receive payment for their energy supply in addition to the payment for their ancillary services capacity. However, suppliers who provide supplemental energy bids will only receive the balance energy payment.

There is no specific penalty on generators or loads for real-time deviations from their final schedules accepted by the ISO. These deviations are settled at the hourly real-time market price. A weighted average of the 10-minute prices is calculated at the end of the hour to settle all uninstructed deviations. The weighted average price is called the hourly ex-post price. The SCs who provide extra supply or have lower than scheduled demand will be paid at this price, while the SCs who provide lower than scheduled supply or have extra demand will pay for this price. In this sense, the real-time power market is the only spot market for energy since all financial settlements are ultimately based on this market.

12.6 RELATIONSHIP BETWEEN POWER MARKETS

There are two forward markets: the day-ahead market for scheduling resources at each hour of the following day, and the hour-ahead market, which is for deviations from the day-ahead schedule. Also, there is the real-time market to balance production and consumption in the real-time. The outcomes the two forward markets and the real-time market are closely related due to their sequential nature in time. This sequential nature in time creates chances for arbitrage by market participants. Market participants can also arbitrate between the PX and ISO markets. The price and quantity relationships between these markets can be figured out and evaluated through analysis of historical data using certain behavior models. These analyses provide useful information for future bidding strategies. In addition to the aforementioned factors, the following factors are to be taken into account in different markets.

- **Impact of price on shifting demand between markets.** Buyers could avoid high prices in PX market by shifting portion of their expected demand into the real-time market.

- **Billing of ancillary services based on scheduled demand.** This ISO practice places an additional cost on the energy purchased in the day-ahead market but is not applied to energy purchased in the real-time market. It gives an incentive for load schedulers to shift some of their demand procurement from the day-ahead to the real-time market in order to not pay the ancillary services costs.

- **Price volatility.** The lower volatility of the day-ahead prices relative to the real-time prices may guarantee a price premium for power purchased in the day-ahead market.

- **Energy supply.** The ISO wants to ensure sufficient generation to supply the load variations through the real-time market. However, because there is no explicit penalty for failure to serve, the ISO cannot assume that the real-time price would lead to enough supply. So, the ISO will have to use replacement reserves to ensure the liquidity of the real-time market. The ISO may not be able to choose the cheapest supply from the real-time market and have to take a spinning reserve bid instead due to some technical limitations such as unit ramp up/down limits.

SPECIAL TOPICS IN INFORMATION SYSTEM

12.6.1 Bidding Strategy

There are several deterministic and stochastic factors that will affect the revenues of power market participants. These factors include accurate forecasting of demand and generation, correct analysis on market power and market elasticity, correct analysis of transmission system operation, and of operation of the whole system, analysis of transmission outage, correct analysis of power generation closely related industries such as oil, gas, coal industries, market relationship, and even the analysis of the weather conditions. A comprehensive bidding process for both energy sellers and buyers is depicted in Figure 12.4.

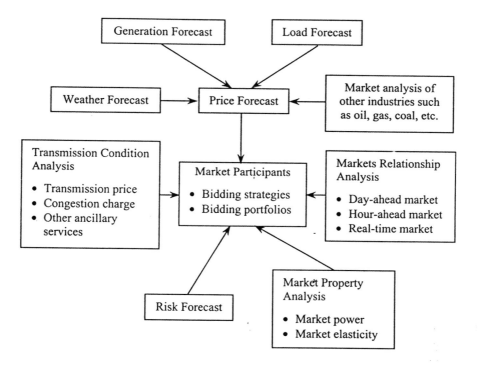

Figure 12.4 Formation of Bidding Strategy

A bidding strategy can be a complicated process, especially within a limited time period. Some useful decision-making support tools and value-added information will have to be provided for market participants by power trading systems.

12.7 GEOGRAPHIC INFORMATION SYSTEM (GIS)

In the 1980s, the geographic information system (GIS) emerged as a new generation of graphics software. GIS has become an indispensable part of a power system information system, and it has received considerable applications in power systems. GIS is a computer-based information system that supports the capturing, management, manipulation, analysis, and modeling of spatially referenced data. GIS has been widely applied to many scientific subjects such as urban planning, utility facility management, transportation management, and gas pipeline management. In GIS, spatial data are linked with the geographic information on a map. As illustrated in Figure 12.5, an entity in a real physical system can be represented on a map by its graphic components (i.e., graphs) and its non-graphic components (i.e., attribute tables). The spatial relationships among entities are reflected in GIS maps.

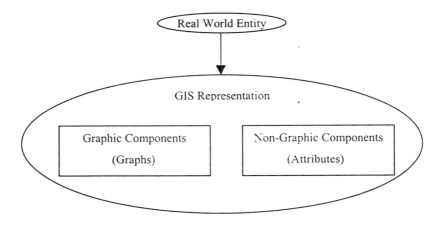

Figure 12.5 Representation of GIS Entities

In the power industry, GIS was first used for the management of distribution system facilities. GIS provides users with a graphic user interface for many applications. Data access routines can be designed to automatically collect graphic and non-graphic data for a specific application. The non-graphic data will be updated in time with the latest computation results and ready for users to view through graphical display. The GIS database provides a very effective means for the managing the information and enhancing the performance of the system operation. For instance, in some applications, such as the trouble call analysis, geographic

SPECIAL TOPICS IN INFORMATION SYSTEM

maps provide more detailed information for the system operator than traditional single-line diagrams.

GIS provides users with an efficient means to solve complicated problems that are usually with spatial features. An electric utility has a large number of facilities that are dispersed in the system, and GIS is a powerful tool for the facility management of an electric utility. For example, there are a large number of pipelines in a power station, most of which are under ground and constitute a complicated pipeline network. This pipeline network may change with the expansion of the power station as time goes on. When a GIS is used to manage this pipeline network, users can very easily find out the information on the geographical distribution of pipelines based on the GIS maps in a computer.

GIS provides a variety of functions such as plotting and editing to manipulate a map file. Users can work on graphs or attribute tables by pointing and clicking on associated function buttons. GIS has a built-in relational database based on various system tables. The schema table, where the entity attributes and symbols are defined, is the master table in the GIS system. The format of the table for graphic and attribute files are determined by users to meet individual requirements. The graphic features such as lines, arcs, and boxes, are stored as coordinate points, while the characteristics of graphic features are stored as attributes which can be queried by the users. For instance, the graphical representation of a distribution line is an arc that connects two corresponding points; the distribution line parameters, such as the type, length, and construction and so on are stored in attribute tables.

12.7.1 GIS Architecture

GIS has taken different architectures at different development stages. It started with a client/server architecture in 1980s. However, with the advent of the Internet, GIS has taken the browser/server architecture. There are two major forms of browser/server GIS: the server-side browser/server GIS and the client-side browser/server GIS. The server-side browser/server GIS uses the common gateway interface technique, which is a common means of interaction between a client browser and a Web server. Users send their queries to the server, and get the replies in JPEG or GIF format from the server. Because the server has to respond to all users queries, the system performance drops by increasing the number of clients. The client-side browser/server GIS adopts plug-in, ActiveX, and Java Applet component techniques. In this architecture, the GIS data and analysis tools are initially stored in the server, and downloaded to the client side the first time they are requested. Since most plug-ins contain executable code and have the capability for local processing, the GIS data analysis and application can be completed more efficiently on the client side. The architectures of the client/server and browser/server of the GIS are depicted in Figure 12.6 and Figure 12.7, respectively. In either architecture, only the server administrator can operate the data in GIS [Ma02].

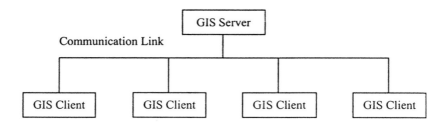

Figure 12.6 Client/Server Architecture

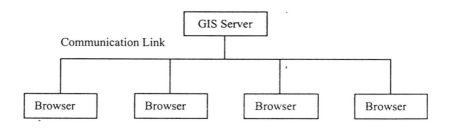

Figure 12.7 Browser/Server Architecture

12.7.2 AM/FM/GIS

The AM/FM (automated mapping/facility management) system, which combined graphic representations and spatial relationships of distribution facilities based on CAD (computer aided drafting) tools, was developed in utilities to facilitate the operation and management of a distribution system. The AM/FM system provides versatile drafting functions and is very efficient in making quality maps. In the AM/FM system, graphic and non-graphic information is utilized to describe the spatial relationships and characteristics of facilities. The graphic information, which includes coordinates, rotation angles, symbols, as well as a non-graphic information pointer, is used for automated mapping. The non-graphic information is used to describe particular attributes of individual facilities.

SPECIAL TOPICS IN INFORMATION SYSTEM

The AM/FM system can be used for work order management, property management, feeder load and voltage analysis, optimal capacitor placement, contingency load transfer, short-circuit calculation and protective coordination, small area load forecasting, substation and primary feeder planning, trouble call analysis, and transformer load management. The major shortcoming of an AM/FM system is that it lacks an implicit data structure and a built-in database for network modeling. Although some CAD tools could provide limited linkage to an external database, this could give rise to other problems such as database normalization and data integrity. A major difference between GIS and AM/FM is that the GIS has a built-in relational database, and a variety of functions for network modeling. The relational database of the GIS can be used for the facility management, and the functions for network modeling can be used to perform network analyses. To make full use of the advantages of the GIS, the AM/FM system is combined with the GIS; the new system is called AM/FM/GIS [Wei95].

The AM/FM/GIS system provides a favorable tool for placing facility elements on a digital map and simultaneously building an associated database that includes graphic information and non-graphic attribute data. In an AM/FM/GIS system, for instance, all components of a distribution network, including feeders, laterals, vaults, transformers, and circuit breakers, can be represented with graphics with its complete data in attribute tables. The information on spatial relationships between facilities is also available as the AM/FM/GIS system specifies where a facility is located on the map and provides detailed information about that facility.

An AM/FM/GIS system supports distribution system planning and design and also provides essential data for many applications for system operation such as transformer load management and related inspections [Chen98]. The AM/FM/GIS system has become a powerful tool for the system operator to operate a distribution system more efficiently. As in GIS, real entities are represented with graphic and non-graphic components in an AM/FM/GIS system. Any change made to graphic files will automatically update the related data in the database and any modification made in the database will be automatically reflected in associated maps.

12.7.3 Integrated SCADA/GIS

The SCADA system helps system operators monitor and control a distribution system in real time. GIS can correlate geographical information with individual facilities for the operation and control of a distribution system. The functions for this purpose include map production, facility inventory, distribution system design, and maintenance service.

As shown in Figure 12.8, GIS and SCADA systems for the same distribution system were usually separated, and data were shared through a local area network. The performance of such architecture is low for a large-scale system, since much information needs to be exchanged between these two systems. To improve the performance of the SCADA system of an increasingly complex distribution system, especially one with distributed generators as shown in Figure 12.9, the SCADA system is integrated with the GIS [Huan02]. An ATM-based network is chosen to connect these two systems, because an ATM-based network is a multi-service network that can meet different service requirements. With the utilization of the ATM network, the data collected from remote terminal units (RTUs) can be transmitted to the SCADA and GIS at the same time. The SCADA and GIS can also share the high-speed information over the ATM network. This integrated SCADA/GIS system not only has more flexibility and better performance than the old stand-alone architecture, but also is suitable for the distributed processing where more than one SCADA/GIS is connected though an ATM network.

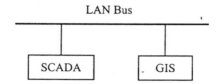

Figure 12.8 Stand alone SCADA and GIS

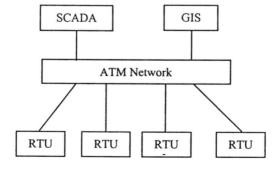

Figure 12.9 Integrated SCADA and GIS

SPECIAL TOPICS IN INFORMATION SYSTEM

We believe the performance of the SCADA system can be improved using GIS functionalities. The same combination of SCADA and GIS can also be implemented for transmission systems.

12.7.4 GIS for Online Security Assessment

Security assessment is an indispensable and effective means to ensure the security of power system operation. In general, security assessment is based on the results of outage power flow computation and transient security analysis. It is hard for the system operator to link the security assessment results to a physical transformation of the power system. However, with the aid of AM/FM/GIS system, the online security assessment results can be visualized. When a power system is depicted with maps for the geographical distribution of system elements, GIS will display security assessment results on these maps and thus provide an intuitive way for system operators to monitor and operate a power system in real time.

When GIS is used for the online security assessment, system operators can choose an outage element directly from GIS maps with a mouse. The outage of a transmission line can be represented directly on the system map, while the outage of a generator or transformer in most cases would need to be represented on the map of the power plant or substation associated to that generator or transformer. The color of the selected element will change in such cases to represent the outage. Actually, when an element is chosen for the contingency power flow computation, the "state" attribute of this element in the GIS attribute table is changed from "true" to "false." The security assessment will be performed based on this new status of the chosen element, and the computation results will be put into the corresponding attribute tables of GIS immediately after the security assessment [Liu98a, Liu01b].

The system operator can check any part of security assessment results, such as the overloaded power flow on a transmission line, on GIS maps. To show the security assessment results vividly, as shown in Figure 12.10, columns and pie charts are used to describe bus voltage magnitudes and phase angles, respectively. Arrows are usually used to represent directions of power flow on a transmission line. Different colors and sizes of graphic elements are used to describe properties of elements that they represent. For instances, green and purple can be used to represent active and reactive power flow respectively, a yellow line without arrow represents a transmission line on outage, a red line represents an overloaded transmission line, and a limit-violated bus will flash with light red. The thickness of a line represents different operation conditions. For instance, a thicker red line represents the overload on that transmission line is higher. Likewise, the performance indexes for contingency selection can be written to the attribute tables of the GIS and displayed on the maps when the system operator wants to seem then. The values of performance indexes can be represented by the thickness of lines. For instance, in the map a thicker line represents a higher value of performance index. In such a way, if a transmission line is taken off for security assessment, the system operator not only can very easily and quickly find out the violated elements on the maps but also can immediately find out which element is affected worst in the system.

(a) Column for Voltage Magnitude   (b) Pie Chart for Phase Angle

Figure 12.10 Representations of Voltage Magnitude and Phase Angle

Correspondingly, GIS can be used as an effective visualization technique for a power system's online security assessment and control. GIS can help improve the online response of system operators to changes in system operating conditions, since with the aid of GIS the system operator would not need to spend much time and energy for analyzing the results of online security assessment. Most important, the useful geographic information, such as the position of a faulted transmission line, which can be only provided by GIS, can offer great help to system operators in decision-making especially under emergency circumstances.

12.7.5 GIS for Planning and Online Equipment Monitoring

A GIS based decision-making support system has been proved very helpful to distribution network planning [Yu00]. The so-called rolling planning scheme is utilized to help seek a practical solution when there is too much inaccuracy and uncertainty in the data for distribution network planning. However, it is hard for system operators to monitor a large number of geographically dispersed equipment in real time. Recently, a GIS-based system was developed for online monitoring of the insulation status of electrical equipment of distribution and transmission systems.

12.7.6 GIS for Distributed Processing

The distributed computation and control of a large-scale power system is supposed to be based on a WAN, which is composed of computers at ISO/RTO and SACCs. Unlike parallel processing, where the computing system is fixed and there are rarely communication faults among processor, the distributed computing and control system based on WAN is much more vulnerable and is subject to machine outages and communication link faults. It is perceived that GIS provides a very useful tool for ISO/RTO about the system architecture before it starts the distributed computation and control.

As shown in Figure 12.11, with the aid of the GIS, the architecture of the distributed computing system can be displayed in real time. Suppose that different colors are used to represent the different operating status of communication links between computers in a distributed computing system. The data corresponding to the performance of each computer on the WAN and features of communication links will be stored in the associated attributes tables.

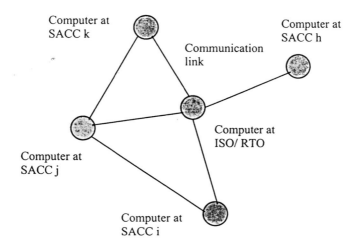

Figure 12.11 Distributed System with GIS

Using this information along with the support of a GPS for timing, the ISO/RTO can use the distributed computing system just like using a parallel machine at hand. Moreover, with the help of GIS, the ISO/RTO can even make online dynamic task allocation for the distributed computing and control system in the case of a computer break-down.

12.7.7 GIS for Congestion Management

In general, transmission congestion is a certain kind of local property since it will primarily affect the market participants within the limited range of the congested line. As shown in Figure 12.12, GIS can greatly facilitate the impact of transmission congestion if used by the ISO/RTO. Correspondingly, the ISO/RTO will focus on the affected area by seeking the most effective local measures for mitigating the congestion. For instance, the ISO/RTO may limit the impact of congestion by utilizing adjustment bids from the regional participants.

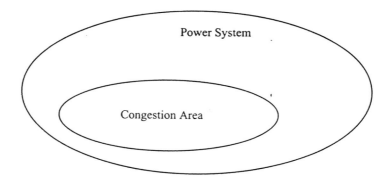

Figure 12.12 Congestion Areas in GIS

In addition, when the regional ALMPs (average locational market prices) are displayed on the screen before the ISO/RTO, as illustrated in Figure 12.13, it will cue the ISO/RTO on how congestion charges and credits are distributed in the system. Hence, the ISO/RTO will have a better handle on the system and be better able to operate the system if congestion occurs. For instance, the ISO/RTO can easily identify GENCOs that are subject to congestion charges and credits, and thus make GENCO's increment or decrement adjustments to mitigate the congestion.

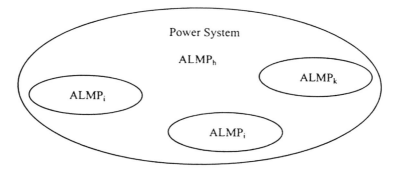

Figure 12.13 Areas with Different LMPs

12.8 GLOBAL POSITIONING SYSTEM (GPS)

As a worldwide navigation and positioning resource, the NAVSTAR (Navigation Satellite Timing and Ranging) GPS was initiated, developed, launched, and maintained by the U.S. Department of Defense in 1978 and promised to provide worldwide coverage and year-round navigation and positioning data for both military and civilian applications.

GPS consists of 24 satellites in six different 12-hour orbital paths, which are deployed in space in such a way that at least five satellites can be viewed from any location on the earth. These satellites continuously transmit navigation data to the earth. Five monitoring stations and four ground antennas located around the world gather data on the satellites' exact position, and relay this information to the master control station at Schriever Air Force Base in Colorado, which provides overall coordination for these satellites and transmits the correction data to the satellites. The GPS signals are transmitted to the ground in two L-band frequencies that are known as L1 and L2. L1 is 1.5754 GHz and carries pseudorandom codes and the status message of the satellites. L2 is 1.2276 GHz and carries more precise military pseudorandom codes. There exist two pseudorandom codes: the coarse acquisition and the precise codes. The GPS transmission is made at a very low power level; the signal strength at the point of reception is about 90 to 120 dbm. A specially designed GPS receiver can receive the signals from four or more satellites at the same time.

A GPS receiver is a miniature device that can be carried along or installed anywhere. A GPS receiver can be used to determine a satellite's location as well as the distance between the receiver and a satellite. With four or more satellites in use, a GPS receiver can determine the user's latitude, longitude, and altitude. Once it has the user's three-dimensional data, the receiver can calculate more information such as speed, and sunrise and sunset time. To get an accurate fix on a moving object, GPS determines the amount of time that it takes for the satellite signal to reach a receiver. Supposing that the signals are synchronous, GPS determine the signal travel time by calculating the number of pseudorandom codes of each satellite. When multiplied by the light speed, the results turn out to be the distance between the satellite and the receiver. The typical GPS receiver has an accuracy of 20 m to 100 m, which is suitable for most applications. Although sophisticated instruments that compare the relative speeds of two timing signals can reach the accuracy of half an inch, they are expensive for generic users. Therefore, many methods have been developed to increase the accuracy of a single, autonomous GPS receiver. Two cost-effective approaches that can greatly enhance the accuracy of GPS are DGPS (differential GPS) and AGPS (assisted GPS).

Because GPS is free of charge and accessible worldwide, it has rapidly become a universal utility with decremental integration costs. The GPS technique can be easily incorporated with vehicles, machinery, computers, and even cellular phones. GPS can provide extremely accurate location information for mobile objects and people — far superior to earlier tracking techniques. Today, GPS has wide range applications, including tracking package delivery, trucking and transportation, mobile commerce, emergency response, exploration, surveying, law enforcement, recreation, wildlife tracking, search and rescue, roadside assistance, stolen vehicle recovery, satellite data processing, and resource management.

The most concerned issue about GPS application is its accuracy. There are several factors that affect the accuracy of GPS. A major factor is that the speed of radio signal is constant only in a vacuum. Water vapor and other particles in the atmosphere can slow down signals resulting in propagation delay. Errors due to multi-path fading, which occurs when a signal bounces off a building or terrain before reaching the receiver's antenna, can also decrease the accuracy. Atomic clock discrepancies, receiver noise, and interruptions to ephemeris monitoring can result in minor errors. Another major source of potential error is selective availability (SA), which is an intentional degradation of the civilian GPS signal. SA was originally inserted as a security measure to prevent a hostile force from exploiting the GPS technology. Authorized users can get a special mechanism to decode SA and eliminate the intentional error. Because the GPS satellites are nearly 11,000 miles away from the earth, an error of a few milliseconds in the calculation of signal travel time could cause an error of 200 miles in position. The monitoring stations and ground antennas check the satellites' speed and position frequently and cleanse ephemeris errors that are caused by gravitational pulls from the moon and sun as well as solar radiation pressure. Besides, the enormous benefits to the world community of increasing GPS accuracy led the U.S. government to turn SA off in 2000.

12.8.1 Differential GPS

The Differential GPS (DGPS) makes use of two receivers: the base receiver and the rover receiver. The actual position of the base is known and compared to the readings received at the same base point. With the estimated error, the readings obtained at the rover can be compensated by simple subtraction. The term "direct DGPS" is used to refer to a GPS configuration in which the position and time measurements are available at the rover station. The term "inverse DGPS" refers to a DGPS instrument in which the results are available at the base station. Although a large distance between these two receivers would degrade the accuracy of DGPS, DGPS can improve the GPs accuracy to one meter or better. The only drawback of DGPS is the requirement for a second GPS receiver and corresponding communication equipment between the base and rover instruments.

The Nationwide Differential Global Positioning System (NDGPS) corrections are broadcast at frequency 283.5 to 325 kHz. DGPS messages are modulated onto the low-medium frequency carrier wave by minimum shift keying (MSK). The selected transmission rates for NDGPS signals are 100 and 200 bits per second. Despite these low transmission rates, they can prevent message losses caused by Gaussian noise and thus achieve higher message throughput under impulse noise conditions. The DGPS broadcast information is contained in a relatively narrow bandwidth, which is an important consideration when concerns about possible interference arise. The NDGPS goals include strengthened national security, integration of GPS into nonmilitary applications, encouraging private sector investment in GPS, and promotion of safety and efficiency in transportation and other activities. Numerous public agencies, including the U.S. Coast Guard and Army Corps of Engineers, transmit DGPS corrections from existing radio beacons placed around harbors, waterways, and other locations to facilitate navigation. Subscription transmission services are also available on FM radio station frequencies or via satellite. In the absence of either a radio receiver or a nearby reference receiver, DGPS corrections could also be distributed via the Internet.

Adequate reception and decoding of DGPS correction messages have been made possible according to the factors such as the level and nature of local noise sources, multi-path, receiver design, receiver antenna type, and placement. Presently, the enhanced positioning accuracy provided by the NDGPS network has got wide application in various fields including electric power industries.

12.8.2 Assisted GPS

Although GPS was not initially designed for indoor use or in urban areas, linking mobile receivers to a cellular, or a wireless local area network infrastructure that has a reference receiver with a clear view of the sky can substantially improve the performance of GPS. Assisted GPS (AGPS) makes use of a reference receiver that provides navigation and signal timing data to a local server. The client device preprocesses and returns basic GPS measurements along with statistical measures that characterize the signal environment to the server, which then performs a series of complex calculations on data received from the client to determine the client's position. AGPS can provide better accuracy than standard GPS within less than 50 feet when users are outdoors.

12.8.3 GPS for Phasor Measurement Synchronization

The GPS-based technique for synchronized phasor measurements have been field-tested and utilized in power systems for many years. A phasor measurement unit (PMU) is a quality digitizer-recorder that has an integral GPS satellite receiver, and many PMUs can be used for sampling events in power systems [Bur94]. Figure 12.14 shows the configuration and the installation of PMUs on a substation bus. By way of time signals of GPS, a number of phasor measurements at the buses that are distributed in a wide geographical area can be synchronized with very high precision.

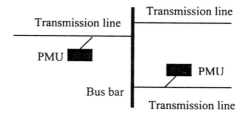

Figure 12.14 Deployments of PMUs

SPECIAL TOPICS IN INFORMATION SYSTEM

Synchronized phasor measurements offer a possibility for tracking system dynamics in real time, and also for enhancing monitoring, protection and control functions of power systems. Previously, the dynamics of electromechanical swings of a power system were only inferred indirectly from a combination of digital and analog oscillographs, using specially designed sensing and monitoring devices such as rotor angle monitors. If phasor measurements can be synchronized precisely, a coherent picture of the whole transient process can then be obtained. The synchronized phasor measurements can be used for observing system dynamic process, because power swings can be recorded via synchronized phasor measurements at a number of buses of the system. The synchronized phasor measurements can give a clear picture of system dynamics and provide useful information for the stability analysis and control. Further, the field measurements of voltage phasors at buses can be used for switching operations.

12.8.4 GPS for Phasor Measurement

The SCADA/EMS is designed and realized in an open-loop centralized control manner as the SCADA system merely captures a quasi-steady state of the system rather than transient processes. With the aide of GPS and PMUs [Pfl92], fast transient processes can be tracked with a high rate of sampling, which results in the possibility of a closed-loop control of power systems, and then power systems could be steered away from certain instable state by executing prompt controls.

When phasor measurements are utilized to help monitor and control the power system, various PMU placement schemes can be adopted, although most advocate the utilization of pilot points which are located at the center of coherent regions of the system [Bal93]. For voltage stability analysis, these coherent regions are required to contain load buses with similar voltage trends, while for transient stability analysis, these coherent regions are required to encompass a group of machines with common slow modes of oscillations. Obviously, this pilot placement scheme has two major drawbacks. First, some systems may not be able to be decomposed into meaningful regions, signifying the necessity to monitor load buses or machine terminals. Second, coherent regions are not stationary but exhibit dynamic behavior; they may either split or coalesce as the system operation conditions change, which lead to the confusion that a pilot placement scheme is optimal for one operating state but can not work well for another.

A PMU placed at a bus can measure the voltage as well as the current phasor of that bus, so it could possible to perform state estimation based on phasor measurements. In this regard, one way to deploy PMUs is to make the system observable, like deploying power measurements for state estimation. However, the cost of this scheme would be very high when all buses are equipped with PMUs. So, to solve the problem, we need to place a minimal set of PMUs that makes the system observable. Because a PMU provides both current and voltage phasor measurements of the bus where it is installed, the system does not need to have at least $2N$-1 PMUs to make the system observable (N is the bus number of the system). A set of less than $2N$-1 PMUs should be able to make the system observable when they are optimally distributed. A minimal set of PMU can be found through a dual search algorithm that combines the bisecting search method and the simulated annealing method [Bal93], where the bisecting search algorithm fixes the number of PMUs while the simulated annealing method looks for a placement set that leads to an observable network for a fixed number of PMUs. To expedite this search process, an initial PMU set can be provided by a graph-theoretic procedure that could generate a spanning measurement subgraph based on the depth-first search. Simulation results have shown that less than one third of the total system buses need to be equipped with PMUs in order to make the system observable.

12.8.5 GPS for Transmission Fault Analysis

The transmission line fault analysis is an indispensable tool at the system control center, and helps the system operator respond quickly to transmission malfunctions. Most transmission line fault analysis approaches utilize expert system to detect and classify faults and to analyze the resulting operations of related relays and circuit breakers. Through combination with an artificial neural network, these approaches could provide even more accurate analysis about the fault detection and classification. However, the crucial part of transmission line fault analysis is the fault locating, an effective approach for fault locating needs to be developed since it could significantly improve the overall performance of transmission line fault analysis. Such an effective fault locating method can be developed through solving transmission line equations at every instant, which requires the sampling of voltages and currents synchronously at both ends of the transmission line. Fortunately, this function can be implemented with GPS [Gal94, Kum98, Moo99].

A digital fault recorder with a GPS satellite receiver could collect the real-time fault data, and this can be realized just by adding a GPS satellite receiver to a standard off-the-shelf digital fault recorder. With these synchronously sampled data, a unique time domain approach for fault analysis can be established, which is extremely fast and accurate and provides robust results even under certain very difficult circumstance such as a time-varying fault resistance. Other fault analysis functions including fault detection and classification can also be implemented by utilizing synchronously sampled data.

A typical application of transmission line fault analysis is locating the fault for HVDC lines. Two approaches based on traveling wave are currently utilized for HVDC fault locating. One is the reflectometry approach, which measures the time difference between the arrival of the first surge and its subsequent reflection at one end of the transmission line. This approach does not need any information from the other end of the transmission line and thus no communication between the two ends of the transmission line is needed. This approach is simple and inexpensive, but it cannot be used for all fault situations. A more reliable approach is to make use of the difference in the arrival times of the fault-generated waves at each end of the transmission line, which conceivably requires a common time reference for the comparison. A dedicated microwave link could be utilized to communicate the arrival times of the surge at each end of the transmission line, but this might be expensive. GPS provides precise time signal and provides a more economical and accurate approach than the microwave link for HVDC fault locating [Dew93]. With the aide of the GPS, fault locating for HVDC lines can be solved with higher precision, since the difference between arrival times of the fault-generated surge at both ends of the transmission line can be calculated more precisely.

12.8.6 GPS for Transmission Capability Calculation

The overhead transmission lines of TRANSCOs form the backbone of the power grid. In an intensely competitive environment, TRANSCOs are under much higher pressure to make optimal use of their facilities. The available transmission capability (ATC) is of more importance in this environment since TRANSCOs make profits through marketing their transmission capacities. To get as much revenue as possible, a TRANSCO needs to calculate accurately the transmission capability of its transmission facilities. The accurate calculation of the transmission capability will not only have obvious financial value but also give the ISO an indication about the operational status of the transmission system.

The transmission capability of an overhead transmission line depends on many factors and these include ambient temperature, wind speed, wind direction, incident solar radiation, conductor characteristics, and conductor configuration. Recent studies have shown that the clearance of an overhead transmission line is also a major factor that could affect the thermal capacity of the overhead transmission line [Men02a]. In this regard, real-time measurements of the transmission line sag can be helpful to the calculation of transmission capability of overhead transmission lines. The sag measurement can be used not only for transmission line planning but also for dynamic line ratings for the real-time monitoring of power grids.

The transmission line sag was previously calculated via indirect measurements, but recently direct measurements of the surface temperature and the tension of the insulator support of the transmission line have been used for the sag calculation. To more rapidly and accurately determine the available transmission capability of the transmission lines, the inverse DGPS is used to measure precisely the position of overhead transmission lines, through which the sag of the related transmission lines can be calculated.

12.8.7 GPS for Synchronizing Multi-agent System

As we showed in Chapter 1 of this book, the power system can be modeled as a system composed of a group of geographically distributed, autonomous, and adaptive intelligent agents that act both competitively and cooperatively for certain goals. Although each agent only has a local view of the system, the whole team of these agents performs control schemes for the entire system through cooperation. In this process, certain kinds of coordination are necessary and important since there could be conflicts among actions of so many autonomous agents, but a key to cooperation among the agents is the time consistency of the information used by each individual agent.

As we saw in Figure 1.12, the reactive layer of MAS (multi-agent system) performs preprogrammed self-healing actions that usually require an immediate response. The coordination layer identifies which triggering event from the reactive layer is urgent based on heuristic knowledge; this layer also analyzes the commands of the top layer and decomposes them into actual control signals, which are sent to reactive agents. The deliberative layer prepares higher-level commands, such as vulnerability assessment and self-healing. The reactive agents may not need much

SPECIAL TOPICS IN INFORMATION SYSTEM

cooperation since in most cases self-healing actions are performed instantly, and there is no time for negotiation. But the coordination layer and the deliberative layer use the same time information, corresponding to reactive agents in the entire system, to coordinate and analyze.

In order to accomplish this task, all reactive agents report their actions to the coordination layer and the deliberative layer by using time tags, which are obtained from GPS. To implement such a scheme, each agent of the system must be equipped with a GPS receiver. Alternatively, when necessary, the coordination layer and the deliberative layer could be preset a time at which reactive agents would submit certain types of information. This scheme is depicted in Figure 12.15. This way, the coordination layer and the deliberative layer obtain an accurate and synchronized time for all agents.

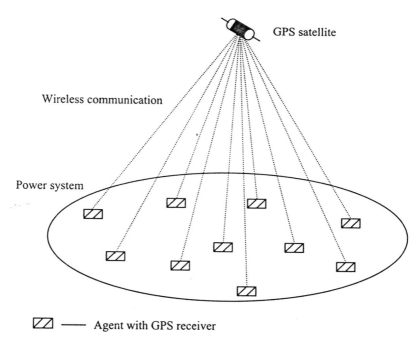

Figure 12.15 Agent Cooperation Based on GPS

In fact, not only the agents of the entire system need to be coordinated for a distinct purpose, the agents of a local area would need to be coordinated to achieve a common goal. For instance, the agent in a certain area of the system would need to coordinate for the congestion elimination just as we saw in Chapter 10.

12.8.8 GPS Synchronization for Distributed Computation and Control

When distributed processing is used for large-scale power system monitoring and control, each SACC collects real-time data from its local control area separately; however, to obtain correct results the distributed computation and control should use the same time data, as that collected by SACCs at the same global time. It the past, due to the lack of a global time reference, it was difficult to maintain the time consistency of all these collected data.

If the data used for distributed processing are not collected at the same time, the results of distributed computation and control could be less degraded. To see this, suppose that as in Figure 12.16 a large-scale power system is divided into three subareas, A, B and C. If the SCADA systems of these three subareas collect their local real-time data within a time discrepancy of Δt, distributed state estimation will not converge to a correct solution after Δt exceeds a certain range. This is obvious because when the data collected at different times in different places are put together, the physical laws with which the system is supposed to comply will be violated.

In fact, the time consistency is an implicit requirement for distributed computation. Distributed control will meet with the same problem. In most cases, to obtain the best control performance, the distributed control commands such as the tertiary voltage/var control command are theoretically supposed to be executed simultaneously. If the execution of the control commend is delayed by a certain SACC, the performance of distributed control will be degraded.

An ideal way to affect distributed computation and control is to get all the measurements of the entire system by taking a "snapshot" of the system, as can be realized with the aide of GPS. Earlier we saw that GPS can provide an accurate global time reference for all SACCs and thus reduce the time discrepancy of SACCs to a negligible extent. As illustrated by Figure 12.16, once the time signal is obtained from the GPS satellite, all SACCs will start to scan their respective areas with their own SCADA systems simultaneously, and all the real time data collected at the same time will be able to form a "snapshot" of the entire system. Moreover, the component computers of the DEMS can start exactly at the same time to execute their distributed applications such as load flow, state estimation, and other various distributed optimization and controls.

SPECIAL TOPICS IN INFORMATION SYSTEM

Further, the utilization of GPS helps improve the convergence performance of distributed algorithms of DEMS, since the discrepancy in the starting time of component computers can exacerbate the asynchronization of distributed computation and thus degrade the performance of distributed computation. The performance of distributed controls of DEMS can also be improved when all control commands are executed simultaneously. In all, through GPS, the performance of distributed computation and control of a DEMS for a large-scale power system can be significantly improved.

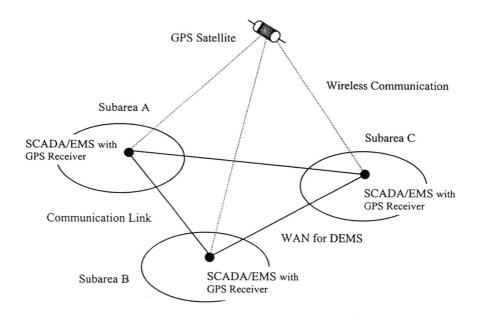

Figure 12.16 GPS for Distributed Processing of DEMS

Appendix A

Example System Data

A.1 PARTITIONING OF THE IEEE 118-BUS SYSTEM

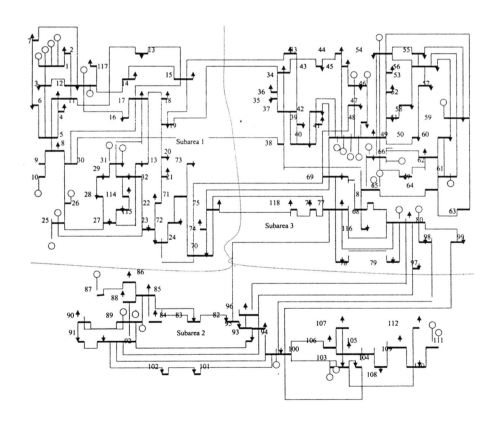

Figure A.1 IEEE 118-bus System Subarea Partitions

A.2 PARAMETERS OF THE IEEE 118-BUS SYSTEM

Table A.1 Line Data of the IEEE 118-Bus System

Line	From Bus	To Bus	R (p.u.)	X (p.u.)	Line	From Bus	To Bus	R (p.u.)	X (p.u.)
1	1	2	0.0303	0.0999	47	35	37	0.011	0.0497
2	1	3	0.0129	0.0424	48	33	37	0.0415	0.142
3	4	5	0.0018	0.008	49	34	36	0.0087	0.0268
4	3	5	0.0241	0.108	50	34	37	0.0026	0.0094
5	5	6	0.0119	0.054	51	38	37	0	0.0375
6	6	7	0.0046	0.0208	52	37	39	0.0321	0.106
7	8	9	0.0024	0.0305	53	37	40	0.0593	0.168
8	8	5	0	0.0267	54	30	38	0.0046	0.054
9	9	10	0.0026	0.0322	55	39	40	0.0184	0.0605
10	4	11	0.0209	0.0688	56	40	41	0.0145	0.0487
11	5	11	0.0203	0.0682	57	40	42	0.0555	0.183
12	11	12	0.006	0.0196	58	41	42	0.041	0.135
13	2	12	0.0187	0.0616	59	43	44	0.0608	0.2454
14	3	12	0.0484	0.16	60	34	43	0.0413	0.1681
15	7	12	0.0086	0.034	61	44	45	0.0224	0.0901
16	11	13	0.0223	0.0731	62	45	46	0.04	0.1356
17	12	14	0.0215	0.0707	63	46	47	0.038	0.127
18	13	15	0.0744	0.2444	64	46	48	0.0601	0.189
19	14	15	0.0595	0.195	65	47	49	0.0191	0.0625
20	12	16	0.0212	0.0834	66	42	49	0.0715	0.323
21	15	17	0.0132	0.0437	67	42	49	0.0715	0.323
22	16	17	0.0454	0.1801	68	45	49	0.0684	0.186
23	17	18	0.0123	0.0505	69	48	49	0.0179	0.0505
24	18	19	0.0112	0.0493	70	49	50	0.0267	0.0752
25	19	20	0.0252	0.117	71	49	51	0.0486	0.137
26	15	19	0.012	0.0394	72	51	52	0.0203	0.0588
27	20	21	0.0183	0.0849	73	52	53	0.0405	0.1635
28	21	22	0.0209	0.097	74	53	54	0.0263	0.122
29	22	23	0.0342	0.159	75	49	54	0.073	0.289
30	23	24	0.0135	0.0492	76	49	54	0.0869	0.291
31	23	25	0.0156	0.08	77	54	55	0.0169	0.0707
32	26	25	0	0.0382	78	54	56	0.0027	0.0095
33	25	27	0.0318	0.163	79	55	56	0.0049	0.0151
34	27	28	0.0191	0.0855	80	56	57	0.0343	0.0966
35	28	29	0.0237	0.0943	81	50	57	0.0474	0.134
36	30	17	0	0.0388	82	56	58	0.0343	0.0966
37	8	30	0.0043	0.0504	83	51	58	0.0255	0.0719
38	26	30	0.008	0.086	84	54	59	0.0503	0.2293
39	17	31	0.0474	0.1563	85	56	59	0.0825	0.251
40	29	31	0.0108	0.0331	86	56	59	0.0803	0.239
41	23	32	0.0317	0.1153	87	55	59	0.0474	0.2158
42	31	32	0.0298	0.0985	88	59	60	0.0317	0.145
43	27	32	0.0229	0.0755	89	59	61	0.0328	0.15
44	15	33	0.038	0.1244	90	60	61	0.0026	0.0135
45	19	34	0.0752	0.247	91	60	62	0.0123	0.0561
46	35	36	0.0022	0.0102	92	61	62	0.0082	0.0376

APPENDIX A

Table A.1 Line Data of the IEEE118-Bus System (Continued)

Line	From Bus	To Bus	R (p.u.)	X (p.u.)	Line	From Bus	To Bus	R (p.u.)	X (p.u.)
93	63	59	0	0.0386	140	90	91	0.0254	0.0836
94	63	64	0.0017	0.02	141	89	92	0.0099	0.0505
95	64	61	0	0.0268	142	89	92	0.0393	0.1581
96	38	65	0.009	0.0986	143	91	92	0.0387	0.1272
97	64	65	0.0027	0.0302	144	92	93	0.0258	0.0848
98	49	66	0.018	0.0919	145	92	94	0.0481	0.158
99	49	66	0.018	0.0919	146	93	94	0.0223	0.0732
100	62	66	0.0482	0.218	147	94	95	0.0132	0.0434
101	62	67	0.0258	0.117	148	80	96	0.0356	0.182
102	65	66	0	0.037	149	82	96	0.0162	0.053
103	66	67	0.0224	0.1015	150	94	96	0.0269	0.0869
104	65	68	0.0014	0.016	151	80	97	0.0183	0.0934
105	47	69	0.0844	0.2778	152	80	98	0.0238	0.108
106	49	69	0.0985	0.324	153	80	99	0.0454	0.206
107	68	69	0	0.037	154	92	100	0.0648	0.295
108	69	70	0.03	0.127	155	94	100	0.0178	0.058
109	24	70	0.0022	0.4115	156	95	96	0.0171	0.0547
110	70	71	0.0088	0.0355	157	96	97	0.0173	0.0885
111	24	72	0.0488	0.196	158	98	100	0.0397	0.179
112	71	72	0.0446	0.18	159	99	100	0.018	0.0813
113	71	73	0.0087	0.0454	160	100	101	0.0277	0.1262
114	70	74	0.0401	0.1323	161	92	102	0.0123	0.0559
115	70	75	0.0428	0.141	162	101	102	0.0246	0.112
116	69	75	0.0405	0.122	163	100	103	0.016	0.0525
117	74	75	0.0123	0.0406	164	100	104	0.0451	0.204
118	76	77	0.0444	0.148	165	103	104	0.0466	0.1584
119	69	77	0.0309	0.101	166	103	105	0.0535	0.1625
120	75	77	0.0601	0.1999	167	100	106	0.0605	0.229
121	77	78	0.0038	0.0124	168	104	105	0.0099	0.0378
122	78	79	0.0055	0.0244	169	105	106	0.014	0.0547
123	77	80	0.017	0.0485	170	105	107	0.053	0.183
124	77	80	0.0294	0.105	171	105	108	0.0261	0.0703
125	79	80	0.0156	0.0704	172	106	107	0.053	0.183
126	68	81	0.0018	0.0202	173	108	109	0.0105	0.0288
127	81	80	0	0.037	174	103	110	0.0391	0.1813
128	77	82	0.0298	0.0853	175	109	110	0.0278	0.0762
129	82	83	0.0112	0.0366	176	110	111	0.022	0.0755
130	83	84	0.0625	0.132	177	110	112	0.0247	0.064
131	83	85	0.043	0.148	178	17	113	0.0091	0.0301
132	84	85	0.0302	0.0641	179	32	113	0.0615	0.203
133	85	86	0.035	0.123	180	32	114	0.0135	0.0612
134	86	87	0.0283	0.2074	181	27	115	0.0164	0.0741
135	85	88	0.02	0.102	182	114	115	0.0023	0.0104
136	85	89	0.0239	0.173	183	68	116	0.0003	0.0041
137	88	89	0.0139	0.0712	184	12	117	0.0329	0.14
138	89	90	0.0518	0.188	185	75	118	0.0145	0.0481
139	89	90	0.0238	0.0997	186	76	118	0.0164	0.0544

A.3 BUS LOAD AND INJECTION DATA OF THE IEEE 118-BUS SYSTEM

Table A.2 Bus Load and Injection Data of the IEEE118-Bus System

Bus	Type	P_d	Q_d	P_g	Q_g	Bus	Type	P_d	Q_d	P_g	Q_g
1	2	51	27	0	0	60	0	78	3	0	0
2	0	20	9	0	0	61	2	0	0	160	0
3	0	39	10	0	0	62	2	77	14	0	0
4	2	30	12	-9	0	63	0	0	0	0	0
5	0	0	0	0	0	64	0	0	0	0	0
6	2	52	22	0	0	65	2	0	0	391	0
7	0	19	2	0	0	66	2	39	18	392	0
8	2	0	0	-28	0	67	0	28	7	0	0
9	0	0	0	0	0	68	0	0	0	0	0
10	2	0	0	450	0	69	3	0	0	516.4	0
11	0	70	23	0	0	70	2	66	20	0	0
12	2	47	10	85	0	71	0	0	0	0	0
13	0	34	16	0	0	72	2	0	0	-12	0
14	0	14	1	0	0	73	2	0	0	-6	0
15	2	90	30	0	0	74	2	68	27	0	0
16	0	25	10	0	0	75	0	47	11	0	0
17	0	11	3	0	0	76	2	68	36	0	0
18	2	60	34	0	0	77	2	61	28	0	0
19	2	45	25	0	0	78	0	71	26	0	0
20	0	18	3	0	0	79	0	39	32	0	0
21	0	14	8	0	0	80	2	130	26	477	0
22	0	10	5	0	0	81	0	0	0	0	0
23	0	7	3	0	0	82	0	54	27	0	0
24	2	0	0	-13	0	83	0	20	10	0	0
25	2	0	0	220	0	84	0	11	7	0	0
26	2	0	0	314	0	85	2	24	15	0	0
27	2	62	13	-9	0	86	0	21	10	0	0
28	0	17	7	0	0	87	2	0	0	4	0
29	0	24	4	0	0	88	0	48	10	0	0
30	0	0	0	0	0	89	2	0	0	607	0
31	2	43	27	7	0	90	2	78	42	-85	0
32	2	59	23	0	0	91	2	0	0	-10	0
33	0	23	9	0	0	92	2	65	10	0	0
34	2	59	26	0	0	93	0	12	7	0	0
35	0	33	9	0	0	94	0	30	16	0	0
36	2	31	17	0	0	95	0	42	31	0	0
37	0	0	0	0	0	96	0	38	15	0	0
38	0	0	0	0	0	97	0	15	9	0	0
39	0	27	11	0	0	98	0	34	8	0	0
40	2	20	23	-46	0	99	2	0	0	-42	0
41	0	37	10	0	0	100	2	37	18	252	0
42	2	37	23	-59	0	101	0	22	15	0	0

APPENDIX A

43	0	18	7	0	0	102	0	5	3	0	0
44	0	16	8	0	0	103	2	23	16	40	0
45	0	53	22	0	0	104	2	38	25	0	0
46	2	28	10	19	0	105	2	31	26	0	0
47	0	34	0	0	0	106	0	43	16	0	0
48	0	20	11	0	0	107	2	28	12	-22	0
49	2	87	30	204	0	108	0	2	1	0	0
50	0	17	4	0	0	109	0	8	3	0	0
51	0	17	8	0	0	110	2	39	30	0	0
52	0	18	5	0	0	111	2	0	0	36	0
53	0	23	11	0	0	112	2	25	13	-43	0
54	2	113	32	48	0	113	2	0	0	-6	0
55	2	63	22	0	0	114	0	8	3	0	0
56	2	84	18	0	0	115	0	22	7	0	0
57	0	12	3	0	0	116	2	0	0	-184	0
58	0	12	3	0	0	117	0	20	8	0	0
59	2	277	113	155	0	118	0	33	15	0	0

Appendix B

Measurement Data for Distributed State Estimation

In the following, each measurement position is represented with the bus number and line number. The measurement types are interpreted as follows.

0: measurement of the reference bus voltage
1: measurement of a bus voltage
2: measurement of an injected active power
3: measurement of an injected reactive power
4: measurement of a line active power
5: measurement of a line reactive power
6: measurement of a line active power at the sending terminal
7: measurement of a line reactive power at the receiving terminal

B.1 MEASUREMENTS OF SUBAREA 1

Table B.1 Measurements of Subarea 1

Position	Type	Value	Weight	Position	Type	Value	Weight
1	4	-0.11815	1	14	6	0.093644	1
1	6	0.118766	1	14	5	-0.00174	1
1	5	0.067954	1	14	7	-0.03734	1
1	7	-0.0912	1	15	4	0.158389	1
2	4	-0.39185	1	15	6	-0.15815	1
2	6	0.394177	1	15	5	-0.06085	1
2	5	0.157595	1	15	7	0.053076	1
2	7	-0.16076	1	16	4	0.342607	1

APPENDIX B

3	4	-1.0175	1	16	6	-0.33995	1
3	6	1.019682	1	16	5	0.018807	1
3	5	-0.44994	1	16	7	-0.02843	1
3	7	0.457705	1	17	4	0.175249	1
4	4	-0.69097	1	17	6	-0.17451	1
4	6	0.702845	1	17	5	-0.07164	1
4	5	0.102281	1	17	7	0.055918	1
4	7	-0.07758	1	18	4	-5.4E-05	1
5	4	0.878097	1	18	6	0.000346	1
5	6	-0.86896	1	18	5	-0.09209	1
5	5	-0.07871	1	18	7	0.031319	1
5	7	0.105838	1	19	4	0.034505	1
6	4	0.34896	1	19	6	-0.03443	1
6	6	-0.34839	1	19	5	-0.03169	1
6	5	-0.05088	1	19	7	-0.01831	1
6	7	0.047962	1	20	4	0.069691	1
7	4	-4.39803	1	20	6	-0.06959	1
7	6	4.445481	1	20	5	-0.02151	1
7	5	-0.90256	1	20	7	0.000532	1
7	7	0.299224	1	21	5	-0.04557	1
8	4	3.359875	1	21	7	0.048632	1
8	6	-3.35988	1	22	4	-0.18041	1
8	5	0.521656	1	22	6	0.181979	1
8	7	-0.22217	1	22	5	-0.06657	1
9	4	-4.44548	1	22	7	0.025459	1
9	6	4.5	1	23	4	0.815562	1
9	5	0.967674	1	23	6	-0.80741	1
9	7	-1.55371	1	23	5	0.129719	1
10	4	0.627506	1	23	7	-0.10945	1
10	6	-0.61911	1	24	4	0.207413	1
10	5	-0.09872	1	24	6	-0.20691	1
10	7	0.108993	1	24	5	-0.05162	1
11	4	0.759245	1	24	7	0.042422	1
11	6	-0.74764	1	25	4	-0.09867	1
11	5	-0.04787	1	25	6	0.099139	1
11	7	0.069493	1	25	5	0.079131	1
12	4	0.324135	1	25	7	-0.10651	1
12	6	-0.32256	1	26	4	0.110981	1
12	5	-0.39831	1	26	6	-0.11082	1
12	7	0.398513	1	26	5	-0.03856	1
13	4	-0.31877	1	26	7	0.02899	1
13	6	0.320699	1	27	4	-0.27914	1
13	5	0.021578	1	27	6	0.280824	1
13	7	-0.03086	1	27	5	0.101717	1
14	4	-0.09321	1	27	7	-0.11505	1
28	4	-0.42083	1	42	6	0.292984	1
28	6	0.424724	1	42	5	0.080243	1
28	5	0.057468	1	42	7	-0.0962	1
28	7	-0.06343	1	43	4	0.114869	1

APPENDIX B

29	4	-0.52472	1	43	6	-0.11454	1
29	6	0.53447	1	43	5	-0.04388	1
29	5	0.045146	1	43	7	0.025704	1
29	7	-0.03993	1	109	4	-0.06927	1
30	4	0.063684	1	109	6	0.069278	1
30	6	-0.06362	1	109	5	-0.04963	1
30	5	-0.00379	1	109	7	-0.05037	1
30	7	-0.04589	1	110	4	0.178104	1
31	4	-1.58505	1	110	6	-0.17778	1
31	6	1.627166	1	110	5	-0.07793	1
31	5	0.401544	1	110	7	0.070458	1
31	7	-0.2721	1	111	4	0.002894	1
32	4	0.840561	1	111	6	-0.00289	1
32	6	-0.84056	1	111	5	-0.02512	1
32	5	1.149157	1	111	7	-0.02368	1
32	7	-1.07779	1	112	4	0.117745	1
33	4	1.413393	1	112	6	-0.11711	1
33	6	-1.34951	1	112	5	-0.04416	1
33	5	-0.19399	1	112	7	0.002249	1
33	7	0.345033	1	113	4	0.060032	1
34	4	0.320787	1	113	6	-0.06	1
34	6	-0.31881	1	113	5	0.006183	1
34	5	-0.02723	1	113	7	-0.01781	1
34	7	0.014544	1	114	4	0.161411	1
35	4	0.148814	1	114	6	-0.16028	1
35	6	-0.14822	1	114	5	-0.06372	1
35	5	-0.06205	1	114	7	0.033781	1
35	7	0.040768	1	115	4	0.000775	1
36	4	2.373445	1	115	6	-0.00078	1
36	6	-2.37345	1	115	5	-0.0191	1
36	5	0.942787	1	115	7	-0.01691	1
36	7	-0.71195	1	117	4	-0.51972	1
37	4	0.758159	1	117	6	0.523363	1
37	6	-0.75558	1	117	5	0.155842	1
37	5	-0.40893	1	117	7	-0.15417	1
37	7	-0.07759	1	178	4	0.038627	1
38	4	2.29944	1	178	6	-0.03618	1
38	6	-2.25718	1	178	5	0.521146	1
38	5	-0.49816	1	178	7	-0.52086	1
38	7	0.040366	1	179	4	0.023864	1
39	4	0.163054	1	179	6	-0.02383	1
39	6	-0.16169	1	179	5	-0.03306	1
39	5	0.034947	1	179	7	-0.01861	1
39	7	-0.07099	1	180	4	0.096991	1
40	4	-0.09178	1	180	6	-0.09684	1
40	6	0.091907	1	180	5	0.037593	1
40	5	-0.0648	1	180	7	-0.0531	1
40	7	0.056924	1	181	4	0.203854	1
41	4	0.91689	1	181	6	-0.20317	1

41	6	-0.8893	1	181	5	0.007067	1
41	5	-0.24053	1	181	7	-0.0236	1
41	7	0.223365	1	182	4	0.016836	1
42	4	-0.29022	1	182	6	-0.01683	1
182	5	0.032547	1	184	6	-0.2	1
182	7	-0.03527	1	184	5	0.033303	1
184	4	0.201421	1	184	7	-0.06258	1

B.2 MEASUREMENTS OF SUBAREA 2

Table B.2 Measurements of Subarea 2

Position	Type	Value	Weight	Position	Type	Value	Weight
82	2	-0.54	1	143	4	-0.09092	1
83	2	-0.19988	1	143	6	0.091274	1
84	2	-0.10999	1	143	5	0.011953	1
85	2	-0.24018	1	143	7	-0.04348	1
86	2	-0.21	1	144	4	0.566297	1
87	2	0.04	1	144	6·	-0.55802	1
88	2	-0.47997	1	144	5	-0.02876	1
89	2	6.046117	1	144	7	0.034436	1
90	2	-1.63535	1	145	4	0.51064	1
91	2	-0.1	1	145	6'	-0.49805	1
92	2	-0.66984	1	145	5	-0.05178	1
93	2	-0.11999	1	145	7	0.053189	1
94	2	-0.29997	1	146	4	0.438031	1
95	2	-0.42	1	146	6	-0.43354	1
96	2	-0.37984	1	146	5	-0.07742	1
100	2	2.150058	1	146	7	0.073912	1
101	2	-0.21998	1	147	4	0.398027	1
102	2	-0.05	1	147	6	-0.39556	1
103	2	0.17	1	147	5	0.145328	1
104	2	-0.38	1	147	7	-0.14783	1
105	2	-0.31	1	149	4	-0.09582	1
106	2	-0.43	1	149	6	0.096	1
107	2	-0.5	1	149	5	0.009123	1
108	2	-0.02	1	149	7	-0.06086	1
109	2	-0.08	1	150	4	0.184969	1
110	2	-0.39	1	150	6	-0.18401	1
111	2	0.36	1	150	5	-0.02791	1
112	2	-0.68	1	150	7	0.008818	1
82	3	-0.27253	1	154	4	0.299188	1
83	3	-0.08415	1	154	6	-0.29322	1
84	3	-0.06716	1	154	5	-0.07507	1
85	3	0.176551	1	154	7	0.055058	1
86	3	-0.1	1	155	4	0.048623	1

APPENDIX B

87	3	-0.01812	1	155	6	-0.04705	1
88	3	-0.09152	1	155	5	-0.31752	1
89	3	-1.14997	1	155	7	0.263206	1
90	3	0.333245	1	156	4	-0.02445	1
91	3	-0.02302	1	156	6	0.024821	1
92	3	0.274957	1	156	5	-0.14914	1
93	3	-0.06278	1	156	7	0.136301	1
94	3	-0.14745	1	160	4	-0.15676	1
95	3	-0.30918	1	160	6	0.157846	1
96	3	-0.13617	1	160	5	0.104844	1
100	3	-0.40471	1	160	7	-0.13234	1
101	3	-0.14627	1	161	4	0.433757	1
102	3	-0.02624	1	161	6	-0.43144	1
103	3	-0.08032	1	161	5	-0.02869	1
104	3	0.032983	1	161	7	0.02465	1
105	3	0.119815	1	162	4	-0.37783	1
106	3	-0.16	1	162	6	0.38144	1
107	3	0.103407	1	162	5	0.016511	1
108	3	-0.01	1	162	7	-0.02904	1
109	3	-0.03	1	163	4	1.195704	1
110	3	-0.05539	1	163	6	-1.17119	1
111	3	-0.11919	1	163	5	-0.34724	1
112	3	0.220369	1	163	7	0.374094	1
129	4	-0.47281	1	164	4	0.579044	1
129	6	0.475506	1	164	6	-0.56355	1
129	5	0.072452	1	164	5	-0.1183	1
129	7	-0.10021	1	164	7	0.1343	1
130	4	-0.25585	1	165	4	0.319589	1
130	6	0.260404	1	165	6	-0.3145	1
130	5	0.058383	1	165	5	-0.10497	1
130	7	-0.07387	1	165	7	0.081577	1
131	4	-0.41954	1	166	4	0.419275	1
131	6	0.427399	1	166	6	-0.40909	1
131	5	0.005292	1	166	5	-0.14129	1
131	7	-0.01246	1	166	7	0.131432	1
132	4	-0.37039	1	167	4	0.607774	1
132	6	0.374648	1	167	6	-0.585	1
132	5	0.025418	1	167	5	-0.11464	1
132	7	-0.02861	1	167	7	0.139265	1
133	4	0.171091	1	168	4	0.498043	1
133	6	-0.17007	1	168	6	-0.49542	1
133	5	-0.00726	1	168	5	-0.13057	1
133	7	-0.01656	1	168	7	0.130679	1
134	4	-0.03994	1	169	4	0.085007	1
134	6	0.04	1	169	6	-0.08474	1
134	5	-0.04787	1	169	5	0.100692	1
134	7	0.004135	1	169	7	-0.1139	1
135	4	-0.49853	1	170	4	0.26796	1
135	6	0.504227	1	170	6	-0.2639	1

135	5	0.177399	1	170	5	-0.0936	1
135	7	-0.1757	1	170	7	0.060435	1
136	4	-0.71479	1	171	4	0.241543	1
136	6	0.72751	1	171	6	-0.23987	1
136	5	0.122158	1	171	5	-0.08407	1
136	7	-0.0771	1	171	7	0.070145	1
137	4	-0.9842	1	172	4	0.239746	1
137	6	0.998075	1	172	6	-0.23611	1
137	5	0.107272	1	172	5	-0.12447	1
137	7	-0.05538	1	172	7	0.090172	1
138	4	0.58074	1	173	4	0.219874	1
138	6	-0.56247	1	173	6	-0.21932	1
138	5	-0.15074	1	173	5	-0.06715	1
138	7	0.164251	1	173	7	0.061077	1
139	4	1.112349	1	174	4	0.602321	1
139	6	-1.08197	1	174	6	-0.5878	1
139	5	-0.25127	1	174	5	-0.11756	1
139	7	0.272549	1	174	7	0.138849	1
140	4	0.009082	1	175	4	0.139322	1
140	6	-0.00908	1	175	6	-0.13866	1
140	5	-0.01346	1	175	5	-0.07721	1
140	7	-0.00794	1	175	7	0.058866	1
141	4	2.000738	1	176	4	-0.35693	1
141	6	-1.9603	1	176	6	0.36	1
141	5	-0.31253	1	176	5	0.099717	1
141	7	0.463988	1	176	7	-0.10919	1
142	4	0.626705	1	177	4	0.69339	1
142	6	-0.61069	1	177	6	-0.68	1
142	5	-0.14228	1	177	5	-0.27867	1
142	7	0.16531	1	177	7	0.25137	1

APPENDIX B

B.3 MEASUREMENTS OF SUBAREA 3

Table B.3 Measurements of Subarea 3

Position	Type	Value	Weight	Position	Type	Value	Weight
46	4	0.012376	1	84	5	0.056177	1
46	6	-0.01236	1	84	7	-0.08945	1
46	5	0.086914	1	85	4	-0.28994	1
46	7	-0.08952	1	85	6	0.297853	1
47	4	-0.34238	1	85	5	0.080182	1
47	6	0.343953	1	85	7	-0.11302	1
47	5	-0.16897	1	86	4	-0.30308	1
47	7	0.162724	1	86	6	0.311537	1
48	4	-0.15744	1	86	5	0.089031	1
48	6	0.158552	1	86	7	-0.11747	1
48	5	-0.06184	1	87	4	-0.37178	1
48	7	0.028557	1	87	6	0.378784	1
49	4	0.298496	1	87	5	0.07014	1
49	6	-0.29764	1	87	7	-0.09468	1
49	5	-0.0984	1	88	4	-0.43581	1
49	7	0.095346	1	88	6	0.442288	1
50	4	-0.9604	1	88	5	0.101266	1
50	6	0.965882	1	88	7	-0.10919	1
50	5	-1.11007	1	89	4	-0.52165	1
50	7	1.120257	1	89	6	0.531189	1
51	4	2.499981	1	89	5	0.117525	1
51	6	-2.49998	1	89	7	-0.1127	1
51	5	1.405055	1	90	4	-1.13516	1
51	7	-1.12993	1	90	6	1.13862	1
52	4	0.571508	1	90	5	0.12438	1
52	6	-0.56128	1	90	7	-0.12124	1
52	5	-0.02866	1	91	4	-0.08712	1
52	7	0.035128	1	91	6	0.087216	1
53	4	0.460088	1	91	5	-0.01194	1
53	6	-0.44761	1	91	7	-0.00229	1
53	5	-0.08698	1	92	4	0.275834	1
53	7	0.079804	1	92	6	-0.27518	1
55	4	0.291281	1	92	5	-0.06378	1
55	6	-0.28946	1	92	7	0.056972	1
55	5	-0.12402	1	93	4	1.586099	1
55	7	0.114499	1	93	6	-1.5861	1
56	4	0.17535	1	93	5	0.588741	1
56	6	-0.17483	1	93	7	-0.48264	1
56	5	0.066886	1	94	4	-1.5861	1
56	7	-0.07728	1	94	6	1.590873	1
57	4	-0.09828	1	94	5	-0.486	1
57	6	0.098872	1	94	7	0.331816	1
57	5	0.007567	1	95	4	0.345643	1
57	7	-0.05223	1	95	6	-0.34564	1
58	4	-0.19517	1	95	5	0.220597	1
58	6	0.196766	1	95	7	-0.21614	1

58	5	0.000326	1	96	4	-1.8628	1
58	7	-0.02927	1	96	6	1.89527	1
59	4	-0.14206	1	96	5	-0.26473	1
59	6	0.143925	1	96	7	-0.41548	1
59	5	0.069055	1	97	4	-1.93652	1
59	7	-0.11972	1	97	6	1.94681	1
60	4	0.038202	1	97	5	-0.26073	1
60	6	-0.03794	1	97	7	-0.00023	1
60	5	0.0483	1	98	4	-1.24583	1
60	7	-0.08895	1	98	6	1.275648	1
61	4	-0.30393	1	98	5	0.31065	1
61	6	0.306305	1	98	7	-0.18322	1
61	5	0.078938	1	99	4	-1.24583	1
61	7	-0.0905	1	99	6	1.275648	1
62	4	-0.35735	1	99	5	0.31065	1
62	6	0.36314	1	99	7	-0.18322	1
62	5	-0.10997	1	100	4	-0.35681	1
62	7	0.097365	1	100	6	0.363377	1
63	4	-0.31149	1	100	5	0.065567	1
63	6	0.315405	1	100	7	-0.09367	1
63	5	0.061285	1	101	4	-0.22523	1
63	7	-0.0799	1	101	6	0.226778	1
64	4	-0.14165	1	101	5	0.08078	1
64	6	0.143199	1	101	7	-0.1046	1
64	5	0.052144	1	102	4	-0.10127	1
64	7	-0.09422	1	102	6	0.101271	1
65	4	-0.08805	1	102	5	2.009679	1
65	6	0.088307	1	102	7	-1.87871	1
65	5	0.066521	1	103	4	0.512721	1
65	7	-0.08178	1	103	6	-0.50678	1
66	4	-0.62801	1	103	5	-0.06295	1
66	6	0.659472	1	103	7	0.063198	1
66	5	0.170561	1	104	4	0.16919	1
66	7	-0.11444	1	104	6	-0.16915	1
67	4	-0.62801	1	104	5	-0.3252	1
67	6	0.659472	1	104	7	-0.31226	1
67	5	0.170561	1	105	4	-0.56736	1
67	7	-0.11444	1	105	6	0.595296	1
68	4	-0.47896	1	105	5	0.069351	1
68	6	0.49577	1	105	7	-0.05106	1
68	5	0.027755	1	106	4	-0.46854	1
68	7	-0.02517	1	106	6	0.490673	1
69	4	-0.3432	1	106	5	0.0309	1
69	6	0.345335	1	106	7	-0.04383	1
69	5	0.013964	1	107	4	-1.18974	1
69	7	-0.02045	1	107	6	1.189738	1
70	4	0.509769	1	107	5	1.017298	1
70	6	-0.50252	1	107	7	-0.93802	1
70	5	-0.11697	1	121	5	0.510757	1
70	7	0.118727	1	121	7	-0.51738	1
71	4	0.631562	1	118	4	-0.6006	1
71	6	-0.61179	1	118	6	0.618629	1

APPENDIX B

71	5	-0.10644	1	118	5	0.194533	1
71	7	0.128467	1	118	7	-0.17124	1
72	4	0.277823	1	119	4	0.678006	1
72	6	-0.27618	1	119	6	-0.66391	1
72	5	-0.04695	1	119	5	0.114768	1
72	7	0.0382	1	119	7	-0.17619	1
73	4	0.096175	1	121	4	0.458193	1
73	6	-0.09571	1	121	6	-0.4564	1
73	5	-0.062	1	122	4	-0.25361	1
73	7	0.024562	1	122	6	0.254369	1
74	4	-0.13429	1	122	5	0.265445	1
74	6	0.134971	1	122	7	-0.26841	1
74	5	-0.09983	1	123	4	-0.90218	1
74	7	0.072428	1	123	6	0.918	1
75	4	0.374238	1	123	5	0.317968	1
75	6	-0.36363	1	123	7	-0.32003	1
75	5	-0.1091	1	124	4	-0.43467	1
75	7	0.077283	1	124	6	0.440745	1
76	4	0.363945	1	124	5	0.122011	1
76	6	-0.35179	1	124	7	-0.1231	1
76	5	-0.12301	1	125	4	-0.64437	1
76	7	0.090729	1	125	6	0.651034	1
77	4	0.068703	1	125	5	-0.03834	1
77	6	-0.06862	1	125	7	0.049969	1
77	5	-0.02634	1	126	4	-0.48182	1
77	7	0.006489	1	126	6	0.483288	1
78	4	0.190348	1	126	5	0.375461	1
78	6	-0.19024	1	126	7	-1.15439	1
78	5	-0.05827	1	127	4	-0.48356	1
78	7	0.051317	1	127	6	0.483562	1
79	4	-0.18961	1	127	5	1.527657	1
79	6	0.1898	1	127	7	-1.44206	1
79	5	0.059737	1	151	4	0.271788	1
79	7	-0.06288	1	151	6	-0.27027	1
80	4	-0.20471	1	151	5	0.083152	1
80	6	0.206578	1	151	7	-0.10045	1
80	5	0.099854	1	152	4	0.296826	1
80	7	-0.11871	1	152	6	-0.29473	1
81	4	0.332521	1	152	5	-0.02033	1
81	6	-0.32658	1	152	7	0.001415	1
81	5	-0.13319	1	153	4	0.203685	1
81	7	0.117056	1	153	6	-0.20173	1
82	4	-0.04237	1	153	5	-0.06753	1
82	6	0.042874	1	153	7	0.021814	1
82	5	0.1022	1	183	4	1.841159	1
82	7	-0.12473	1	183	6	-1.84	1
83	4	0.163964	1	183	5	-0.26182	1
83	6	-0.16288	1	183	7	0.111696	1
83	5	-0.12971	1	186	4	-0.07938	1
83	7	0.115327	1	186	6	0.079769	1
84	4	-0.32903	1	186	5	0.124912	1
84	6	0.334849	1	186	7	-0.13711	1

B.4 MEASUREMENTS ON TIE LINES

Table B.4 Measurements on Tie Lines

Position	Type	Value	Weight	Position	Type	Value	Weight
44	4	0.072774	1	128	4	-0.02695	1
44	6	-0.07256	1	128	6	0.028632	1
44	5	-0.03733	1	128	5	0.195005	1
44	7	0.006103	1	128	7	-0.27038	1
45	4	-0.0336	1	148	4	0.198242	1
45	6	0.033694	1	148	6	-0.19666	1
45	5	-0.0212	1	148	5	0.047253	1
45	7	-0.04169	1	148	7	-0.0876	1
54	4	0.639315	1	157	4	-0.11999	1
54	6	-0.63718	1	157	6	0.120281	1
54	5	0.02533	1	157	5	-0.05327	1
54	7	-0.42039	1	157	7	0.031549	1
108	4	1.103948	1	158	4	-0.04528	1
108	6	-1.06956	1	158	6	0.04538	1
108	5	0.031149	1	158	5	-0.04605	1
108	7	-0.01191	1	158	7	-0.0008	1
116	4	1.126008	1	159	4	-0.21828	1
116	6	-1.07807	1	159	6	0.219182	1
116	5	-0.06915	1	159	5	0.039667	1
116	7	0.085119	1	159	7	-0.05719	1
120	4	-0.32677	1	185	4	0.412239	1
120	6	0.333938	1	185	6	-0.40978	1
120	5	0.086955	1	185	5	-0.00219	1
120	7	-0.1129	1	185	7	-0.00155	1

Appendix C

IEEE-30 Bus System Data

C.1 BUS LOAD AND INJECTION DATA OF THE IEEE 30-BUS SYSTEM

Table C.1 Bus Load and Injection Data of IEEE 30-Bus System

Bus	Load (MW)	Bus	Load (MW)
1	0.0	16	3.5
2	21.7	17	9.0
3	2.4	18	3.2
4	67.6	19	9.5
5	34.2	20	2.2
6	0.0	21	17.5
7	22.8	22	0.0
8	30.0	23	3.2
9	0.0	24	8.7
10	5.8	25	0.0
11	0.0	26	3.5
12	11.2	27	0.0
13	0.0	28	0.0
14	6.2	29	2.4
15	8.2	30	10.6

C.2 REACTIVE POWER LIMITS OF THE IEEE 30-BUS SYSTEM

Table C.2 Reactive power limit of IEEE 30-Bus System

Bus	Qmin (p.u.)	Qmax (p.u.)	Bus	Qmin (p.u.)	Qmax (p.u.)
1	-0.2	0.0	16		
2	-0.2	0.2	17	-0.05	0.05
3			18	0.0	0.055
4			19		
5	-0.15	0.15	20		
6			21		
7			22		
8	-0.15	0.15	23	-0.05	0.055
9			24		
10			25		
11	-0.1	0.1	26		
12			27	-0.055	0.055
13	-0.15	0.15	28		
14			29		
15			30		

C.3 LINE PARAMETERS OF THE IEEE 30-BUS SYSTEM

Table C3 Line Parameter of 30-Bus System

Line	From Bus	To Bus	R (p.u.)	X (p.u.)	Tap Ratio	Rating (p.u.)
1	1	2	0.0192	0.0575		0.300
2	1	3	0.0452	0.1852	0.9610	0.300
3	2	4	0.0570	0.1737	0.9560	0.300
4	3	4	0.0132	0.0379		0.300
5	2	5	0.0472	0.1983		0.300
6	2	6	0.0581	0.1763		0.300
7	4	6	0.0119	0.0414		0.300
8	5	7	0.0460	0.1160		0.300
9	6	7	0.0267	0.0820		0.300
10	6	8	0.0120	0.0420		0.300
11	6	9	0.0000	0.2080		0.300
12	6	10	0.0000	0.5560		0.300
13	9	11	0.0000	0.2080		0.300
14	9	10	0.0000	0.1100	0.9700	0.300
15	4	12	0.0000	0.2560	0.9650	0.650
16	12	13	0.0000	0.1400	0.9635	0.650
17	12	14	0.1231	0.2559		0.320
18	12	15	0.0662	0.1304		0.320
19	12	16	0.0945	0.1987		0.320
20	14	15	0.2210	0.1997		0.160
21	16	17	0.0824	0.1932		0.160
22	15	18	0.1070	0.2185		0.160
23	18	19	0.0639	0.1292	0.9590	0.160
24	19	20	0.0340	0.0680		0.320
25	10	20	0.0936	0.2090		0.320
26	10	17	0.0324	0.0845	0.9850	0.320
27	10	21	0.0348	0.0749		0.300
28	10	22	0.0727	0.1499		0.300
29	21	22	0.0116	0.0236		0.300
30	15	23	0.1000	0.2020		0.160
31	22	24	0.1150	0.1790		0.300
32	23	24	0.1320	0.2700	0.9655	0.160
33	24	25	0.1885	0.3292		0.300
34	25	26	0.2544	0.3800		0.300
35	25	27	0.1093	0.2087		0.300
36	28	27	0.0000	0.3960		0.300
37	27	29	0.2198	0.4153	0.9810	0.300
38	27	30	0.3202	0.6027		0.300
39	29	30	0.2399	0.4533		0.300
40	8	28	0.0636	0.2000	0.9530	0.300
41	6	28	0.0169	0.0599		0.300

Appendix D

Acronyms

A

ACC	Area Control Center
AEP	American Electric Power
AGC	Automatic Generation Control
AGPS	Assisted GPS
ALMP	Average Locational Market Price
API	Application Program Interface
ATM	Asynchronous Transfer Mode
ATC	Available Transmission Capability
AVC	Automatic Voltage Controller

B

BBDM	Bordered Block Diagonal Matrix

C

CAS	Complex Adaptive System
CASM	Common Application Service Model
CCAPI	Control Center Application Program Interface
CCC	Central Control Center
CEMS	Centralized Energy Management System
CHP	Combined Heat and Power
CIS	Component Interface Specification
COW	Cluster of Workstation
CT	Combustion Turbines

D

DBMS	Database Management System
DCC	Dispatching and Control Center
DDET	Dynamic Decision Event Tree
DER	Distributed Energy Resource
DEM	Distributed Memory Environment
DG	Distributed generation
DGPS	Differential GPS
DISTCO	Distribution Company
DOE	Department of Energy
DPS	Distributed Processing System
DNO	Distribution Network Operator

E

EDFA	Erbium Doped Fiber amplifier
ELD	Equivalent Load
EMS-API	Energy Management System Application Program Interface
EPRI	Electric Power Research Institute
ESP	Energy Service Provider

F

FRAD	Frame Relay Access Device

G

GCA	Generation Control Area
GENCO	Generation Company
GIS	Geographic Information System
GOMSFE	Generic Object Models for Substation and Feeder Equipment
GOMSFE	General Object Model for Substation and Field Equipment
GPS	Global positioning system

H

HTTP	Hypertext Transfer Protocol

APPENDIX D

I

IC	Internal Combustion
ICCP	Inter-Control-Center Communications Protocol
ICCS	Integrated Control Center Systems
IDC	Interchange Distribution Calculator
IED	Intelligent Electronic Device
IP	Internet Protocol
IPP	Independent Power Producer
IPS	Initial Preferred Schedule
IT	Information Technology
ISN	Interregional Security Network
ISO	Independent System Operator
IUC	Integrated Utility Communications
IS&R	Information Storage and Retrieval

J

JDBC	Java Database Connectivity

M

MAS	Multi-agent System
MCFC	Molten Carbonate Fuel Cells
MMS	Manufacturing Message Specifications
MPI	Message Passing Interface
MPL	Message Passing Library
MPλS	Multiprotocol Lambda Switching
MPLS	Multiprotocol Label Switching
MSK	Minimum Shift Keying

N

NTP	Network Time Protocol
NERC	North American Electric Reliability Council
NDGPS	Nationwide Differential Global Positioning System

O

OADM	Optical Add/Drop Multiplexer
OASIS	Open Access Same-time Information System

OSI	Open Systems Interconnection
OMS	Outage Management System
OXC	Optical Cross-Connects

P

PAFC	Phosphoric Acid Fuel Cells
PDF	Power Distribution Factors
PEM	Proton Exchange Membrane
PID	Proportional-Integral-Differential
PLC	Power Line Carrier
PSE	Purchasing-Selling Entity
PU	Processing Unit
PURPA	Public Utilities Regulatory Policy Act
PV	Photovoltaic
PVM	Parallel Virtual Machine

R

RDF	Resource Description Framework
RPX	Retail Power Exchanges
RPC	Remote Procedure Call
RRC	Regional Reliability Council
RTO	Regional Transmission Organization
RTP	Real-Time Protocol

S

SACC	Subarea Control Center
SAN	System Area Network
SC	Schedule Coordinator
SCC	Satellite Control Center
SCC	Security Coordination Center
SCP	Standards and Communications Protocols
SDX	System Data Exchange
SGI	Silicon Graphic Interface
SMP	Shared Memory Processing
SOA	Semiconductor Optical Amplifier
SOFC	Solid Oxide Fuel Cell
SONET/SDH	Synchronous Optical Network/Synchronous Digital Hierarchy

APPENDIX D

SQL	Structured Query Language
SVC	Static Voltage Controller

T

TCP	Transmission Control Protocol
TMS	Transaction Management System
TIS	Transaction Information System
TRANSCO	Transmission Company
TASE	Telecontrol Application Service Element Number
TDM	Time-Division Multiplexing
TCP/IP	Transmission Control Protocol/Internet Protocol

U

UCA	Utility Communications Architecture
UCT	Universal Coordinated Time
UDP	User Datagram Protocol
UFPC	Unified Power Flow Controller
UML	Unified Modeling Language
ULTC	Transformer Under Load Tap Changers

W

WAN	Wide Area Network
WDM	Wavelength Division Multiplexed
W3C	World Wide Web Consortium
WPX	Wholesale Power Exchange

X

XML	Extensible Markup Language

Bibliography

[Abb88] Abbasy, N., Shahidehpour, S.M., "An Optimal Set of Measurements for the Estimation of System States in Large-Scale Power Networks," <u>Electric Machines and Power Systems</u>, Vol. 15, pp. 311-332, 1988.

[Abu91] Abur, A., Celik, M.K., "A Fast Algorithm for the Weighted Least Absolute Value State Estimation," <u>IEEE Transactions on Power Systems</u>, Vol. 6, No. 1, pp. 1-8, February 1991.

[Ada98] Adamiak, M., Redfern, M., "Communications Systems for Protective Relaying," <u>IEEE Computer Applications in Power</u>, Vol. 11, No.3, pp. 14-18, July 1998.

[Ada99] Adamiak, M., Premerlani, W., "Data Communications in a Restructured Environment," <u>IEEE Computer Applications in Power</u>, Vol.12, No.3, pp. 36-29, July 1999.

[Alo96] Aloisio, G., Bochicchio, M.A., La Scala, M., Trovato, M., "A Metacomputing Approach for Real-time Transient Stability Analysis," 8th Mediterranean Electrotechnical Conference, Vol.2, pp. 893-896, 1996.

[Alv98] Alves, A.B., Monticelli, A., "Static Security Analysis Using Pipeline Decomposition," <u>IEE Proceedings of Generation, Transmission and Distribution</u>, Vol. 145, No.2, pp. 105-110, March 1998.

[Ama96] Amano, M., Zecevic, A.I., Siljak, D.D., "An Improved Block Parallel Newton Method via Epsilon Decompositions for Load Flow Calculations," <u>IEEE Transactions on Power Systems</u>, Vol.11, No.3, pp. 1519 -1527, August 1996.

[Amb01] Ambriz-Perez, H., Acha, E., Fuerte-Esquivel, C.R., "Closure to Discussion of Advanced SVC Models for Newton-Raphson Load Flow and Newton Optimal Power Flow Studies," <u>IEEE Transactions on Power Systems</u>, Vol.16, No. 4, pp. 947-948, November 2001.

[Ami01] Amin, M., "Toward Self-healing Energy Infrastructure Systems," IEEE Computer Applications in Power, Vol. 14, No.1, pp. 20-28, January 2001.

[Ara81] Arapostathis, A., Sastry, S., Varaiya, P., "Analysis of Power Flow Equation," Electrical Power and Energy Systems, Vol. 3, July 1981.

[Asc77] Aschmoneit, F.C., Peterson, N., Adrian, E. C., "State Estimation with Constraints," Proceeding of the Tenth Power Industry Computer Application Conference, Toronto, pp. 427-430, May 1977.

[Aug01] Augugliaro, A., Dusonchet, L., Ippolito, M.G., Sanseverino, E.R, "An Efficient Iterative Method for Load Flow Solution in Radial Distribution Networks," IEEE Porto Power Tech Proceedings, Vol.3, 2001.

[Awd01] Awduche, D., Rekhter, Y., "Multi-Protocol Lambda Switching: Combining MPLS Traffic Engineering Control with Optical Crossconnects," IEEE Communication Magazine, Vol.39, No.3, pp.111-116, March 2001.

[Baj02] Bajaj, R., Ranaweera, S.L., Agrawal, D.P., "GPS: location-tracking technology," IEEE Computer Magazine, Vol.35, No.3, pp. 92-94, March 2002.

[Bal93] Baldwin, T.L., Mili, L., Boisen, M.B., Jr., Adapa, R., "Power System observability with minimal phasor measurement placement," IEEE Transactions on Power Systems, Vol. 8, No. 2, pp.707-715, May 1993.

[Bal99] Baldick, R., Kim, B.H., Chase, C., Luo, Y., "A Fast Distributed Implementation of Optimal Power Flow," IEEE Transactions on Power Systems, Vol. 14, No. 3, pp. 858-864, August 1999.

[Bar95] Baran, B., Kaszkurewicz, E., Falcao, D.M., "Team Algorithms in Distributed Load Flow Computations," IEE Proceedings of Generation, Transmission and Distribution, Vol. 142, No.6, pp. 583 -588, November 1995.

[Bar00] Barker, P.P., De Mello, R.W., "Determining the Impact of Distributed Generation on Power Systems. I. Radial Distribution Systems," Proceedings of IEEE Power Engineering Society Winter Meeting, Vol. 3, pp. 1645 -1656, 2000.

BIBLIOGRAPHY

[Bec00] Becker, D., Falk, H., Gillerman, J., Mauser, S., Podmore, R., Schneberger, L. "Standards-Based Approach Integrates Utility Applications," IEEE Computer Applications in Power, Vol. 13, No. 4, pp.13 -20, October 2000.

[Beg01] Begovic, M., Pregelj, A., Rohatgi, A., Novosel, D., "Impact of Renewable Distributed Generation on Power Systems," Proceedings of the 34th Annual Hawaii International Conference on System Sciences, pp. 654 -663, 2001.

[Ber89] Bertsekas, D. P., Tsitsiklis, J. N., Parallel and Distributed Computation: Numerical Methods, Prentice Hall Inc., 1989.

[Ber00] Berry, T. "Standards for Energy Management System Application Program Interfaces," Proceedings of International Conference on Electric Utility Deregulation and Restructuring and Power Technologies, pp.156 -161, 2000.

[Bla90] Blackman, J.M., Hissey, T.W, "Impact of Local and Wide Area Networks on SCADA and SCADA/EMS Systems," IEE Colloquium on Advanced SCADA and Energy Management Systems, pp. 8/1 –815, Dec 1990.

[Bra00] Braz, L.M.C., Castro, C.A., Murati, C.A.F., "A Critical Evaluation of Step Size Optimization Based Load Flow Methods," IEEE Transactions on Power Systems, Vol. 15, No. 1, pp. 202 -207,February 2000.

[Bri82] Brice, C.W., Cavin, R.K., "Multiprocessor Static State Estimation," IEEE Transactions on Power Apparatus and Systems, Vol. 101, No. 2, pp. 302-308, February 1982.

[Bri91] Britton, J. "Utility Control Centers Open to Change," IEEE Computer Applications in Power, Vol. 4 No. 4, pp. 35-39, Oct. 91.

[Bur94] Burnett, R.O., Butts, M.M., Cease, T.W., Centeno, V., Michel, G., Murphy, R.J., Phadke, A.G, "Synchronized Phasor Measurements of a Power System Event," IEEE Transactions on Power Systems, Vol. 9, No.3, pp.1643-1650, August 1994.

[Can01] Canever, D., Dudgeon, G.J.W., Massucco, S., McDonald, J.R., Silvestro, F., "Model Validation and Coordinated Operation of a Photovoltaic Array and a Diesel Power Plant for Distributed Generation," Proceedings of IEEE Power Engineering Society Summer Meeting, Vol. 1, pp. 626 -631, 2001.

[CAP01] "Fundamentals of Utilities Communication Architecture," IEEE Computer Applications in Power, Vol.14, No.3, pp.15 -21, July 2001

[Car98] Carvalho, J.B., Barbosa, F.M., " Parallel and Distributed Processing in State Estimation of Power System Energy," 9th Mediterranean Electrotechnical Conference, 1998. Vol. 2, pp. 969-973, 1998.

[Cau96] Cauley, G., Hirsch, P., Vojdani, A., Saxton, T., Cleveland, F., "Information Network Supports Open Access," IEEE Computer Applications in Power, Vol. 9, No. 3, pp.12-19, Jul 1996.

[Che93] Chen, H.M., Berry, F.C., "Parallel Load-Flow Algorithm Using a Decomposition Method for Space-Based Power Systems," IEEE Transactions on Aerospace and Electronic Systems, Vol. 29, No.3, pp.1024-1030, July 1993.

[Che98] Chen, T., Cherng, J., "Design of a TLM Application Program Based on An AM/FM/GIS System," IEEE Transactions on Power Systems, Vol. 13, No. 3, pp. 904-909, August 1998.

[Che02] Chen, J., Thorp, J.S., "A Reliability Study of Transmission System Protection Via a Hidden Failure DC Load Flow Model," Proceedings of Fifth International Conference on Power System Management and Control, pp. 384 -389, 2002.

[CIG92] CIGRE TF 39/02, "Voltage and Reactive Power Control," CIGRE, Report 39-203, Paris 1992.

[Cle72] Clements, K. A., Denison, O. J., Ringlee, R. J., "A Multi-Area Approach to State Estimation in Power System Networks," Proceedings of IEEE/PES Summer Meeting, paper C72-465-3, July 1972.

[Dal01] Daly, P.A., Morrison, J., "Understanding the Potential Benefits of Distributed Generation on Power Delivery Systems," Rural Electric Power Conference, pp. A2/1 -A213, 2001.

[Cle99] Clermont, S. "Lessons Learned: Interfacing Existing External Systems to New EMS," Proceedings of the 21st IEEE International Conference Power Industry Computer Applications, pp. 227-231, 1999.

[Cle00] Cleveland, F. M., "Information Exchange Modeling (IEM) and Extensible Markup Language (XML) Technologies,"

Proceedings IEEE Power Engineering Society Winter Meeting, Vol.1, pp. 592 -595, 2000.

[Con99] Contreras, J., Wu, F.F, "Coalition Formation in Transmission Expansion Planning," IEEE Transactions on Power Systems, Vol. 14, No.3, pp.1144 -1152, August 1999.

[Con02] Contreras, J., Losi, A., Russo, M., Wu, F.F., "Simulation and Evaluation of Optimization Problem Solutions in Distributed Energy Management Systems," IEEE Transactions on Power Systems, Vol. 17, No.1, pp. 57 -62, February 2002.

[Cor95] Corsi, S., Marannino, P., Losignore, N., Moreschini G., Piccini, G., "Coordination Between the Reactive Power Scheduling Function and the Hierarchical Voltage Control of the EHV ENEL System," IEEE Transactions on Power Systems, Vol.10, No.2, pp. 686-694, May 1995.

[Cos81] Simoes-Costa, A., Quintana, V.H., "A Robust Numerical Technique for Power System State Estimation," IEEE Transactions on Power Apparatus and Systems, Vol. PAS--100, No. 2, pp. 691--698, Feb. 1981.

[Cut81] Cutsem, T. V., Horward, J. L., Ribben-Pavella, M., "A Two-Level Static State Estimator for Electric Power Systems," IEEE Transactions on Power Apparatus and Systems, Vol.100, No. 8, pp. 3722-3732, August 1981.

[Cut83] Cutsem, T. V., Ribben-Pavella, M., "Critical Survey of Hierarchical Methods for State Estimation of Electric Power Systems," IEEE Transactions on Power Apparatus and Systems, Vol. 102, No. 10, pp. 3415-3424, October 1983.

[Das94] Das, D., Nagi, H.S., Kothari, D.P., "Novel Method for Solving Radial Distribution Networks," IEE Proceedings of Generation, Transmission and Distribution, Vol. 141, No.4, pp. 291 -298, July 1994.

[Dav02] David, A.K., Lin, X., "Dynamic Security Enhancement in Power-Market Systems," IEEE Transactions on Power Systems, Vol. 17, No. 2, pp. 431 -438, May 2002.

[Dee88] Deeb, N., Shahidehpour, S.M., "An Efficient Technique for Reactive Power Dispatch Using a Revised Linear Programming Approach," Electric Power Systems Research Journal, Vol. 15, No.2, pp. 121-134, 1988.

[Dee89a] Deeb, N., Shahidehpour, S.M., "Linear Reactive Power Optimization in a Large Power Network Using the Decomposition Approach", IEEE Transactions on power Systems, pp. 665-673, 1989.

[Dee89b] Deeb, N., Shahidehpour, S.M., "Economic Allocation of Reactive Power Supply in An Electric Power Network," Proceeding of IEEE International Symposium on Circuits and Systems, pp. 1859-1862, 1989.

[Dee90] Deeb, N., Shahidehpour, S.M., "Lost Minimization in Allocation of Reactive Power Sources," International Journal of Electric Power and Energy Systems, Vol. 12, No. 4, pp. 263-270, 1990.

[Dee91] Deeb, N., Shahidehpour, S.M., "Decomposition Approach for Minimizing Real Power Losses in Power Systems," IEE Proceedings-C, Vol. 138, No. 1, pp. 27-38, January 1991.

[Del01] Del Monaco, J., "The Role of Distributed Generation in the Critical Electric Power Infrastructure," IEEE Power Engineering Society Winter Meeting, Vol. 1, pp. 144-145, 2001.

[Del02] De la Villa Jaen, A., Gomez-Exposito, A., "Implicitly Constrained Substation Model for State Estimation," IEEE Transactions on Power Systems, Vol.17, No.3, pp. 850-856, August 2002.

[Den88] Denzel, D., Edwin, K., Graf, F., Glavitsch, H., "Optimal Power Flow and Its Real-Time Application at the RWE Energy Control Center," CIGRE, Report 39-19, Paris, 1988.

[deV01] de Vos, A., Rowbotham, C.T., "Knowledge Representation for Power System Modeling," 22nd IEEE Power Engineering Society International Conference on Power Industry Computer Applications, pp. 50-56 2001.

[Dew93] Dewe, M.B., Sankar, S., Arrillaga, J., "The Application of Satellite Time References to HVDC Fault Location," IEEE Transactions on Power Delivery, Vol.8, No. 3, pp.1295-1302, July 1993.

[Din01] Ding X, Girgis, A., "Optimal Load Shedding Strategy in Power Systems With Distributed Generation," Proceedings of IEEE Power Engineering Society Winter Meeting, Vol. 2, pp. 788-793, 2001.

BIBLIOGRAPHY

[Doi99] Doi, H., Serizawa, Y., Tode H., Ikeda, H., "Simulation Study of Qos Guaranteed ATM Transmission for Future Power System Communication," IEEE Transactions on Power Delivery, Vol. 14, No. 2, pp. 342-348, April 1999.

[Dop70a] Dopazo, J., Klitin, O., Stagg, G., Van Slyck, L., "State Calculation of Power Systems from Line Flow Measurements: Part I," IEEE Transactions on Power Apparatus and Systems, Vol. PAS-89, No. 7, pp. 1698--1708, September/October 1970.

[Dop70b] Dopazo, J, Klitin, O., Van Slyck, L., "State Calculation of Power Systems from Line Flow Measurements: Part II," IEEE Transactions on Power Apparatus and Systems, No. 7, pp. 145-151.

[Dor96] Dorigo, M., Maniezzo, V., Colorni, A., "Ant System: Optimization By a Colony of Cooperating Agents," IEEE Transactions on Systems, Man and Cybernetics, Part B, Vol.26, No.1, pp. 29-41 February 1996.

[Dug00] Dugan, R.C., Price, S. K., "Issues for Distributed Generation in the US," Proceedings of IEEE Power Engineering Society Winter Meeting, Vol. 1, pp.121-126, 2002.

[Dyl94] Dy-Liacco, T.E., "Modern Control Centers and Computer Networking," IEEE Computer Applications in Power, Vol. 7 No. 4, Oct 1994

Page(s): 17 -22[Ebr00] Ebrahimian, R., Baldick, R., "State Estimation Distributed Processing," IEEE Transactions on Power Systems, Vol. 15, No.4, pp. 1240-1246, November 2000.

[Edr98] Edris, A., Mehraban, A.S., Rahman, M., Gyugyi, L., Arabi, S., Reitman, T., "Controlling the Flow of Real and Reactive Power" IEEE Computer Applications in Power, Vol. 11, No. 1, pp. 20-25, January 1998.

[EPR96] Electric Power Research Institute, Guidelines for Control Center Application Program Interfaces, EPRI TR-106324 3654-01, Final Report, June 1996.

[EPR01a] Electric Power Research Institute, Grid Operations & Planning, March 2001.

[EPR01b] "The benefits of Integrating Information Systems Across the Energy Enterprise," Electric Power Research Institute Report 10013242001.

[Eri97] Ericsson, G., Johnsson, A., "Examination of ELCOM-90, TASE.1, and ICCP/TASE.2 for Inter-Control Center Communication," IEEE Transactions on Power Delivery, Vol.12, No.2, pp. 607 -615, April 1997

[Exp02] Exposito, A., Ramos, E., "Augmented Rectangular Load Flow Model," IEEE Transactions on Power Systems, Vol. 17, No. 2, pp. 271 -276, May 2002.

[Fal95] Falcao, D., Wu, F., Murphy, L., "Parallel and Distributed State Estimation," IEEE Transactions on Power Systems, Vol.10, No.2, pp. 724 -730, May 1995.

[Fer01] Ferreira, C., Dias Pinto, J., Barbosa, F., "Dynamic Security Analysis of An Electric Power System Using a Combined Monte Carlo-Hybrid Transient Stability Approach," 2001 IEEE Porto Power Tech Proceedings, Vol. 2, 2001.

[FuC99] Fu, C., Bose, A., "Contingency Ranking Based on Severity Indices in Dynamic Security Analysis," IEEE Transactions on Power Systems, Vol. 14, No.3, pp. 980 -985, August 1999.

[FuY01] Fu, Y., Groeman, J., van der Geest, H., Zee, H., " a Case Study on the Alternative Solutions for the Load Flow Problem in a 50 Kv Cable Grid ," Seventh International Conference on AC-DC Power Transmission, pp. 311 -315, 2001.

[Gal94] Gale, P., "The Use of GPS for Precise Time Tagging of Power System Disturbances and in Overhead Line Fault Location," Developments in the Use of Global Positioning Systems, pp. 5/1 -5/2, 1994.

[Gil98] Gilbert, G., Bouchard, D., Chikhani, A., "A Comparison of Load Flow Analysis Using Distflow, Gauss-Seidel, and Optimal Load Flow Algorithms," Proceedings of 1998 IEEE Canadian Conference on Electrical and Computer Engineering, Vol.2, pp. 850-853, 1998.

[Gil01] Gilbert, S., "The Nations Largest Fuel Cell Project, a 1 MW Fuel Cell Power Plant Deployed As a Distributed Generation Resource, Anchorage, Alaska Project Dedication August 9, 2000," Proceedings of Rural Electric Power Conference, pp. A4/1 -A4/8, 2001.

[Gje85] Gjelsvik, A., Aam, S., Holten, L., "Hachtel"s Augmented Matrix Method--A Rapid Method for Improving Numerical Stability in Power System Static State Estimation," IEEE Transactions on Power Apparatus & Systems, Vol. 104, No.11, pp. 2987-2993, November 1985.

[Gla90] Glavitsch, H., Asal, H., Schaffer, G., "Experiences and New Concepts in Reactive Power and Voltage Control in Interconnected Power Systems," CIGRE, Report 38/39-08, Paris, 1990

[Gla98] Glavitsch, H., Alvarado, F., "Management of Multiple Congested Conditions in Unbundled Operation of a Power System," IEEE Transactions on Power Systems, Vol. 13, No.3, pp. 1013-1019, August 1998.

[Gle01] Glendinning, P. "Potential Solutions to Voltage Control Issues for Distribution Networks Containing Independent Generators," 16th International Conference and Exhibition on Electricity Distribution, pp. 240 -240, 2001.

[Goo99] Goodrich, M., "Using the IEEE SDF Import Format and the CIM for Model Exchange," Proceedings of IEEE Power Engineering Society Summer Meeting, Vol. 2, pp. 876 -878, 1999.

[Gow95] Gowar, J., Optical Communication Systems, Prentice Hall, 1995.

[Gre92] Green, T.A.; Bose, A., "Open systems benefit energy control centers," IEEE Computer Applications in Power, Vol. 5, No. 3 , pp. 60-64, Jul 1992

[GuJ83] Gu, J., Clements, K., Krumpholz, K., Davis, P., "The Solution of U-Conditioned Power Systems via the Method of Peters and Wilkinson," Proceedings of the Thirteenth Power Industry Computer Application, pp. 239-246, 1983.

[Ham01] Hamoud, G., Bradley, L., "Assessment of Transmission Congestion Cost and Locational Marginal Pricing in a Competitive Electricity Market," Proceedings of Power Engineering Society Summer Meeting, Vol. 3, pp. 1468 -1472, 2001.

[Har00] Harp, S., Brignone, S., Wollenberg, B., Samad, T., "Sepia. A Simulator for Electric Power Industry Agents," IEEE Control Systems Magazine, Vol. 20, No. 4, pp. 53 -69, August 2000.

[Hat93] Hatziargyriou, N., Karakatsanis, T., Papadopoulos, M., "Probabilistic Load Flow in Distribution Systems Containing Dispersed Wind Power Generation," IEEE Transactions on Power Systems, Vol. 8, No. 1, pp. 159-165, February 1993

[Hey01] Heydt, G., Liu, C., Phadke, A., Vittal, V., "Solutions for the Crisis in Electric Power Supply," IEEE Computer Applications in Power, Vol.14, No. 3, pp. 22-30, July 2001.

[Hir99] Hirsch, P.M., "Common Information Model: Panel Session," Proceedings of IEEE Power Engineering Society Summer Meeting, Vol. 2, pp. 860-860, 1999.

[Hol88] Holten, L., Gjelsvik, A., Aam, S., Wu, F., Liu, W., "Comparison of Different Methods for State Estimation" IEEE Transactions on Power Systems, Vol. 3, No. 4, pp. 1798-1806, November 1988.

[Hor96] Horiike, S., Okazaki, Y. "Modeling and Simulation for Performance Estimation of Open Distributed Energy Management Systems," IEEE Transactions on Power Systems, Vol.11, No. 1, pp. 463-468, February 1996.

[Hos02] Hossack, J., McArthur, S., McDonald, J., Stokoe, J., Cumming, T., "A Multi-Agent Approach to Power System Disturbance Diagnosis," Fifth International Conference on Power System Management and Control, pp. 317-322, 2002.

[Hua02] Huang, S., Lin, C., "Application of ATM-Based Network for An Integrated Distribution SCADA-GIS System," IEEE Transactions on Power Systems, Vol.17, No. 1, pp. 80-86, February 2002.

[IEC01] Draft IEC 61970: Energy Management System Application Program Interface (EMS-API)—Part 301: Common Information Model (CIM) Core.

[IEC02] Draft IEC 61970: Energy Management System Application Program Interface (EMS-API) - Part 302: Common Information Model (CIM) Financial, Energy Scheduling, and Reservation.

[IEC03] Draft IEC 61970: Energy Management System Application Program Interface (EMS-API) - Part 303: Common Information Model (CIM)

[IEE73] IEEE Committee Report, "Common Format for Exchange of Solved Load Flow Data," IEEE Transactions on Power

BIBLIOGRAPHY 513

Apparatus and Systems, vol. 92, pp.1916- 1925, July/December 1973.

[IEE86] IEEE Committee Report, "Proposed Data Structure for Exchange of Power System Analytical Data," IEEE Transactions on Power Systems, Vol.PWRS-1, pp. 8-16, May 1986.

[IEE95] IEEE Communication Protocol Tutorial, February 1995.

[IEE01] "Cap Tutorial-Fundamentals of Utilities Communication Architecture," IEEE Computer Applications in Power, Vol.14, No. 3, pp. 15-21, July 2001.

[IEE88] IEEE Tutorial Course on Distribution Automation, Report88EH0280@-8-PWR, 1988.

[Ili95] Ilic, M., Liu, X., Leung, G., Athans, M., "Improved Secondary and New Tertiary Voltage Control," IEEE Transactions on Power Systems, Vol.10, No.4, November 1995

[Ili99] Ilic, M., Galiana, F., Fink, L., "Power System Restructuring -- Engineering and Economics: Bulk-Power Reliability and Commercial Implications of Distributed Resources," Kluwer Academic Publishers, 1998.

[Irv78]} Irving, M., Owen, R., Sterling, M., "Power System State Estimation Using Linear Programming," Proceedings of IEE, Vol. 125, No. 9, pp. 879-885, Sept. 1978.

[Kat92] Kato, K., Fudeh, H., "Performance Simulation of Distributed Energy Management Systems," IEEE Transactions on Power Systems, Vol.7, No. 2, pp. 820-827, May 1992.

[Keb01] Kebaili, S., Adjeroud, F., Zehar, K., "Extension of the Modified Newton Method for Radial Distribution Systems Load Flow," Proceedings of Canadian Conference on Electrical and Computer Engineering, Vol. 2, pp. 781 -784, 2001.

[Kei91] El-Keib, A., Nieplocha, J., Singh, H., Maratukulam, D., "A Decomposed State Estimation Technique Suitable for Parallel Processor Implementation," Proceedings of IEEE/PES Summer Meeting, San Diego, CA, July 1991.

[Kez96] Kezunovic, M., Perunicic, B., "Automated Transmission Line Fault Analysis Using Synchronized Sampling at Two Ends,"

IEEE Transactions on Power Systems, Vol. 11, No. 1, pp. 441-447, February 1996.

[Kha01] Khadem, M., "Progress Report on CIM XML for Model Exchange Interoperability Tests," Proceedings of IEEE Power Engineering Society Summer Meeting, Vol. 2, pp. 840 -841, 2001.

[Kho02] Khosla, R., Li, Q., "Multi-Layered Multi-Agent Architecture With Fuzzy Application in Electrical Power Systems," Proceedings of the 2002 IEEE International Conference on Fuzzy Systems, Vol. 1, pp. 209 -214, 2002.

[Kim97] Kim, B., Baldick, R., " Coarse Grained Distributed Optimal Power Flow," IEEE Transactions on Power Systems, Vol. 12, No. 2, pp. 932-939, May 1997.

[Kir94] Kirkham, H., Johnson, A. R., "Design Consideration for a Fiber Optic Communications Network for Power Systems," IEEE Transactions on Power Delivery, Vol. 9, No. 1, pp. 510 -518, January 1994.

[Kav96] Kavicky, J., and Shahidehpour, M., "Parallel Path Aspects of Transmission Modeling," IEEE Transactions on Power Systems, Vol. 11, No. 3, pp. 1180-1190, 1996.

[Kri98] Krishna, V., Ramesh, V., "Intelligent Agents for Negotiations in Market Games - 2 Application," IEEE Transactions on Power Systems, Vol.13, No. 3, pp. 1109-1114, August. 1998

[Kob74] Kobayashi, H., Narita, S., Hammam, A., "Model Coordination Method Applied to Power System Control and Estimation Problems," Proceedings of the forth International Conference on Digital Computer Applications to Process Control, pp. 114-128, 1974.

[Kot82]} Kotiuga, W., Vidyasagar, M., "Bad Data Rejection Properties of Weighted Least Absolute Value Techniques Applied to Static State Estimation," IEEE Transactions on Power Apparatus and Systems, Vol. PAS--101, No. 4, pp. 844--853, April 1982.

[Kum97] Kumar, A.B.R., Brandwajn, A., Ipakchi, A., Adapa, R., "Integrated Framework for Dynamic Security Analysis," 20th Power Industry Computer Applications Conference, pp. 260 - 265,1997.

BIBLIOGRAPHY

[Kum98] Kumar, A.B.R., Brandwajn, V., Ipakchi, A., Adapa, R., "Integrated Framework for Dynamic Security Analysis," <u>IEEE Transactions on Power Systems</u>, Vol. 13, No.3, pp. 816 -821, August 1998.

[Kwo92] Kwon, W. H., Chung, B. J., "Real-Time Fiber Optic Network for An Integrated Digital Protection and Control System," <u>IEEE Transactions on Power Delivery</u>, Vol.7, No.1, pp. 160-166, January 1992.

[Lag89] Lagonotte, P., Sabonnadiere, J.C., Leost, J.-Y., Paul, J.-P, "Structural Analysis of the Electrical System: Application to Secondary Voltage Control in France" <u>IEEE Transactions on Power Systems</u>, Vol. 4, No.2, pp. 479-486, May 1989.

[Lan93] Langhorne, C., Carlson, C., Chowdhury, S., "Wide Area Network Performance Modeling of Distributed Energy Management Systems," <u>Power Industry Computer Application Conference</u>, pp. 27 -32. 1993.

[Lan94] Langhorne, C., Carlson, C., Chowdhury, S., "Wide Area Network Performance Modeling of Distributed Energy Management Systems," <u>IEEE Transactions on Power Systems</u>, Vol. 9, No. 2, pp. 730 -735, May 1994.

[Lar70a] Larson, R., Tinney, W., Peschon, J., "State Estimation in Power Systems, Part I: Theory and Feasibility," <u>IEEE Transactions on Power Apparatus and Systems</u>, Vol. PAS-89, No. 3, pp. 345-352, March 1970.

[Lar70b] Larson, R., Tinney, W., Hajdu, L., Piercy, D., "State Estimation in Power Systems, Part II: Implementation and Applications,"" <u>IEEE Transactions on Power Apparatus and Systems</u>, Vol. PAS-89, No. 3, pp. 352-363, March 1970.

[Lee99] Lee, S., "The EPRI Common Information Model for Operation and Planning," <u>Proceedings of IEEE Power Engineering Society Summer Meeting</u>, Vol. 2, pp. 866 -871 1999.

[Lee02] Lee, S., "Simulation Model Explores Alternative Wholesale Power Market Structures," <u>IEEE Computer Applications in Power</u>, Vol. 15, No. 2, pp. 28 -35, April 2002.

[Lia02] Liang, J., Green, T., Weiss, G., Zhong, Q., "Evaluation of Repetitive Control for Power Quality Improvement of Distributed Generation," <u>Proceedings of IEEE 33rd Annual</u>

Power Electronics Specialists Conference, Vol. 4, pp. 1803 - 1808, 2002.

[Lin92] Lin, S., "A Distributed State Estimator for Electric Power Systems," IEEE Transactions on Power Systems, Vol. 7, No. 2, pp. 551-557, May 1992.

[Lin94] Lin S., Lin, C., "An Implementable Distributed State Estimator and Distributed Bad Data Processing Schemes for Electric Power Systems," IEEE Transactions on Power Systems, Vol. 9, No. 3, pp. 1277-1284., August 1994.

[Liu85] Liu, C., Wu, F., "Analysis of Small-disturbance Stability Region of Power System Models With Real and Reactive Power Flows," Large Scale Systems, No.9, 1985.

[Liu98a] Liu, Y., Qiu, J., "Visualization of Power System Static Security Assessment Based on GIS," Proceedings of International Conference on Power System Technology, Vol.2, pp.1266-1270, 1998.

[Liu98b] Liu J., Yu X., "Open Communication and Control Architecture for EMS in Electrical Integrated Information System," Proceedings of International Conference on Power System Technology, Vol. 2, pp. 1285 -1289, 1998.

[Liu00] Liu, C., Heydt, G., Phadke, A., et al, "The Strategic Power Infrastructure Defense (SPID) System-A conceptual design," IEEE Control System Magazine, Vol.20, No. 4, pp. 40 -52, August 2000.

[Liu01a] Liu, C., "Adaptation of Multi-Agent Systems for Power Infrastructure Defense," IEEE Power Engineering Society Winter Meeting, 2001, Vol. 1, 2001.

[Liu01b] Liu, L., Sun, C., Zhou, Q., Gu, L., Deng, Q., "A Novel Electrical Equipment on-Line Monitoring System Based on Geographic Information System," Proceedings of International Symposium on Electrical Insulating Materials, pp. 205-208, 2001.

[LoK86] Lo, K., Mahmoud, Y., "A Decoupled Linear Programming Technique for Power System State Estimation," IEEE Transactions on Power Systems, Vol. PWRS-1, No. 1, pp. 154-160, Feb. 1986.

BIBLIOGRAPHY

[Los00] Losi, A., Russo, M., "An Object Oriented Approach to Load Flow in Distribution Systems," IEEE Power Engineering Society Winter Meeting, Vol.4, pp. 2332 -2337, 2000.

[LuC97] Lu, C., Chen, K., Lin, M., Lin, Y., "Performance Assessment of An Integrated Distribution SCADA-AM/FM System," IEEE Transactions on Power Delivery, Vol. 12, No. 2, pp.971-978, April 1997.

[Luk00] Lukic, M., Nakashima, T., Niimura, T., "Voltage Security Analysis of Power Systems Under Deregulation," Canadian Conference on Electrical and Computer Engineering, Vol.1, pp. 133 -137, 2000.

[Lun93] Lun, S.-M., Lo, T., Wu, F., Murohy, L., Sen, A, "LANSIM and Its Application to Distributed EMS," Proceedings of Power Industry Computer Application Conference, pp. 20-26, 1993.

[Lyn97] Lynch, N. A., Distributed Algorithms, Morgan Kaufmann Publishers, Inc. 1997

[McD02] McDermott, T.E., Dugan, R.C., "Distributed Generation Impact on Reliability and Power Quality Indices," Proceedings of IEEE Rural Electric Power Conference, pp. D3 -1-7, 2002

[Mah94] Mahadev, P., Christie, R., "Envisioning Power System Data: Vulnerability and Severity Representations for Static Security Assessment," IEEE Transactions on Power Systems, Vol. 11, No. 3, pp. 1915-1920, 1994.

[Mak96] Mak, S., Radford, D., "Communication System Requirements for Implementation of Large Scale Demand Side Management and Distribution Automation," IEEE Transactions on Power Delivery, Vol. 11, No. 2, pp. 683 -689, April 1996.

[Mar00] Marmiroli, H., Suzuki, H. "Web-based Framework for Electricity Market," Electric Utility and Restructuring and Power Technologies, DRPT 2000.

[MaS02] Ma, S., Qi, L., Liu, W., Ma, W., "Power Station GIS Design and Implementation," IEEE Computer Applications in Power, Vol. 15, No. 2, pp. 41-45, April 2002.

[Mau02] Mauser, S., "Status of EPRI CCAPI Common Graphics Exchange," Proceedings of IEEE Power Engineering Society Winter Meeting, Vol.1, pp. 614 -615, 2002.

[Mek01] Mekhamer, S., Soliman, S., Mostafa, M., El-Hawary, M., "Load Flow Solution of Radial Distribution Feeders: a New Approach," IEEE Porto Power Tech Proceedings, Vol.3, 2001.

[Men02a] Mensah-Bonsu, C., Krekeler, U., Heydt, G., Hoverson, Y., Schilleci, J., Agrawal, B., "Application of the Global Positioning System to the Measurement of Overhead Power Transmission Conductor Sag," IEEE Transactions on Power Delivery, Vol. 17, No. 1, pp. 273-278, January 2002.

[Men02b] Meng Y., Schlueter, R., "Classification of Types, Classes, and Agents for Power System Bifurcations," IEEE Power Engineering Review, Vol. 22, No.8, pp. 55 -57, August 2002.

[Mer71] Merrill H., Schweppe, F., "Bad Data Suppression in Power System Static State Estimation," IEEE Transactions on Power Apparatus and Systems, Vol. PAS-90, pp. 2718-2725, Nov/Dec. 1971.

[Meu98] Meulen, R. "New Technologies for Distributed Measurement and Control," Paper No. VFB. 6/3 98005, ibid, pp. 117-124.

[Mil02] Miller, S., Staron, J., Oatts, M., Enns, M., "A Dictionary and Self Defining Protocol for Exchanging Power System Information," IEEE Transactions on Power Systems, Vol.17, No.2, pp. 337-341, May 2002.

[Mon84] Monticelli, A., Wu, F., "A Method That Combines Internal State Estimation and External Network Modeling," Proceedings of IEEE PES Winter Meeting, Dallas, TX, USA, January 1984.

[Moo99] Moore, P., Crossley, P., "GPS Applications in Power Systems. I. Introduction to GPS," Power Engineering Journal, Vol. 13, No. 1, pp. 33 -39, February 1999.

[Nag01] Nagata, T., Sasaki, H., "A Multi-Agent System for Power System Restoration," Proceedings of IEEE Power Engineering Society Winter Meeting, Vol. 3, pp. 1359 -1364, 2001.

[Nak01] Naka, S., Genji, T., Fukuyama, Y., "Practical Equipment Models for Fast Distribution Power Flow Considering Interconnection of Distributed Generators," Proceedings of Power Engineering Society Summer Meeting, Vol. 2, pp.1007 -1012, 2001.

BIBLIOGRAPHY 519

[Nam02] Nambiar, U., Lacroix, Z., Bressan, S., Lee, M., Li, Y., "Current Approaches to XML Management," IEEE Internet Computing, Vol. 6, No. 4, pp. 43 -51, July-August 2002.

[Nan98] Nanda, J., Kothari, M., Srinivas, M., "on Some Aspects of Distribution Load Flow," IEEE Region 10 International Conference on Global Connectivity in Energy, Computer, Communication and Control, Vol. 2, pp. 510 -513, 1998.

[Nan00] Nanda, J., Srinivas, M., Sharma, M., Dey, S., Lai, L., "New Findings on Radial Distribution System Load Flow Algorithms," Proceedings of IEEE Power Engineering Society Winter Meeting, Vol.2, pp. 1157 -1161, 2000.

[Neu01] Neumann, S., "CIM Extensions for Electrical Distribution," Proceedings of IEEE Power Engineering Society Winter Meeting, Vol. 2, pp. 904 -907, 2001.

[Nie89] Nielsen, E.K., Ackerman, W.J., Barrie, D., Bucciero, J.M., Fowler, N.V., Koehler, J.E., Millar, P.W., Stockard, P.M., Traynor, P.J., Willson, J.D., "Backup Control Centers: Justification, Requirements, Emergency Planning, and Drills," IEEE Transactions on Power Systems, Vol. 4, No. 1, pp. 248 – 256, Feb 1989

[Nie97] Niemeyer, R.E., "Using Web Technologies in Two MLS Environments: a Security Analysis," Proceedings of 13th Annual Computer Security Applications Conference, pp. 205 - 214, 1997.

[Nii01] Niioka, S., Kozu, A., Yokoyama, R., Okada, N., "Supply Reliability Evaluation Taking Account of Transmission Congestion," IEEE PortoPower Tech Proceedings, Vol.1, 2001.

[Niu02] Niu, Y., Cong, Y.L., Niimura, T., "Transmission Congestion Relief Solutions By Load Management," Proceedings of IEEE Canadian Conference on Electrical and Computer Engineering, Vol. 1, pp. 18 -23, 2002.

[Nwa91a] Nwankpa, C., and Shahidehpour, M., "Analysis of Small Disturbances of Transmission Lines in Dynamic Stability Studies," Int. J. Systems Sciences, Vol. 22, No. 5, pp. 845-872, Apr. 1991.

[Nwa91b] Nwankpa, C., and Shahidehpour, M., "Stochastic Model for Power System Planning Studies," IEE Proceedings, Part C, Vol. 138, No. 4, pp. 307-320, July 1991.

[Nwa92] Nwankpa, M. Shahidehpour and Z. Schuss, "A Stochastic Approach to Small Signal Stability Analysis," IEEE Transactions on Power Systems, Vol. 7, No. 3, pp. 1519-1528, Nov. 1992.

[Ong02] Ong, E., "MPI Ruby: Scripting in a Parallel Environment," Computing in Science & Engineering, Vol. 4, No. 4, pp. 78 -82, July-August 2002.

[Osa97] Osano, M., Capretz, M.A.M., "A Distributed Method for Solving Nonlinear Equations Applying the Power Load Flow Calculation," Proceedings of the Thirtieth Hwaii International Conference on System Sciences, Vol. 5, pp. 676 -680, 1997.

[Ove97] Overbye, T. J., et al, "Visualization Power System Operation in An Open Market," IEEE Computer Applications in Power, Vol. 10, No.1, pp.51-58, January 1997.

[Pac97] Pacheco, P.S., Parallel Programming with MPI, Morgan Kaufmann Publishers, Inc. 1997.

[Pal96] Palmer, E. W., Ledwich, G., "Optimal Placement of Angle Transducers in Power Systems," IEEE Transactions on Power Systems, Vol. 11, No. 2, pp.788-793, May 1996.

[Pau87] Paul, J. P., Leost, J. Y., Tesseron, J. M., "Survey of the Secondary Voltage Control in France: Present Realization and Investigations," IEEE Transactions on Power Systems, Vol.2, No.2, pp. 505-511, May 1987.

[Pfl92] Pflieger, K., Enge, P.K., Clements, K.A., "Improving Power Network State Estimation Using GPS Time Transfer," IEEE Position Location and Navigation Symposium, pp. 188-193, 1992.

[Pim02] Pimpa, C., Premrudeepreechacharn, S., "Voltage Control in Power System Using Expert System Based on SCADA System," Proceedings of IEEE Power Engineering Society Winter Meeting, Vol.2, pp.1282 -1286, 2002.

[Pir92] Piret, I. P., Antoine, I. P., Stubbe, M., Ianssens, N., Delince, I. M., "The Study of a Centralized Voltage Control Method Applicable to the Belgian Systems," CIGRE, Report 39-201, Paris 1992.

[Pod99a] Podmore, R., "Building a Real-Time CIM for the Eastern Interconnection," Proceedings of IEEE Power Engineering Society Summer Meeting, Vol.2, pp. 861 -865, 1999.

[Pod99b] Podmore, R., Becker, D., Fairchild, R., Robinson, M., "Common Information Model-A Developer's Perspective," Proceedings of the 32nd Annual Hawaii International Conference on Systems Sciences, 1999.

[Rob95] Robinson, J.T., Saxton, T., Vojdani, A., Ambrose, D., Schimmel, G., Blaesing, R.R., Larson, R., "Development of the Intercontrol Center Communications Protocol (ICCP)." Proceedings of Power Industry Computer Application Conference, pp. 449 –455, May 1995.

[Ram01] Ramakumar, R., "Role of Distributed Generation in Reinforcing the Critical Electric Power Infrastructure," Proceedings of IEEE Power Engineering Society Winter Meeting, Vol.1, pp. 139-139, 2001.

[Sal01] Salman, S.K., Rida, I.M., Stirling, R., Zhang, R., Hughes, J., "Development of automatic voltage control using object-oriented programming technique," IEEE Porto Power Tech Proceedings, Vol.3, 2001.

[Sim81] Simoes-costa, A., Quintana, V. H., "An Orthogonal Row Processing Algorithm for Power System Sequential State Estimation," IEEE Transactions on Power Apparatus and Systems, Vol. 100, pp.3791-3800, May 1992.

[Sko02] Skopp, A.R., "Implementing Back-up Control Centers Proceedings of IEEE Power Engineering Society Summer Meeting, Vol. 1, p. 374, 2002.

[Qiu99] Qiu, B., Gooi, H.B., "Internet-based SCADA Display Systems (WSDS) for Access via Internet," IEEE Transaction on Power System, Vol.15, No.2, pp 681 -686, May 2000.

[Qui02] Quintela, A.S., Castro, C.A., "Improved Branch-Based Voltage Stability Proximity Indices. II. Application in Security Analysis," Large Engineering Systems Conference on Power Engineering, pp. 115 -119, 2002.

[Raj94] Rajagopal, S., Rafian-Naini, M., Lake, J., Velo, E., Turke, A., Sigari, P., Silverman, S., Henry, L., Hamlin, R., Bednarik, R.A., "Workstation Based Advanced Operator Training Simulator for Consolidated Edison," IEEE Transactions on Power Systems, Vol. 9, No. 4, Page(s): 1980 –1986, Nov. 1994

[Rau02] Rau, N.S., "Transmission Congestion and Expansion Under Regional Transmission Organizations," IEEE Power

Engineering Review, Vol. 22, No. 9, pp. 47 -49, September 2002.

[Rob95] Robinson, J.T., Saxton, T., Vojdani, A., Ambrose, D., Schimmel, G., Blaesing, R.R., Larson, R., "Development of the Intercontrol Center Communications Protocol (ICCP)," Proceedings of Power Industry Computer Application Conference, pp. 449 -455, 1995

[Roy93] Roytelman, I., Shahidehpour, S. M., "State Estimation for Electric Power Distribution Systems in Quasi Real-Time Conditions" IEEE Transactions on Power Delivery, Vol. 8, No. 4, pp. 2009-2015, October 1993.

[Sab96] Sabir, S., Mahoney, H., "Building a Backbone for Integrated Business Communications," IEEE Computer Application in Power, Vol. 9, No.1, pp. 38-43, February 1996.

[Sak02] Meliopoulos, S., "State Estimation for Mega RTOs," Proceedings of IEEE Power Engineering Society Winter Meeting, Vol.2, pp. 1449 -1454. 2002

[Sch70] Schweppe, F.C., Wildes, J., "Power System Static-State Estimation, Part 1: Exact Model," IEEE Transactions on Power Apparatus and Systems, Vol. 89, No. 1, pp.120-125, January 1970.

[Sch74] Schweppe, F.C., Handschin, E., "Static State Estimation in Electric Power Systems," Proceedings of IEEE, Vol. 62, No. 7, pp. 972--982, July 1974.

[Sch93] Schellstede, G., Schroppel, W., "A Distributed Energy Management System Based on Open Architecture Rules," Proceedings of Athens Power Tech, Vol. 1, pp. 114 -119, 1993.

[Sch99] Schlumberger, Y., Lebrevelec, C., De Pasquale, M.," Power Systems Security Analysis-New Approaches Used At EDF," Proceedings of IEEE Power Engineering Society Summer Meeting, 1999. , Vol. 1, pp. 147 -151, 1999.

[Sch02] Schaffner, C., "An Internet-Based Load Flow Visualization Software for Education in Power Engineering," IEEE Power Engineering Society Winter Meeting, Vol. 2, pp.1415 -1420, 2002.

[Seri99] Serizawa, Y., Shimizu K., Fujikawa, F., "ATM Transmissions of Microprocessor-Based Current Differential Teleprotection

Signals," IEEE Transactions on Power Delivery, Vol. 14, No. 2, pp. 335-341, April 1999.

[Sha96] Shahidehpour, M., Ramesh, V., "Nonlinear Programming Algorithms and Decomposition Strategies for OPF," chapter in IEEE/PES Tutorial on Optimal Power Flow, 1996.

[Sha99a] Shahidehpour, M., Ferrero, M., "Electricity Supply Industry," chapter in Wiley Encyclopedia of Electrical and Electronics Engineers, 1999.

[Sha99b] Shahidehpour, M., Yamin, H., "Risk Management using Game Theory in Transmission Constrained Unit Commitment within a Deregulated Power Market," chapter in IEEE/PES Tutorial on Applications of Gaming Methods to Power System Operation, 1999.

[Sha99c] Shahidehpour, M., Alomoush, M., "Decision Making in a Deregulated Power Environment based on Fuzzy Sets," Chapter in Modern Optimization Techniques in Power Systems, edited by Y.H. Song, 1999.

[Sha00] Shahidehpour, M., Marwali, M., Maintenance Scheduling in Restructured Power Systems, Kluwer Academic Publishers, May 2000.

[Sha01] Shahidehpour, M., Alomoush, M., Restructured Electric Power Systems, Marcel Dekker Publishers, May 2001.

[Sha02] Shahidehpour, M., Yamin, H., Li, Z., Market Operations in Electric Power Systems, John Wiley and Sons, May 2002.

[She91] Sheble, G.B, "Distributed Distribution Energy Management System (DDEMS) Exploratory Experiments," IEEE Computer Applications in Power, Vol. 4, No. 1, pp. 37-42, January 1991.

[Shi02] Shih, K., Huang, S., "Application of a Robust Algorithm for Dynamic State Estimation of a Power System," IEEE Transactions on Power Systems, Vol. 17, No. 1, pp. 141 - 147,February 2002.

[Sid00] Sidhu, T.S., Cui, L., "Contingency Screening for Steady-State Security Analysis By Using FFT and Artificial Neural Networks," IEEE Transactions on Power Systems, Vol. 15, No. 1, pp. 421-426, February 2000.

[Sil02] Silva, J.M. "Evaluation of the Potential for Power Line Noise to Degrade Real Time Differential GPS Messages Broadcast At

283.5-325 Khz," <u>IEEE Transactions on Power Delivery</u>, Vol.17, No. 2, pp. 326 -333, April 2002.

[Sin98] Singh, H., Hao, S., Papalexopoulos, A., "Transmission Congestion Management in Competitive Electricity Markets," <u>IEEE Transactions on Power Systems</u>, Vol. 13, No. 2, pp.672-680, May 1998.

[Sin01] Singh, J. "XML for Power Market Data Exchange," <u>Proceedings of IEEE Power Engineering Society Winter Meeting</u>, Vol. 2, pp. 755 -756, 2001.

[Sko02] Skoulidas, C., Vournas, C., Papavassilopoulos, G., "Adaptive Game Modeling of Deregulated Power Markets," IEEE Power Engineering Review, Vol. 22, No.9, pp. 42-45, September 2002.

[Sod02] Sodan, A.C., "Towards Asynchronous Metacomputing in MPI," <u>Proceedings of 16th Annual International Symposium on High Performance Computing Systems and Applications</u>, pp. 221-228, 2002.

[Sri00] Srinivas, M.S., "Distribution Load Flows: a Brief Review," <u>Proceedings of IEEE Power Engineering Society Winter Meeting</u>, Vol.2, pp. 942 -945, 2000.

[Sta91] Stankovic, A., Ilic, M., Maratukulam, D. "Recent Results in Secondary Voltage Control of Power Systems," <u>IEEE Transactions on Power Systems</u>, Vol. 6, No.1, pp. 94-101, February 1991.

[Sta02] Stankovic, N., Zhang, K., "A Distributed Parallel Programming Framework," <u>IEEE Transactions on Software Engineering</u>, Vol. 28, No. 5, pp. 478-493, May 2002.

[Su99a] Su, C.-L., Lu, C.-N., Lin, M.-C., "Migration Path Study of a Distribution SCADA System," <u>IEE Proceedings of Generation, Transmission and Distribution</u>, Vol.146, No.3, pp. 313-317, May 1999.

[Su99b] Su, C.L., Lu, C.N., Lin, M.C., "Wide Area Network Performance Study of a Distribution Management System," <u>IEEE Transmission and Distribution Conference</u>, Vol.1, pp. 136 -141, 1999.

[Sum98] Sumic, Z., "The Role of Internet in the Re-Regulated Utility Industry," <u>Proceedings of Distributech 98 Conference and</u>

BIBLIOGRAPHY

Exhibition, Tampa, Florida, Paper No. VFB. 6/2 98033, pp.108-116, 1998.

[Tal94] Talukdar, S., Ramesh, V.C., "A Multi-Agent Technique for Contingency Constrained Optimal Power Flows," IEEE Transactions on Power Systems, Vol.9, No. 2, pp. 855 -861, May 1994.

[Tho86] Thorp, J.S., Ilic-spong, M., Varghese, M., "An Optimal Secondary Voltage/Var Control Technique," Automatica, Vol. 22, No.2, pp. 217-221, 1986.

[Tho01] Thornley, V.P., Hiscock, N.J., "Improved Voltage Quality Through Advances in Voltage Control Techniques," Proceedings of Seventh International Conference on Developments in Power System Protection, pp. 355 -358, 2001.

[Tol01] Tolbert, L.M., Hairong Qi, Peng, F.Z., "Scalable Multi-Agent System for Real-Time Electric Power Management," Proceedings of Power Engineering Society Summer Meeting, Vol. 3, pp. 1676 -1679, 2001.

[Tov01] Tovar-Hernandez, J.H., et. al., "Advanced SVC Models for Newton-Raphson Load Flow and Newton Optimal Power Flow Studies," IEEE Transactions on Power Systems, Vol.16, No. 4, pp. 946 -948, November 2001

[Tsa93] Tsay, M.T., Lu, C.N., Lin, W.M., Chen, C.S., "Data Extraction from a Geographic Information System for Power System Applications," Proceedings of Power Industry Computer Application Conference, pp. 154 -161, 1993.

[Tyl91] Tylavsky, D. J., et al "Parallel Processing in Power System Computation," Proceedings of IEEE Power Engineering Society Summer Meeting, San Diego, July 1991.

[Vaa01] Vaahedi, E., Chang, A.Y., Mokhtari, S., Muller, N., Irisarri, G, "A Future Application Environment for BC Hydro's EMS," IEEE Transactions on Power Systems, Vol. 16, No.1, pp.9-14, February 2001.

[Vee01] Veeraraghavan, M., Karri, R., "Architecture and Protocols That Enable New Applications on Optical Network," IEEE Communication Magazine, March 2001. Vol.39, No.3, pp. 118 -127.

[Vet02]   Vetter, J.S., Mueller, F. "Communication Characteristics of Large-Scale Scientific Applications for Contemporary Cluster Architectures," <u>Proceedings of International Parallel and Distributed Processing Symposium</u>, pp. 272-281, 2002.

[Vis02]   Visudhiphan, P., Ilic, M.D., Mladjan, M., "on the Complexity of Market Power Assessment in the Electricity Spot Markets," <u>Proceedings of IEEE Power Engineering Society Winter Meeting</u>, Vol. 1, pp. 440-446, 2002.

[Vla01]   Vlachogiannis, J.G., "Fuzzy Logic Application in Load Flow Studies,"<u>IEE Proceedings of Generation, Transmission and Distribution</u>, Vol. 148, No.1, pp. 34-40, January 2001.

[Wag99]   Wagner, I.A., Lindenbaum, M., Bruckstein, A.M, "Distributed Covering by Ant-Robots Using Evaporating Traces," <u>IEEE Transactions on Robotics and Automation</u>, Vol.15, No.5, pp. 918-933, October 1999.

[Wan91]   Wang, Y., Hachtel's Augmented Matrix Method for Power System State Estimation and Its Implementation, <u>MS Thesis</u>, Tianjin University, February 1991.

[Wan94]   Wang, Y, A Study on Distributed Asynchronous Iterative Algorithms of Distributed Energy Management Systems (DEMS) of Power System, <u>PhD Dissertation,</u> Tianjin University, August 1994.

[Wan96]   Wang, Y., Yu, Y., "A Distributed Asynchronous State Estimation Algorithm for Distributed Energy Management Systems," <u>Proceedings of the Chinese Society for Electrical Engineering</u>, Vol.16, No.5, May 1996.

[Wan97]   Wang, Y., Electric Power System State Estimation engineering Application Program Design and Hierarchical and Distributed Optimal and Close-Looped Voltage/VAR Control, Final Research Report, Tsinghua University, Beijing China, March, 1997.

[Wan00]   Wang, X., Schulz, N.N, "Development of Three-Phase Distribution Power Flow Using Common Information Model," <u>Proceedings of IEEE Power Engineering Society Summer Meeting</u>, Vol. 4, pp. 2320-2325, 2000.

[Wan01]   Wang, H.F., "Multi-Agent Co-ordination for the Secondary Voltage Control in Power-System Contingencies," <u>IEE</u>

Proceedings of Generation, Transmission and Distribution, Vol.148, No.1, pp. 61 -66, January 2001.

[Web01] http://www.caiso.com: Sponsored by the California Independent System operator.

[Web02] http://www.calpx.com: Sponsored by the California Power Exchange.

[Web03] http://www.cnie.com: Sponsored by the Committee for the National Institute for Environment.

[Web04] http://www.eroc.som: Sponsored by Division of Electric Reliability Council of Texas.

[Web05] http://www.ferc.fed.us: Sponsored by Federal Energy Regulation Commission.

[Web06] http://www.idaopower.com: Sponsored by Independent Grid Operator.

[Web07] http://www.iso-ne.com: Sponsored by New England Inc.

[Web08] http://www.mcs.anl.gov/Projects/mpi/standard.html: Sponsored by Argonne National Laboratory.

[Web09] http://www.mhpcc.edu: Sponsored by Maui High Performance Computing Center

[Web10] http://www.midwestiso.org: Sponsored by Midwest Independent Transmission System Operator, LLC.

[Web11] http://www.nerc.com: Sponsored by North America Electric Reliability Council.

[Web12] http://www.nypowerpool.com: Sponsored by New York Independent System Operator.

[Web13] http://www.ornl.com: Sponsored by Oak Ridge National Laboratory.

[Web14] http://www.pjm.com: Sponsored by Pennsylvania-New Jersey-Maryland interconnection.

[Web15] http://www.w3.org/XML: Sponsored by W3C Consortium.

[Web16] Extensible markup language (XML) 1.0, available at http://www.w3.org/TR/REC-xml.

[Wei95] Wei, X. G., Sumic, Z., Venkata, S.S., "ADSM-An Automated Distribution System Modeling Tool for Engineering Analyses,"

	IEEE Transactions on Power Systems, Vol. 10, No. 1, pp.393-399, February 1995.
[Wid99]	Widergren, S., deVos, A., Jun Zhu, "XML for Data Exchange," Proceedings of IEEE Power Engineering Society Summer Meeting, Vol.2, pp. 840 -842, 1999
[Wil97]	Wildberger, A. M., "Autonomous Adaptive Agents for Distributed Control of the Electric Power Grid in a Competitive Electric Power Industry," 1997 first International Conference on Knowledge-Based Intelligence Electronic Systems, Adelaide, Australia, pp. 2-11, May 1997.
[Wil02]	Williamson, S. "Recent Advances in Energy Management Systems for a Deregulated Market Environment," Internal Report, Department of Electrical and Computer Engineering Department, Illinois Institute of Technology, 2002.
[WuF71]	Wu, F.F., "Theoretical Study of the Convergence of the Fast Decoupled Load Flow," IEEE Transactions on Power Apparatus and Systems, Vol. 96, No.1, January/February 1971.
[WuF85]	Wu, F.F., Neyer, A, F., "Asynchronous Distributed State Estimation for Power Distribution System," Proceedings of the 9th Power System Computation Conference, pp. 439-446, 1985.
[WuF88]	Wu, F.F., "Real-time Network Security Monitoring, Assessment and Optimization," International Journal of Electrical Power & Energy Systems, Vol. 10, No. 2, April 1988.
[WuF92]	Wu, F.F., and et al, "Parallel Processing in Power Systems Computation," IEEE Transactions on Power Systems, Vol. 7, No. 2, pp. 629 -638, May 1992.
[WuF98]	Wu, F., Yeung, C., Poon, A., Yen, J., "A Multi-Agent Approach to the Deregulation and Restructuring of Power Industry," Proceedings of the Thirty-First Hawaii International Conference on System Sciences, Vol. 3, pp. 122 -131, 1998.
[WuJ85]	Wu, J., Static Security Analysis of Electric Power Systems, Shanghai Communication University Press, 1985.
[Yos00]	Yoshikawa, T., Matsuoka, H., "Optical Interconnections for Parallel and Distributed Computing," Proceedings of the IEEE, Vol. 88, No. 6, pp. 849 -855, June 2000.
[Yu93]	Yu, Y., Wang, Y., "An Asynchronously Coordinating Load Flow Calculation Algorithm for Distributed Energy

BIBLIOGRAPHY

Management Systems," Proceedings of IEEE TENCON"93 on Computer, Communication, Control and Power Engineering, Beijing, China, October 19-21, Vol. 1, pp. 483-486, 1993.

[Yan01] Yan, P., "Modified Distributed Slack Bus Load Flow Algorithm for Determining Economic Dispatch in Deregulated Power Systems," Proceedings of IEEE Power Engineering Society Winter Meeting, Vol. 3, pp. 1226-1231, 2001.

[Yu00] Yu, Y.X., Wang, C.S., Xiao, J., Yan, X.F., Ge, S.Y., Huang, C. H., "A Decision Making Support System for Urban Distribution Planning: Models, Methods and Experiences," Proceedings of International Conference on Power System Technology, Vol.1, pp. 485-490, 2000.

[Zab80] Zaborsky, J., Whang, K. W., Prasad, K. V., "Ultra Fast State Estimation for the Large Electric Power System," IEEE Transactions on Automatic Control, Vol. 25, No. 4, pp. 839-84, August 1980.

[Zur01] Zurlo, L.S., Mercado, P.E., de la Vega, C.E., "Parallelization of the Linear Load Flow Equations," IEEE Porto Power Tech Proceedings, Vol.3, 2001.

[Zwi90] Zwibel, H.S., Peasley, C.E., Malachesky, P.A., Simburger, E.J., "GPS Electrical Power System Computer Simulation," Proceedings of the 25th Intersociety Energy Conversion Engineering Conference, Vol. 2, pp. 90-95, 1990.

Index

A

Agent, 17, 34-38, 44, 349, 352, 361-363, 367, 368, 371-374, 380-382, 384-389
Architecture, 103, 112, 117, 138, 170, 171, 173, 457, 458
Area control center, 2, 20, 32, 33
Asynchronization, 51, 55, 73
Asynchronous, 8, 44, 179, 188-190, 192-195, 197-202, 205-207, 211, 218, 219, 223-225, 232, 234, 259, 261, 263
ATM, 6, 8, 9, 86, 460
Attribute, 141, 142, 147, 162

B

Bid, 27, 358, 359, 371, 374, 375, 377, 386, 388, 436
Bilateral, 24, 369, 382, 403
Bulk, 4, 24, 356, 403, 411, 429, 437

C

Catastrophic failures, 3, 14-16, 19, 38
CCAPI, 135, 137
CIM, 135-150, 155, 157, 162-164, 166-168, 172, 173, 175
Client/server, 96
Communication, 102-106, 110-119, 121-127, 131, 133, 137, 184, 349, 351, 355, 373, 446
Communications system, 4, 6, 13-15, 17, 19, 22, 23, 33, 40
Computer network, 81, 83
Congestion, 177, 349, 357, 365, 367, 368, 375, 378, 385, 387, 392, 395, 403, 417, 430, 434, 436-438, 448-451, 453, 464, 473
Control, 1-10, 12-20, 22-27, 29-39, 41-45, 177-179, 183, 203, 391, 417-419, 426, 428, 429
Control Center, 107, 136, 175, 209, 229, 235, 238, 253, 254, 266, 268, 272, 277, 293, 295, 299, 300, 304, 305, 308, 329, 333, 334, 339, 346, 347, 349
Convergence, 48, 51, 63, 72, 73, 74, 178, 183, 187, 190-193, 197-200, 205-207, 210, 211, 216, 218, 221-223, 225, 227, 228, 232, 239, 260-263

D

Data communication, 48-51, 53, 60, 63, 66, 69, 72, 73, 82-86, 89, 96-98
DEMS, 1, 14, 22, 23, 27, 44, 53, 72, 76, 78, 80, 83, 178, 179, 190
Decentralized, 325
DISTCO, 24, 28, 82, 99
Distributed database, 98, 99
Distributed generation, 44, 45, 391
Distributed load flow, 209, 250, 258, 260, 261, 263
Distributed processing, 1, 2, 10, 11, 13, 14, 16, 17, 32, 34, 43-45, 47-50, 52, 53, 55, 56, 58, 60, 62, 70, 72, 75, 76, 78-84, 87, 89, 92, 93, 96-98, 164, 166, 168, 460, 474
Distribution system, 74, 82, 209, 210, 219, 228, 391, 395, 402-406, 409, 410, 412-416, 418-420, 422-428, 430, 432, 435, 437
DMS, 395, 412, 414-416, 419, 422, 423

E

E-commerce, 42, 45, 439, 440-444, 448
Electronic tagging, 103, 129, 130
EMS, 135-138, 142, 166-168, 173, 175, 177, 178, 238, 248, 256, 266, 267, 293, 300, 301, 308, 312
Energy trading, 82
EPRI, 135, 137, 354

F

Fast decoupled, 177, 199-202
Feeder, 154, 212-214, 216, 217, 220-222, 412, 420, 427
FERC, 268
Fiber optical technique, 7
Firewall, 103, 106

G

GENCO, 24, 28, 48, 81, 92, 99, 358, 361, 364, 365, 368, 369, 370, 371, 373, 374, 376-386
GIS, 45, 439, 456-464
GPS, 6, 17, 19, 34, 45, 439, 464-475

H

Heterogeneous, 349

I

ICCP, 103, 114, 115, 121-123, 126-129
ICCS, 102-116, 125-130, 132-134
Information, 102, 103, 106-108, 110, 114, 117-119, 121-124, 126, 127, 130-138, 140-142, 147, 149, 151-159, 161-164, 166-173, 175
Information system, 101, 110, 111
Integration, 44, 107, 117, 121, 123, 135, 164, 165, 168, 170, 172-174, 391, 409, 410, 443, 445, 446, 466, 467
Interface, 102, 108, 110, 111, 114, 116, 123
Internet, 47, 48, 53, 54, 82-86, 96, 97, 99, 102, 111, 112, 114-116, 119, 126-128, 133, 134,

175, 439, 440-444, 446-448, 457, 467
IP, 7-9, 12, 15
ISO, 2, 24-31, 43, 47, 48, 72, 79, 80, 82-84, 87, 92, 98, 101-116, 119, 123, 127-130, 132-134, 177, 178, 267, 268, 270, 272, 274-276, 293-304, 349, 357-361, 363-366, 371, 373-377, 379-385, 387-390, 440, 443, 444, 449, 451-454, 463, 464, 471

J

Java language, 99

K

Kirchhoff, 236

L

LAN, 47, 53, 69, 83, 96, 97, 102-106, 115, 118-120, 124, 127, 134
Load flow, 405, 416, 417, 419-423

M

Middleware, 135, 165-167
MPI, 51, 53, 62, 83, 87-95, 175

N

NERC, 102, 111, 112, 114, 119, 127-130
Network, 102, 112, 114, 116-119, 121-124, 126, 128, 134, 209-212, 214, 216, 219, 220, 223, 224, 230, 232, 439, 441, 443, 457, 459, 460, 462, 468, 470

O

OASIS, 12, 19, 24, 27, 28, 30
Object-oriented, 34, 35, 41, 138, 142
Over-priced, 18, 161, 162
OXC, 8

P

Parallel and distributed processing, 2, 43, 47, 50
Parallel load flow, 178, 183
Parallel processing, 47, 49-53, 56, 60, 67, 463
Power grid, 4-6, 13-15, 34, 36, 38
Protocol, 8, 9, 12, 18, 42, 102, 165

Q

Query, 13, 100, 131, 168, 169, 172

R

Restructuring, 1, 3, 4, 14, 24, 25, 28, 40, 400, 403, 406-409, 415, 429, 440, 444
Restructured, 349, 372
RTO, 2, 29, 30, 31, 43, 267-270, 272, 276, 293, 295, 296, 298-300, 302-304
Reliability, 101-103, 105, 112, 115

S

Satellite, 6, 13, 19, 20, 27, 31
SCADA, 18, 20-22, 32, 33, 41, 123, 128, 137, 140, 141, 164, 173, 235, 245, 249, 253, 256, 266, 299, 395, 405, 412, 414-

416, 418, 422, 428, 459-461, 469, 474
Security, 177, 178, 265-267, 298-301, 307, 311, 325, 349, 354, 356, 410, 413-416, 420, 443, 445, 448, 461, 462, 466, 467
Self-healing, 14, 15, 17, 34, 37, 39-41
State estimation, 238, 247, 416, 422
Synchronous, 8, 179, 187-191, 197, 199, 204, 205, 207, 217, 218, 225, 232, 234, 259, 263, 264
Synchronization, 55, 63, 64, 66-69, 71, 73, 91, 92

T

Task allocation, 58, 59, 69
TCP/IP, 84, 102, 119, 126, 132
TRANSCO, 24, 28, 30, 361, 363, 372-375, 471
Transmission system, 75, 210-212, 225, 357, 403, 404

U

UCA, 41, 103, 117-124
Utilities, 4, 6, 7, 9, 10, 15, 20, 24

V

Vertically integrated, 1, 2, 4, 19, 24, 27
Vulnerable, 14

W

WAN, 1, 9, 10, 14, 17, 22, 43, 47, 54, 62, 64, 69, 83, 98, 102, 105, 112, 113, 115, 118, 119, 124, 238, 463

X

XML, 7, 10-13, 15, 42

Z

Zone, 31, 357, 358, 365, 366, 371, 378, 379, 383, 386, 387